Sheep Production in the Tropics and Sub-Tropics

TROPICAL AGRICULTURE SERIES

The Tropical Agriculture Series, of which this volume forms part, is published under the editorship of Gordon Wrigley

ALREADY PUBLISHED

Tobacco *B. C. Akehurst*
Sugar-cane *Frank Blackburn*
Tropical Grassland Husbandry *L. V. Crowder and H. R. Chheda*
Sorghum *H. Doggett*
Tea *T. Eden*
Rice *D. H. Grist*
The Oil Palm *C. W. S. Hartley*
Cattle Production in the Tropics Volume 1 *W. J. A. Payne*
Spices Vols 1 & 2 *J. W. Purseglove* et al.
Tropical Fruits *J. A. Samson.*
Bananas *N. W. Simmonds*
Agriculture in the Tropics *C. C. Webster and P. N. Wilson*
Tropical Oilseed Crops *E. A. Weiss*
An Introduction to Animal Husbandry in the Tropics
    G. Williamson and W. J. A. Payne
Cocoa *G. A. R. Wood and R. A. Lass*

Sheep Production in the Tropics and Sub-Tropics

Ruth M. Gatenby

Longman
London and New York

Longman Group Limited
Longman House, Burnt Mill, Harlow
Essex CM20 2JE, England
Associated companies throughout the world

*Published in the United States of America
by Longman Inc., New York*

© Longman Group Limited 1986

All rights reserved; no part of this publication may be reproduced, stored in a retrieval system, or transmitted in any form or by any means, electronic, mechanical, photocopying, recording, or otherwise, without the prior written permission of the Publishers.

First published 1986

British Library Cataloguing in Publication Data
Gatenby, Ruth M.
  Sheep production in the tropics and
  sub-tropics.—(Tropical agriculture series)
  1. Sheep—Tropics
  I. Title   II. Series
  636.3'00913   SF375.5.T76

ISBN 0 582 40404 5

Library of Congress Cataloguing in Publication Data
Gatenby, Ruth M., 1955–
  Sheep production in the tropics and sub-tropics.

  (Tropical agriculture series)
  Bibliography: p.
  Includes index.
  1. Sheep—Tropics. I. Title. II. Series.
  SF375.5.T76G38 1986   636.3'00913   85–4272
ISBN 0-582-40404-5

Produced by Longman Singapore Publishers (Pte) Ltd.
Printed in Singapore.

Contents

FOREWORD

PREFACE

ACKNOWLEDGEMENTS

1	Introduction	1
2	Management (by R. T. Wilson)	17
3	Health	40
4	Nutrition	54
5	Feeding	78
6	Reproduction	121
7	Growth and meat production	145
8	Wool production (by Marca Burns)	166
9	Milk and other products	202
10	Sociology and economics	226
11	Improvement	248

APPENDIX I Glossary of technical terms — 267

APPENDIX II Breeds of sheep in the tropics and sub-tropics — 269

REFERENCES — 293

INDEX — 336

Foreword

It is often said that the tropical sheep does not receive its fair share of attention in respect of research and in the funds devoted to promoting the species as a provider of edible and other products. It is, indeed, less fortunate in these respects than its temperate counterpart. Yet sheep farming requires less capital to start and maintain than farming with beef or dairy cattle – a consideration of great importance in Third World countries. The paradox is heightened when one considers that if a domestic animal species for the peasant farmer were to be engineered from scratch, the sheep would match many of the design criteria. It has a high reproductive rate compared with cattle or buffaloes; and has a much greater potential for exploitation of this trait than these other species. This attribute, coupled with a short generation interval, gives the sheep good potential for a high rate of genetic improvement in a number of performance traits. Its size, particularly in the tropics, makes it easy to handle. The sheep or lamb across the shoulders of a shepherd is a familiar image in the tropics. And lastly, along with the goat, it is the exploiter *par excellence* of marginal land.

Natural and artifical selection has resulted in a great diversity of types of sheep and of products, and the tropical sheep exemplifies this diversity well. The products range from the milk of an Awassi ewe to the carcass of a fat-rumped sheep and to skins used for apparel, tents and floor coverings. For the productivity of a domestic species to be improved, and for the species to be used more effectively within the agricultural economy as a whole, it is necessary for the most important information on the species to be readily available to those who have influence on and/or responsibility for agricultural development, and to people working at the extension, advisory and production levels. It is mainly at such groups that Dr Gatenby has directed her book. It treats tropical sheep production, not in isolation, but in the context of its social and economic milieu.

While the biological and economic foundations of animal production in tropical and temperate areas are the same, the constraints imposed by the social and physical environment and by economic circumstance in the tropics are very different from those of temperate, developed countries, and such constraints are fully taken account of in the book.

I hope that this book will achieve two things. Firstly, that it will assist people working with sheep in tropical areas to get better performance and economic efficiency out of their animals, and secondly, that it will help the newcomer to the tropics to realise that improved sheep production in the tropics is not just a matter of transporting Western technology.

Commonwealth Bureau of Animal Breeding and Genetics, Edinburgh

J. D. Turton,
Director,

Preface

This book is written for students of agriculture and veterinary science at degree and postgraduate levels. Its aim is to give a broad view of sheep production in the tropics and sub-tropics, and to review the growing mass of relevant scientific literature. There must be omissions and errors in the book: I apologise for these, and would be grateful for constructive criticisms from readers.

I received financial support to write the book from the International Livestock Centre for Africa (ILCA) and the Commonwealth Development Corporation. The Commonwealth Bureau of Animal Breeding and Genetics (CBABG), the University of Edinburgh Centre for Tropical Veterinary Medicine, the Animal Breeding Research Organisation and ILCA provided facilities. To these organizations I am very grateful. I would particularly like to thank Mr J. D. Turton (Director), Mrs. Alison Barfield and Ms Helen Dorrian of the CBABG for their help in many ways. Dr M. Burns encouraged me to write the book, and contributed Chapter 8. Mr. R. T. Wilson contributed Chapter 2. I also thank the following people who read and commented on various parts of the manuscript: Prof. D. W. Brocklesby, Dr D. Fielding, Mr T. W. Fison, Dr J. A. Hammond, Dr C. E. Hinks, Dr R. H. F. Hunter, Mr I. L. Mason, Dr J. C. Mathers, Mr R. W. Matthewman, Dr M. L. Ryder, Mr J. D. Turton, Mr H. R. Wagstaff and Dr G. Wiener. Finally, I thank Ruth Paynter and other friends who helped me in many ways during my year of writing.

April 1985   Ruth M. Gatenby

Acknowledgements

We are grateful to the following for permission to reproduce copyright material:

Commonwealth Agricultural Bureaux for our table 4.6 from table 3 p 15 (Hansard 1983); Food & Agriculture Organization of the United Nations for our fig 10.1 from fig 6 p 124, tables 10.4, 10.5, 10.6 from tables 10, 11, 9 pp 27/28 (UNDP/FAD 1976b); the author, P J Hogarth for our fig 6.4 from fig 7.3 p 108 (Hogarth 1978); Indian Council of Agricultural Research for our fig 7.2 from fig 1 p 26 (Malik & Acharya 1972); Institut d'Elevage et de Médecine Vétérinaire des Pays Tropicaux for our figs 2.4 from fig 6 p 18 (Baron 1955), 6.3 from graph 1 p 289 (Galliard 1979); International Livestock Centre for Africa for our table 2.5 (Wilson et al 1983); Journal of Range Management for our fig 2.7 from fig 1 p 297 (Morag 1972); Longman Group Ltd for our figs 1.14 from fig 9.5 (Ryder 1983b), 4.3 from fig 8.4 p 141 & table 4.8 from table 5.1 p 56 (Macdonald et al 1981), table 2.6 from table 2 p 26 (Wilson 1983); Small Ruminant CRSP for our table 2.4 from table 2 p 20 (De Boer 1983a); South African Journal of Animal Science for our fig 7.3 from fig 1 p 27 (Reyneke & Fair 1972); the authors, Dr A Thahar & R J Petheram and Applied Science Publishers Ltd for our table 2.2 (Thahar & Petheram 1983).

We are grateful to the following for permission to reproduce copyright photographs:

Cambridge University Press for our figs 8.3 from Plate 17a p 332 (Burns, Clarkson 1949), 8.4, 8.5 from Plates 1 and 2 p 172 (Burns 1966); the author and Gerald Duckworth and Co. Ltd. for our fig 9.1 from fig 5.7 p 219 (Ryder 1983a); Mr J E Sumberg for our fig 2.1 from fig 2 p 49 (Sumberg 1983).

Chapter 1
Introduction

Sheep convert vegetation into products required by man. They thrive in a wide variety of environments in the tropics and sub-tropics. Sheep are particularly well adapted to semi-arid areas which they share with goats and camels. Breeds of sheep in the dry tropics are able to live on the low-quality vegetation and to withstand seasonal shortages of food and water. In the humid tropics sheep usually form only a small part of the agricultural economy. In the humid and sub-humid tropics, sheep must be tolerant of disease. For instance, tsetse flies infest approximately one-third of Africa (ILCA 1979a) so that in these tsetse-infested areas small trypanotolerant breeds of sheep and goats are found.

A flock of sheep can provide a family with food each day in the form of milk and blood, but only in limited parts of the world are sheep milked or bled. In all parts of the tropics sheep meat is eaten. In contrast to large ruminants, sheep are small enough to be totally consumed on the day of slaughter, thus avoiding the need for storage which is very difficult in a hot climate. Sheep also produce non-edible products: manure, skins and, in some cases, wool and pelts. Sheep are a form of investment and this function is particularly important in societies where there are few banking facilities. Pastoralists cannot own communal grazing land, and many small farmers in less developed countries are tenants. As a form of investment, sheep have many advantages. One or more animals can be readily sold for cash. The flock requires few inputs other than labour, and gives a good rate of return. The reproductive rate of sheep is rapid compared with that of large ruminants so that if there is a disaster (such as a drought or disease epidemic) the number of animals in the flock can build up again quickly.

Geographical distribution of sheep

Figure 1.1 shows the distribution of sheep in the tropics and sub-tropics. The density in each country was calculated from its total number of sheep and its total land area, with the effect that variations in the density

2 *Introduction*

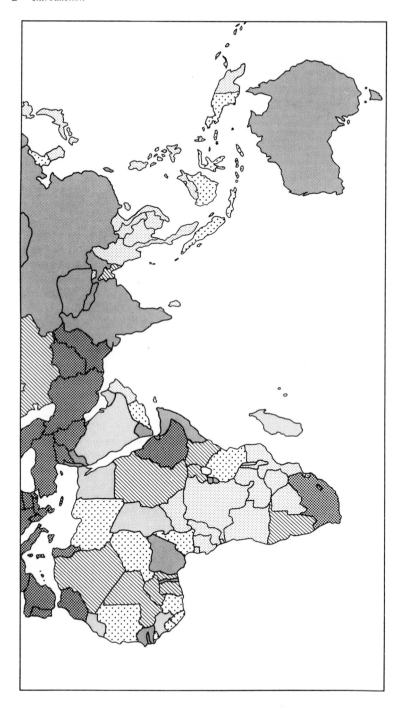

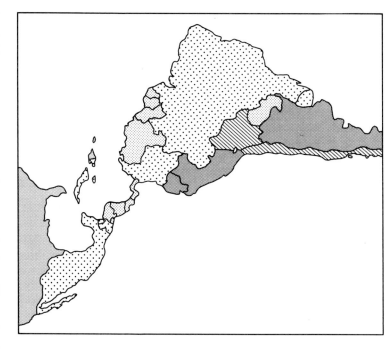

Fig. 1.1 Geographical distribution of sheep in the tropics and sub-tropics (Data from FAO 1983a)

of sheep within the country are not shown. The highest densities of sheep are found in the semi-arid parts of the tropics and sub-tropics.

In Africa, the most important sheep-rearing areas (apart from South Africa) are those surrounding the Sahara – Senegal, Upper Volta, Morocco, Algeria, Tunisia and northern Nigeria – and several countries in East and Northeast Africa – Sudan, Ethiopia, Somalia, Uganda and Kenya. In Asia, there are many sheep in the Near East and in India, but relatively few in the humid environment of Southeast Asia. In tropical America the highest densities of sheep are found in Peru, Ecuador and Bolivia.

Breeds of sheep in the tropics

There are numerous breeds of sheep in the tropics. A list of over 200 breeds is given in Appendix II (p. 269), and in addition there are very many more localised indigenous types, imported breeds and crossbred sheep in tropical areas. Breeds in different areas have developed in response to two selection pressures. Firstly, natural selection favours animals which thrive and reproduce, i.e. those which are adapted to all aspects of the environment (Rendel 1981). Secondly, man has selected animals with those characteristics he desires.

Sheep and goats evolved about 2.5 million years ago (Ryder 1983a), and were the first ruminants to be domesticated by man, some time between 10 000 and 6000 BC (Zeuner 1963). It appears that there were three prototypes of domestic sheep (*Ovis aries*) in southern and western Asia: the Mouflon (*O. musimon*), the Argali (*O. ammon*) and the Urial (*O. orientalis*). There are no indigenous domesticated sheep in Central and South America. The so-called indigenous sheep in Africa are derived from Asia or Europe.

Breeds indigenous to the tropics can be grouped according to their type of tail (fat tail, fat rump or thin tail) and their pelage (hair or wool). Thus there are five categories: fat-tailed hair (Figs **1.2** and **1.3***), fat-tailed wool, fat-rumped hair (Figs **1.4** and 1.5), thin-tailed hair (Figs **1.6**, **1.7**, 1.8, 1.9, 1.10, **1.11, 1.12**) and thin-tailed wool (Fig. 1.13) types. The geographical distribution of these types in Africa and Asia is shown in Fig. 1.14.

In general, the sheep indigenous to the humid parts of the tropics are hair types. Wool sheep predominate in the Mediterranean littoral of Africa, the Near East and northern India. A fat tail or rump is an adaptive advantage to animals living in a hot dry climate as it acts as a food store for the long dry season. Breeds in the Near East and eastern and southern Africa are mostly fat-tailed or fat-rumped. However, in

---

*Figure numbers in bold type are colour illustrations in a separate section between p. 6 and 7.

Geographical distribution of sheep

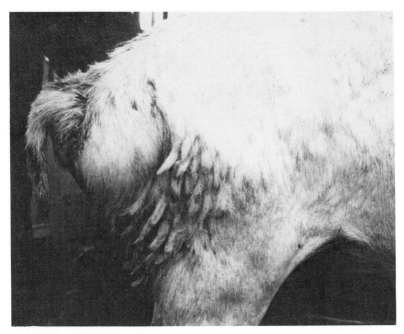

Fig. 1.5 Fat rump of sheep (Somali type), Kenya

the hot dry climates in West and Central Africa and India, sheep have thin tails. As a generalisation, breeds in dry areas tend to be long-legged and larger than breeds in humid areas.

Africa

There are many different types of sheep in Africa, and numerous breeds. Apart from some breeds found close to the Mediterranean, all the breeds indigenous to Africa are hair breeds, i.e. they do not produce wool. The African Long-legged type predominates around the Sahara. This type is tall, has a thin tail and is well adapted to a migratory existence. It includes the West African Long-legged breeds (Fulani, Maure, Tuareg), the Dongola, Northern Sudanese, Southern Sudanese and Zaghawa, and at the southern extent of its range, the Zaïre Long-legged. In East Africa, the East African Fat-tailed type predominiates. Breeds of this type are the Abyssinian, East African Blackheaded and Masai. Further south, sheep are of the African Long-fat-tailed type: Africander, Madagascar, Malawi, Nguni, Zimbabwe, Tanzania Long-tailed and Tswana.

In South Africa, many of the sheep are derived from wool breeds imported from temperate countries. Particularly important is the South African Merino. Other breeds for the harsher areas of South Africa have

Fig. 1.8 Nellore ewe (with tassels), Andhra Pradesh, India

been developed from African Hair sheep. For instance, the Blackhead Persian in South Africa is a fat-rumped hair breed which has originated from the Somali. Crosses between wool and hair sheep have been developed into recognised breeds such as the Dorper (**Dor**set x Blackhead **Per**sian).

In the south of West Africa, where the climate is humid or sub-humid, the West African Dwarf breed is found. Unlike all the African breeds already mentioned, West African Dwarf sheep are small, prolific and have substantial resistance to many diseases including trypanosomiasis. The Atlantic Coast type of sheep are found on the northwestern coast of Africa. Further east along the north coast of Africa and extending into Asia, are the Near East Fat-tailed type which includes the Barbary, Ossimi, Barki, Fellahi and Rahmani breeds.

Asia

In the hotter parts of the Near East the indigenous sheep are fat-tailed. The best-known breed is the Awassi. In Israel, the Israeli Improved

Fig. 1.2 Abyssinian sheep (Adal type), Ethiopia

Fig. 1.3 Fat-tailed sheep, Yemen (Photo: M. J. Nicholson)

Fig. 1.4 Somali ram, Somalia (Photo: R. M. Edelstein)

Fig. 1.6 Mandya ewe, Andhra Pradesh, India

Fig. 1.7 Madras Red ewes, Tamil Nadu, India

Fig. 1.11 West African Dwarf ram, Ghana (Photo: R. M. Edelstein)

Fig. 1.12 Barbados Blackbelly ewe, Barbados (Photo: N. W. Simmonds)

Fig. 1.15 Dorset rams, Yucatan, Mexico (Photo: S. M. Broom)

Geographical distribution of sheep 7

Fig. 1.9 Yankasa flock grazing with cattle, Bauchi State, Nigeria (Photo: M. M. H. Sewell)

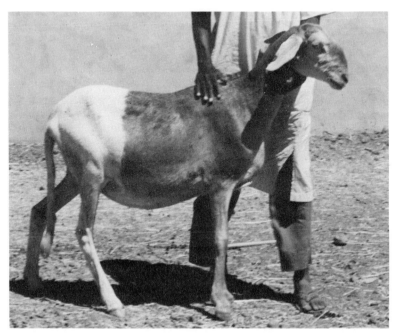

Fig. 1.10 Uda ewe, Katsina, Nigeria (Photo: M. Burns)

Awassi and the Assaf have been developed from the Awassi for milk production.

In the Indian sub-continent there are several distinct geographical regions and each has a distinct type of sheep. In northwestern India and lowland Pakistan, the sheep are small, coarsewooled and kept on transhumant agro-pastoral systems. Many of these breeds are categorised as the Bikaneri type. In peninsular India, sheep are either coarsewooled (e.g. Deccani) or are of the South India Hair type (e.g. Mandya). The Jaffna breed in Sri Lanka is similar to the South India Hair type. In lowland northeastern India and Bangladesh, the sheep are of a small hair type, but they are relatively unimportant in the intensive farming system. Finally, in the Himalayan region (which is outside the scope of this book) there are wooled breeds.

In Indonesia there are three main breeds of sheep: the Javanese Thin-tailed, Priangan and East Java Fat-tailed. The Javanese Thin-tailed is the breed indigenous to West Java, the Priangan is said to be the result of crossing indigenous sheep with sheep imported from South Africa, and the East Java Fat-tailed was probably introduced from the Near East by Arab traders. In Malaysia and Thailand, the indigenous sheep are similar to the Javanese Thin-tailed and are known as Kelantan.

America

There are no domesticated sheep truly indigenous to the Americas. The first sheep imported to South America, probably during the sixteenth century, were wooled animals, probably from Spain (Mason 1980b).

Fig. 1.13 Jaisalmeri flock (with goats), Rajasthan, India

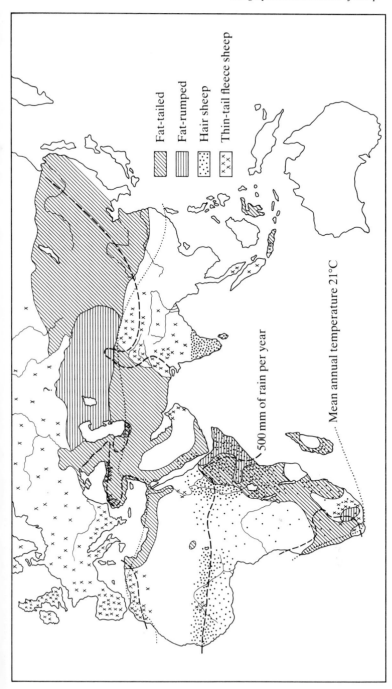

Fig. 1.14 Distribution of sheep types in Africa and Asia (From Ryder 1983b)

These gave rise to the wooled South American Criollo (of diverse appearance) and the Navajo in the USA. During the seventeenth and eighteenth centuries when slaves were transported from West and Central Africa to America, hair sheep were transported too. These sheep, of the West African Dwarf type gave rise to the American Hair breeds such as the Barbados Blackbelly, Morada Nova, Pelibuey and West African. The Blackhead Persian of Central America must be descended from the Somali breed of northeastern Africa. A fat-tailed breed in Brazil, the Rabo Largo, must also have originated from fat-tail sheep imported from Africa or Asia. Importation of the Bergamasca from Italy, and subsequent crossing with hair sheep has resulted in the Santa Inês breed in Brazil.

Oceania

The sheep industry in Australia is based largely on the Australian Merino (Belschner 1959). However, today there are relatively few sheep in the tropical (i.e. northern) parts of Australia. On the tropical Pacific islands, too, there are no indigenous and few imported sheep.

Some tropical breeds such as the Barbados Blackbelly are well described, others have some documentation, but for most tropical breeds even the description of appearance is not complete. Division of animal types into distinct breeds is fraught with problems. There may be a gradation of sheep type corresponding with a geographical gradation, so that two areas have distinctly different animals, but there is no sharp geographical or morphological division between the two types. Thus one breed name may be used for a variety of types of animals. On the other hand, breeds may be named after the locality in which they are found, so that several breed names may be used for one type of sheep. Further problems arise with different languages. In an attempt to minimise this confusion and standardise the presentation of scientific information, Mason (1969) compiled a dictionary of livestock breeds. With a few exceptions, the names of African breeds and types used in the present book correspond to those recommended by Mason. For tropical America, classification is based on Mason (1980a,b) and Fitzhugh and Bradford (1983a). Indian breeds are listed by Bhat *et al.* (1980) and are described by Acharya (1982).

The productivity of tropical breeds is even less well documented than their appearance. What information is available often comes from research stations where sheep are subjected to management which is very different from that in the surrounding area. Gathering useful information on breed performance in the field is not easy. Methods of evaluating sheep breeds are discussed by Turner (1982). The Society for Advancement in Animal Breeding Researches in Asia and Oceania (SABRAO) are establishing data banks for Asian breeds. A form on which the characteristics of sheep breeds can be documented is given by SABRAO (1980).

Most indigenous breeds in the tropics have not been subjected to such intense selection as has been practised for the last 100 years in temperate areas, even though there has obviously been natural selection favouring adapted animals, particularly in dry areas. This lack of selection suggests that for productive traits the phenotypic variation within the population may be very large compared with that observed for temperate breeds.

There are many records of the introduction of temperate breeds into the tropics. In some cases where the environment is satisfactory, such as in the highlands of East Africa, these exotic breeds – particularly Corriedale and Romney Marsh – and their crosses are successfully established. Similarly, the wooled sheep in tropical and sub-tropical South America are descended from sheep imported from southern Europe. Merino breeds, based largely on stock derived from Spanish Merinos, have been successfully established in many parts of the world, notably Australia, the USSR, South Africa, South America, the USA and other parts of Europe, but Merinos are less successful in the hot tropics. The Karakul, a fur-bearing sheep native to Central Asia, was introduced into South West Africa at the beginning of the twentieth century, and became successfully established.

In other cases, the descendants of introduced animals remain in only small numbers today; for example the Macina in Mali and the Nilgiri in South India are wooled sheep found in limited areas surrounded by large populations of hair sheep. Dorset sheep in the Yucatan, Mexico are shown in Figs **1.15**. Many introductions of temperate breeds into the tropics have been unsuccessful, resulting in high mortality and low productivity (e.g. Holmes and Leche 1977; Wilson 1981). Nevertheless, temperate breeds continue to be imported into less developed countries in the tropics by national and international development and aid programmes. For instance, Soviet Merinos, Rambouillets, Dorset Horns and Suffolks have been introduced into India (India, CSWRI 1983) in an attempt to improve the productivity of Indian flocks. Often exotic breeds in the tropics appear to be very different from the ancestral breed; the original importations consisted of only a few animals, and subsequent inbreeding and adaptation to the environment have given rise to smaller, less productive animals.

Genetics

The characteristics of a sheep depend both on its genotype and its environment. Genetic characters can be roughly divided into two categories: those known as quantitative characteristics, which are controlled by many genes (polygenes), and those which are controlled by genes at only one or a few loci. Most production traits – growth-rate, litter size, wool production, etc. – come into the first category. The second category includes a diverse collection of characteristics which, despite being known as characteristics controlled by major genes, are usually of less economic significance.

The study of quantitative genetics is a complex subject which is covered in detail in specialised texts, such as Johansson and Rendel (1968), Turner and Young (1969) and Falconer (1981). The extent to which phenotypes are determined by the genes transmitted from the parents is known as heritability, and is given the symbol h^2. More specifically, h^2 is defined as the proportion of the phenotypic variance which can be attributed to additive genetic effects. The h^2s of fleece weight and quality are quite high, ranging from 0.3 to 0.6. Milk yield and post-weaning growth also have reasonably high heritabilities. For pre-weaning growth, h^2 is 0.1–0.4, and for reproductive traits (litter size, barrenness, lamb survival), h^2 is less than 0.2 (Rae 1982).

Another useful measure is the repeatability of a trait. This is defined as the correlation between repeated measurements of the same trait on animals. It is estimated from either: (i) the product moment correlation between two measurements; (ii) the regression of the second measurement on the first; or (iii) the ratio of the variance between measurements on individuals to the variance between individuals. The advantage of the third method is that it can be used for more than two measurements per animal.

The possibilities for improvement of production traits by breeding are discussed in Chapter 11. The rate of genetic improvement by selection for any genetic trait in a flock depends on the heritability of the trait, the superiority of the individuals selected as parents and the generation interval. Characteristics in a population can also be changed by crossbreeding with a different genotype. Care must be taken to ensure that the potential advantages of the crossbred progeny are not outweighed by serious defects such as a high mortality rate. Although exotic breeds may appear to have superiority over indigenous tropical breeds in terms of growth potential or wool yield, these characteristics are not realised in an inhospitable environment.

The offspring of two sheep of different genotypes may be better than would be predicted from the mean of the parents. This effect is known as heterosis or hybrid vigour and can be substantial, particularly for reproductive traits. Few studies in the tropics give enough information to allow the effects of heterosis on reproduction rate to be estimated. In Egypt, Aboul-Naga (1975) found that for native Ossimi, Merino and F_1 hybrid ewes the numbers of lambs weaned per ewe available for mating were 0.81, 0.80 and 1.08, respectively, indicating a 34% advantage of the hybrids over the pure breeds. In the same experiment Merino × Barki ewes showed a 20% advantage. It is not clear what proportion of the F_1 superiority can be attributed to heterosis and what proportion to additive genetic effects and the suitability of the intermediate genotype to the environment.

Characteristics of sheep controlled by major genes include sex, coat colour, birthcoat type, the presence of horns and tassels, and haemoglobin and other blood characteristics (Ricordeau 1982). The nature of qualitative characteristics may depend on genes at several genetic loci. For instance coat colour is determined by eight loci, and one of these loci has several possible alles (Adalsteinsson 1982).

Although the inheritance of litter size (S) has long been known to be polygenic, there is now evidence that S may also be controlled by one or more major genes. There is evidence that Australian Merinos of the Booroola strain carry a gene for high litter size at birth (Piper et al. 1982), and there have recently been suggestions that similar major genes control prolificacy in other breeds.

Two erythrocyte polymorphic characters – potassium and haemoglobin types – affect the adaptation and productivity of sheep (Agar et al. 1972). In hot dry environments in the Sudan, northern Nigeria or India, sheep with haemoglobin type B predominate (Khattab 1968; Veen and Folaranmi 1978; Arora and Acharya 1972), and their productivity (particularly reproductive rate) is slightly higher than those sheep with types A and AB. In contrast, for West African Dwarf sheep in a more humid environment in southern Nigeria, haemoglobin type A is more common than the other types. It has been suggested that sheep with haemoglobin type A are more resistant to helminths (Evans et al. 1963) and thus are suited to the humid tropics. Sheep in temperate areas have haemoglobin types A, AB and B so that it has been suggested that screening of individuals for haemoglobin type before importation to the tropics would be beneficial.

Within breeds of sheep, the potassium concentrations in erythrocytes and the whole blood of sheep fall into two categories: high concentrations (HK) and low concentrations (LK), and few, if any, animals have intermediate potassium concentrations (Agar et al. 1969). Erythrocyte potassium concentration is controlled by a single locus with two alleles, the LK being recessive. The LK types are thought to have an adaptive advantage with respect to water deprivation and semen production in a hot dry climate (More et al. 1980; 1981). Erythrocyte potassium concentration also appears to be correlated with wool quality. For LK sheep there is a positive relationship between mean fibre diameter and potassium concentration for wool sheep in India (Taneja et al. 1969). Although both haemoglobin type and potassium concentration appear to be related to adaptation and productivity, there is no direct relationship between them.

Transferrins are proteins which facilitate the transport of iron ions in blood (Johansson and Rendel 1968). There are definite genetic differences in the type of transferrins in sheep sera, and transferrin type is thought to affect the growth of lambs (Rahman and Konuk 1977).

Sheep have fifty-four chromosomes and goats have sixty, and the two species do not normally interbreed. However, there are reports of successful crossings giving rise to 'shoats'. There is even a report (Bunch et al. 1976) that a female shoat was successfully reared, was mated by a ram and gave birth to twins. Recently there have been reports of experimental chimerism between sheep and goats (Fehilly et al. 1984; Meinecke-Tillmann and Meinecke 1984).

Environmental physiology

The environment affects sheep in two ways. Firstly it affects the vegetation, and thus the supplies of food and water to sheep, and the disease pattern. These are known as the indirect effects. Secondly, the environment affects sheep directly, mainly through the effects of temperature, radiation and humidity. A study of these direct interactions is known as environmental physics, and the study of the subsequent effects on the sheep is known as environmental physiology.

Of all the direct effects of the environment on sheep in the tropics, temperature stress is usually the most serious. If the sheep is too hot its chronic response is to reduce its food intake, and thus its productivity (measured in terms of reproductive rate, growth rate, milk yield, etc.) is impaired. Much detailed research has been conducted on the physical causes of heat stress and physiological responses of sheep (see Monteith 1973; Mount 1979). For the present, a brief discussion of the causes and effects of heat stress will suffice to allow an understanding of methods of alleviating deleterious effects.

An animal is too hot if, over an extended period, the sum of its rate of internal heat production plus radiant heat absorbed is greater than the rate at which heat is lost from its body. Heat is lost as sensible heat (convection, long-wave radiation and conduction), and by the evaporation of water from the skin and the respiratory tract, or is stored in the body and is seen as a rise in body temperature. Four factors contribute to heat stress:
1. High ambient temperature which reduces the potential for sensible heat loss from the body;
2. Solar radiation, which imposes an external heat load on the animal;
3. A high metabolic rate, which gives a high rate of internal heat production. Productive sheep on a good diet have high metabolic rates, but the metabolic rate of any sheep is temporarily increased by exercise; and
4. High ambient humidity, which limits the rate at which water can evaporate from the animal.

Thus a heavily pregnant or lactating ewe running in the sun on a hot humid day will be likely to suffer heat stress. Such a ewe will attempt to cool herself by behavioural thermoresulation, i.e. she will attempt to lie down or stand quietly in the shade.

Characteristics of sheep which are beneficial in preventing heat stress include a coat which reflects a large proportion of solar radiation (thus sheep with light-coloured sleek coats absorb less solar radiation than those with dark, rough or woolly coats, although the differing insulations of wool and hair coats complicates this simple picture), the ability to sweat rapidly when necessary, and the ability to tolerate a diurnal fluctuation in body temperature. These characteristics are displayed by sheep which are native to tropical areas. Sheep imported from temperate areas are usually less able to withstand heat. Thus in India, Rambouillets are more susceptible to heat stress than the native Malpuras or Choklas (Singh et al. 1980b).

Ambient humidity affects the rate of evaporation from an animal, and if the absolute humidity is high it limits the rate of evaporative heat loss. In addition to its effect on the heat budget, relative humidity is a primary factor determining the survival of disease micro-organisms. A humid environment favours the survival and spread of disease, so that the incidence of many diseases is higher in the rainy season than in the dry season.

The water intake of sheep is closely related to their thermal environment. Under heat stress, the rate of water loss from the body is increased as the sheep sweats and pants (e.g. Quartermain 1964; Gatenby et al. 1983). The requirement of sheep for water is discussed in Chapter 4.

Even in the tropics, cold stress arises when sheep are subjected to low temperatures together with rain. Under these circumstances energy is used to keep the body warm. Cold stress can be fatal to young lambs born in the rainy season because the rate at which they can generate heat is insufficient to maintain body temperature at a level which allows the lamb to move and suckle its dam. Good management – ensuring that the lamb receives colostrum soon after birth, is not separated from its mother and is given shelter – is the best way of reducing the effects of cold stress in lambs.

Environmental stress can have very serious consequences when sheep are transported. Confinement of sheep in an enclosed space where ventilation is inadequate generates a warm, moist environment. Under these conditions, diseases spread rapidly. Many sheep are exported each year by ship from Australia to Singapore. For this journey taking only 8 days, the mortality rate is on average 0.75% (Gardiner and Craig 1970; Australian Bureau of Animal Health 1981). For the longer journey from Australia to the Near East, average mortality is about 4%. Most deaths are thought to result from heat stress and salmonellosis.

An animal can do much to combat thermal stress by behavioural thermoregulation. In a hot environment sheep may seek out shade at midday (Fig. 1.16), and if shade is not available, wooled sheep can shade their heads (which are the most sensitive parts of their bodies, yet which have minimal thermal insulation) by standing in a group with their

Fig. 1.16 West African Dwarf sheep in the shade of a tree, Nigeria (Photo: M. M. H. Sewell)

heads down. In the rainy season ewes select sheltered places to give birth, so that the young lamb is protected against cold stress. The chronic response of sheep to heat stress is a reduction in food intake, and it is primarily this response which lowers productivity in a hot environment.

Another aspect of the physical environment which has a direct effect on sheep is visible radiation or light. In addition to its heating effects, light affects sheep through:

1. *Length of day. or 'photoperiod'.* Breeds of sheep which have evolved in temperate parts of the world are affected by photoperiod. Ewes exhibit oestrous cycles in response to decreasing photoperiod. This effect is discussed in Chapter 6. In addition, food intake on *ad libitum* diets is higher when daylength is long. Some scientists therefore advocate increasing the growth of finishing lambs by providing lights at night.
2. *Intensity of light.* Dim light reduces aggression in animals kept in close confinement. However, too little light can cause vitamin D deficiency in young animals. This is rarely a problem with lambs.
3. *Quality of light.* Ultraviolet light causes photosensitisation of sheep which have eaten certain plants including *Lantana* spp. and *Tribulis terrestris* (Beasley 1967; Sastry and Singh 1979).

Chapter 2

Management

In the tropics and sub-tropics, the vast majority of sheep are managed in traditional ways. There is normally little of what would be understood to be modern methods of management, which might include fencing, dipping and veterinary treatment. None the less, many traditional owners are good managers by any standards and make use of age-old concepts to produce the best out of their animals from what are often very limited environmental resources. Breed types, ownership patterns, flock structures and risk-avoidance strategies related to grazing and water, are all reflected in this management ability.

Systems of management

The principal factor affecting the strategy of management is the climate. While there are minor differences in strategy throughout the tropics, it is generally true to say that the drier the area the more movement is involved. In wetter areas, management systems tend to be sedentary. Table 2.1 shows how management depends on rainfall in the drier areas of Africa. Nomadic systems are generally defined as those in which no fixed base is recognised; transhumant systems as those involving seasonal movement but returning to a settled base for at least part of the year; and sedentary systems as those in which no seasonal or annual movements normally take place, the animals remaining near the village or cultivation area throughout the year.

The word 'transhumance' originally had a specific meaning, relating to the seasonal movement of goats and sheep from the European lowlands where they wintered, to the high Alpine pastures where they spent the summer. This particular type of management (which has been called 'vertical migration') is relatively rare in the tropics and sub-tropics. It does, however, occur in the northern parts of India (Acharya 1982) where flocks transhume to the 'alpine' regions of the Himalayas to avoid the dryness and heat of the summer. At this time they graze on the alpine pastures, returning in the autumn to graze on harvested fields and on fallow lands, remaining in the vicinity of the homestead until the

Table 2.1 Ecology and management of sheep in the drier areas of Africa

Climatic regime (rainfall) (mm)	'Macro'-management	Country/Ethnic group	'Micro'-management Day	'Micro'-management Night	Size of flock/herding group
Arid (200)	Nomadic	Mauritania/Moor		Open camp	100– 350
		Ethiopia/Afar		Penned	
		Sudan/Kababish		Open camp	
(300)	Transhumant	Mali/Tuareg	Loose flock		
		Niger/Tuareg			
		Chad/Zhagawa			200– 250
		Kenya/Turkana		Penned	
		Ethiopia/Afar		Penned	
Semi-arid (400)	Semi-sedentary	Sudan/Baggara	Tight flock	Penned	50– 150
		Mali/Fulani	Loose flock	Open camp	20– 60
(500)		Kenya/Masai	Tight flock	Penned	200– 500
	Sedentary	Sudan/Daju, etc.	Tight flock	Penned	20– 80
(600)		Mali/Bambara	Tight flock	Penned/tied	5– 10
	Stall feeding	West Africa/'Mouton de case'	Tied	Tied	0– 10
					1– 5
Highlands	Ranching	Kenya/large-scale farms	Extensive paddocks		500–1000

early winter. Later in the winter and during the spring, the animals move to the foothills where they graze or browse on the leaves and pods of trees. In the hilly parts of Uttar Pradesh some 87% of flocks follow this migratory system of management and the remaining 13% are stationary. Similarly, in the hilly parts of eastern India, a substantial proportion of flocks follow a vertical migration. The stated reasons for transhumance in the East being to avoid the rains, leeches and poisonous plants in the plains and lower hills (Acharya 1982).

The arid and semi-arid regions of northwestern India comprise Punjab, Haryana, Rajasthan, Gujarat and the plains of Uttar Pradesh and Madhya Pradesh. In much of this region, the majority of flocks are totally sedentary, but in the drier parts of Rajasthan a substantial proportion are migratory, following a 'horizontal' pattern. The owners of totally nomadic flocks usually have no permanent rights to any land. On migration, the sheep move in groups of four to ten flocks, usually together with other domestic species particularly camels and goats. These flocks graze open stubbles after crops have been harvested, or they may be more intensively folded on specific fields in which case the farmer may compensate the flock owners in cash or in kind.

In the Near East, classic forms of horizontal nomadism are practised by the Bedouin. Unlike the examples just given for India, the whole Bedouin family, averaging about eleven persons, usually moves with the flock (Bhattacharya and Harb 1973). Migration routes may be followed regularly, although in areas where the rainfall is very low and erratic, the migration pattern is often varied to exploit areas of ephemeral grazing which occur due to sporadic precipitation. Following their age-old routes of migration, the Bedouins pay scant heed to modern political boundaries. For instance, owners who spend much of the spring and summer months in the coastal plains of Lebanon where the grazing is good only from April to July, spend the winter months in neighbouring Syria. In Syria, the livestock owners may pay rent to farmers for stubble grazing, this being the opposite of the custom in India. Many owners with small or medium-sized flocks now migrate by lorry or pick-up truck preferring, evidently, the use of fossil energy to that of their sheep and themselves.

The Awassi and related breeds of sheep in the Near East are fat-tailed. They are able, when temperatures are not too high and grazing is green, to go for extended periods without drinking. In times of sparse grazing they are said to live largely on their fat. Non-migratory flocks owned by sedentary farmers have a daily grazing orbit of 6–8 km. Migratory flocks on good desert grazing may also cover this distance, but on moving from one area to another, 16 km is more usual. When pressed, they can be made to travel up to 35 km in 24 hours, although as sheep are generally slow movers, some goats are often kept in the flock to encourage the sheep to move faster (Williamson 1949).

In most nomadic and transhumant systems the aim of movement is to avoid the very low-quality vegetation which is the only food available in arid areas in the dry season. Sometimes, however, movement avoids floods or, as in the case of the Baggara in western Sudan, mud and flies (Wilson and Clarke 1975). The inundation zone of the River Niger in West Africa is flooded to varying depths from July to November causing the Fulani of Central Niger to move their wooled Macina sheep to areas of dry ground where there is rainfed grazing (Wilson 1983). Although a few Fulani move their animals as far as 300 km, comparable with the distances trekked by the Bedouin, most move only the minimum distance to get their sheep out of the flood area.

In humid areas there is usually much less seasonality in the rainfall and herbage growth, and sedentary management is the norm. Sheep managed in sedentary systems are often much more closely integrated on a permanent basis with cropped lands than are sheep in migratory systems. This applies even where the shepherds are not farmers. In Sri Lanka, for example, where sheep are kept primarily as producers of manure, they are grazed during the day either on paddy stubbles or on common grazing land (Buvanendran 1978). The flock owners are paid by the farmers for this service. When crops are actually growing and sheep are penned away from fields or gardens, the manure is collected and sold.

In the humid zones of West Africa, a similar system to that described for Sri Lanka is developing. The traditional practice is for sheep (and goats) to wander freely around the villages scavenging for food (Matthewman 1980). Many farmers are now adopting a more sophisticated system (Sumbers 1983) in which sheep are confined to small areas by means of hedges, often of a leguminous browse species such as leucaena or gliricidia, with the animals wearing a collar to prevent them escaping (Fig. 2.1).

Integration of sheep with permanent tree crops is also a feature of sedentary management. This system is common under coconuts and cashew in coastal Tanzania, is being introduced with oil-palms in the Ivory Coast, and is also gaining popularity in Malaysia under rubber. In the last-named example, sheep were reported not to damage the rubber trees but ate fallen seeds and leaves in addition to grasses, legumes and other plants (Vanselow 1982).

In much of Central and tropical South America, sedentary sheep are generally managed as an adjunct to other livestock enterprises. In Mexico and Colombia, cattle ranchers keep a few sheep, but they are rarely given much attention (Fitzhugh and Bradford 1983a). Sheep are regarded as an easy source of petty cash and as a meat supply for festivals or for workers' rations. In parts of Northeast Brazil where skins provide an additional source of income, management is similar, sheep grazing on unimproved rangeland (known as caatinga) and even in bad years not being given any supplementary feed (Figueiredo et al. 1983).

Almost all sedentary flocks, whatever their daytime management, are confined at night. When sheep are confined during the day also, they are said to be stall-fed. The Kikuyu of Kenya are thought to have practised stall-feeding, cutting and carting all forage, for centuries (Lyne-Watt 1942). In the Indonesian humid zones management practices range from total confinement of sheep in the upland areas (where 95% of farmers feed native grasses, 28% banana leaves, 28% maize tops and 18% cassava leaves), to grazing on rice stubbles or under rubber trees in lowland areas (Mathius *et al.* 1983). In West Java, sixteen different systems have recently been identified (Table 2.2) with marked seasonal differences occurring in types of feeding (Thahar and Petheram 1983).

In areas where there are many landless people or where pressure on land is high, tethering of animals on roadside verges, on steep hillsides or on any other available location is common. This is the practice in much of the West Indies and, where the nutrition from natural vegetation is considered inadequate, supplementary feed in the form of household scraps, banana skins, sugar-cane by-products, etc. may be offered.

Some reflection will quickly lead to the conclusion that these different

Fig. 2.1 A locally made collar and a live hedge to restrain sheep in a confined area (From Sumberg 1983)

Table 2.2 *Feeding practices in West Java, Indonesia. The figures show the percentage of respondents using each feeding practice in the wet and dry seasons. The total number of respondents in each season was 350. The largest component of a combination system is the first shown – for example, hand feeding plus herding involves over 50% hand feeding*

Feeding system	West season	Dry season
Full hand feeding	32.6	23.4
Grazing		
Full herding	13.7	30.6
Tethering	2.5	4.9
Free range	0.6	5.4
Herding and tethering	0.0	1.1
Tethering and herding	0.3	0.0
Tethering and free range	0.0	0.3
Total	17.1	42.3
Combinations, mainly hand feeding		
Hand feeding and herding	16.0	6.0
Hand feeding and tethering	1.4	1.4
Hand feeding and free range	1.4	0.0
Hand feeding and herding and tethering	0.3	0.0
Total	19.1	7.4
Combinations, mainly grazing		
Herding and hand feeding	23.1	21.4
Tethering and hand feeding	6.9	4.3
Free range and hand feeding	0.6	0.3
Herding and hand feeding and tethering	0.3	0.6
Herding and tethering and hand feeding	0.3	0.3
Total	31.2	26.9

Source: After Thahar and Petheram (1983).

strategies are adapted to the resources available. Nomadism is a sophisticated management response to a resource base which is always seasonally and often almost totally deficient. Stall-feeding is equally a response to the avilability of nutrients in a particular environment and to a demand, often very strictly defined in time and space, for a convenient quantity of meat. These two examples of management options are the extremes of a wide range which forms a continuum in terms of food availability from the almost totally unendowed very arid areas to the much more favourable humid regions. Where irrigation possibilities exist, even in the arid zones, there are opportunities for relatively sophisticated interventions. These may lead to greatly increased output even where only native stock and locally grown foods are available. In Mali for example, stall-fed sheep gain weight at twice the rate ($117 \, g \, d^{-1}$) of those managed extensively ($60 \, g \, d^{-1}$) and weigh almost 40 kg at 9 months of age (Kolff and Wilson 1984).

Ownership patterns

Patterns of ownership are extremely varied and, for an outsider, difficult to establish and understand, especially in Africa. The ramifications of many African kinship systems, the complicated systems of 'stock friends', loans and flock splitting, the herding-out procedures involving professional shepherds, often of a different ethnic group, all lead to a rather fluid idea of ownership. This often involves many displacements of an animal over its lifetime. Although the system of sharing is perhaps best described from Africa, it is also known in Java and Sumatra (Knipscheer et al. 1983) and occurs elsewhere as well.

Despite these difficulties in defining an individual flock, it is none the less true to say that the drier the area the larger the average flock size. This applies equally in Africa (Table 2.1), Southeast Asia (Mathius et al. 1983), the Near East (Bhattacharya and Harb 1973) and Brazil (De Boer 1983a).

In Northeast Brazil, 58% of landowners own sheep, many of these (39% of all landowners) not owning goats, with an average flock size of 142 animals (Table 2.4). However, many groups other than landowners own sheep, including manager-sharecroppers, sharecroppers, settlement scheme farmers, cash renters, communal owners and employees.

In the lowlands of northern India the sedentary flocks tend to be small and rarely composed of more than 5 animals, while the migratory flocks average 22 head. Several flocks are combined when animals transhume so that management units of 100 to 200 head are controlled by one, often professional, shepherd (Acharya 1982). In Sri Lanka, the average flock size is reported to be about 100 Sheep (Buvanendran 1978). This is a surprisingly large flock size for this climatic zone and can probably be attributed to the ownership of sheep by people who are not farming and the commercial use of sheep as producers of manure.

In the humid zone of Java, the numbers of sheep per farmer are between 3 and 5 animals (Mason 1978) and in the forest zone of West Africa (where 60% of farmers own small ruminants) the average sheep holding is only 2.5 head (Matthewman 1980).

The reduction in flock size as one moves towards the wetter areas is possibly only indirectly influenced by climate. This is because a larger proportion of income is generated from crops in what have been described as agro-pastoral than in pastoral systems (Wilson et al. 1983). In agro-pastoral systems, livestock assume the role of a reserve in case of crop failure, rather than being the principal source of subsistence. In Kenya, for example, the pastoral Masai own on average 44.0 sheep, while the agro-pastoral Pokot own only 5.3 (Peacock, personal communication). Similarly in Chad, the pastoral Zioud own an average of 43.5 sheep while the agro-pastoral Salamat own only 2.0 (Dumas 1977). In all four cases goats and cattle are owned in addition to sheep.

Flock structures

'Prestige' and 'perverse supply' were once catchwords used to typify the attitudes of traditional livestock owners in the tropics. While traditional owners are undoubtedly conservative, it is doubtful if they are any more so than their modern peers in Europe, Australia or North America. Reasons for keeping a particular animal are rarely irrational. Almost invariably, if an animal is in the flock it has an economic use, although this does not entirely preclude a social use. Flock structures support this contention and the early offtake of males either for home consumption or for sale is normal practice throughout the world. Only where males or castrates have an economic use such as for wool, hair or the production of excessive fat, are they retained.

In Africa, a generalised flock structure has about 75% of the flock as females, with 55% of the flock being females over 10 months of age and capable of breeding. Exceptions to this general pattern can be explained, as can be seen from Table 2.3, by the different management objectives. Contrary to another popular misconception, culling practices are also of a fairly high standard. Parturition intervals in most flocks average 250–279 days; females with long intervals are usually culled. Older females are also culled under normal circumstances such that broken-mouthed ewes are usually less than 5% of the total in large pastoral flocks, and rarely exceed 10% in small agro-pastoral ones.

In India, flock structures, while showing a considerable preponderance of females, vary by geographical region (Acharya 1982). In the arid and semi-arid Northwest, structures are similar to the generalised African type with about 75% of the total flock being females. In the southern peninsular region, where some 80–95% of flocks are sedentary and are generally small in size, females in the flocks average more than 85%. A fair proportion of flocks in the peninsular

Table 2.3 *African examples of management objectives related to flock structures (per cent of total animals)*

Country/Ethnic group	Use	Females Total	'Breeding'*	Males Total	Castrates
Ethiopia/Afar	Milk	92.2	61.4	7.8	0.0
Niger/Tuareg	Milk (meat)	81.8	63.9	18.2	4.7
Mauritania/Moor	Meat/hair	78.1	58.6	21.9	6.2
Sudan/Baggara	Meat	77.8	57.7	22.2	0.0
Mali/Fulani	Meat/wool	74.5	55.9	25.5	11.3
Chad/Arab	Meat/milk	73.3	53.7	26.7	'few'
Upper Volta/Fulani	Meat/milk	70.5	47.6	29.5	?
Kenya/Masai	Meat (fat)	68.6	54.2	31.4	15.4

* Over 10 months of age.

area (up to 23% in some ethnic groupings) do not contain any males; these must evidently resort to other flocks on communal grazing lands in order to ensure that fertilisation of ewes takes place. Females in the flocks in the eastern areas of India are in the range 60–70%, among the lowest ratios reported for all traditional societies. Sri Lankan flocks are said to comprise 90–95% females (Buvanendran 1978).

Ewes comprise 67% of the Bedouin flocks of the Near East, these being covered by two rams (Bhattacharya and Harb 1973). That these flocks are kept primarily for milk production is reflected in the flock structure which resembles that of the milk pastoralists of Africa shown in Table 2.3.

Flock structures in Brazil (Figueiredo *et al.* 1983) appear to be similar to those of other tropical areas. Breeding ewes over 12 months of age account for about 55% of the flock, with all females contributing 71% of all animals. Breeding males over 2 years of age amount to about 2.5% of the total.

Daily management

The macro-management strategies of husbandry (leading to nomadism, transhumance or sedentary systems) are determined primarily by the climatic and nutritional regimes. Indirectly these two factors also impose certain micro-management tactics such as flock sizes, daily grazing cycles and methods of watering.

Pastoralists, who rely mainly on livestock for their livelihood, tend to be better managers than agro-pastoralists who derive a large part of their income from crops. There is, of course, a considerable range of individual management ability within systems. In Mali, for instance, management has been shown to account for differences in flock output of the order of 10 : 1 between best and worst flocks in the same environment (Wilson *et al.* 1983).

Herding practices

In extensive management systems in arid and semi-arid areas, the daily grazing period is of long duration, animals being away from the camp for 16–18 hours or even longer in some cases. Along the southern fringe of the Sahara, the pastoral Tuareg do not return to their temporary camp until 21.00 and leave again before dawn at 04.00 or 05.00. Animals are watered only every 2 days or at longer intervals in the cold season. Sheep are physiologically and behaviourally adapted to this regime which demands of them less energy than daily trekking long distances to water. Similarly, there is less energy demand on their herders who not only have to trek with the sheep but need to clean out silted hand-dug wells and lift water in small containers from depths of up to 20 m. At depths greater than this, some form of animal power is usually employed to draw water.

Where predators are not a problem, flocks are often allowed to roam more or less freely at night, and are shepherded by children during the day. These children often make pets of favourite ewes (Fig. 2.2). This leaves the adult men free to herd camels or go about the more serious business of smoking, drinking tea and philosophising. In the lowlands of northwestern India, a similar type of daily management is in vogue, with animals usually being watered at midday, and in the hot summer months they have a resting period during the afternoon and then spend most of the night grazing (Acharya 1982).

Where milk is the major economic product, regular watering is more important: animals return to the camp or village at fixed intervals but may still be grazed at night. This is the case for both fellahin (sedentary or transhumant) and Bedouin (nomadic) flocks in the Near East (Epstein 1982). Men are responsible for herding (often accompanied by

Fig. 2.2 Baggara child with pet ewe, Southern Darfur, Sudan (Photo: R. T. Wilson)

dogs), care of sick animals, shearing and tying up ewes for milking. The actual milking, which is done while the ewes are tied in series to a long rope, is performed by women.

In areas of higher rainfall and greater potential, there are different considerations. Animals have to be protected from theft and perhaps from inclement weather. On the other hand, crops have to be protected from the animals. Night grazing is therefore a much less common practice.

In some agricultural societies, herding is viewed as a low-status occupation, particularly in the highlands of North Thailand (Falvey 1979). Daily management, then, is often the work of women and children. Crop protection is a prime consideration and, as it is unusual to confine the animals in enclosures during the day, fences may be built around the crops. These are seldom completely effective and herding is the rule, except during the non-growing season when animals may be allowed to roam freely. Where tethering is practised, problems of this nature normally do not arise.

The small size of individual holdings in agro-pastoral systems usually leads to a number of owners combining their flocks into a single herding unit. This flock is then taken out to graze either by a member of one of the owning families or is looked after by a professional shepherd. The latter are remunerated in cash, kind (milk, or a part of the lamb or wool crop) or a combination of both. Professional herding is probably on the increase, especially in Africa where urban dwellers are investing more and more in livestock and where, concurrently, droughts have led to a reduction in pastoralists' flocks.

In many agro-pastoral societies, a factor contributing to low productivity is the shortness of the period allowed for grazing, which may be as little as 5 or 6 $h\,d^{-1}$ compared with 10.5 $h\,d^{-1}$ for Bedouin sheep (Bhattacharya and Harb 1973). The reasons for this may be practical from the point of view of the farmer and his family who have other occupations such as tilling the fields, collecting firewood and preparing food, which they consider more important than herding. Other reasons often quoted (which may have a rational basis) are problems with ticks on wet grass and the deleterious effects on general health of grazing in the early morning.

Night management

Confinement at night varies from a simple brush enclosure, through attachment to a stake, to a completely closed shed.

In northwestern arid and semi-arid India, as well as in the peninsular region, some 60% of flocks are penned in fields away from the house, the remainder being penned in open enclosures made of thorn or earth. In the hilly parts of eastern India sheep, along with other livestock, may be kept in the lower parts of domestic dwellings. In the hilly regions of northern India the common practice is to pen sheep near the house, but a small number of owners keep them underground (Acharya 1982).

In Sri Lanka, as already noted, following grazing on stubbles and common land during the day, sheep are penned in branch and stick enclosures at night on market-garden plots (Buvanendran 1978).

In the nomadic systems of the Near East, sheep remain in open camp at night as they do in most of the African pastoral systems. On the South American ranches, where sheep are of only minor importance compared with cattle, they may come back to the vicinity of the homestead at night, particularly if the only water on the ranch is to be found there (Mason 1980b).

In the agro-pastoral areas of Africa, the most common practice is to put sheep in a thorn enclosure, although they are sometimes tied individually to pegs. In this area goats are tied to pegs much more often than sheep, undoubtedly a reflection of their more independent nature. In some areas, for example among the Mossi of Upper Volta (Fig. 2.3), sheep are kept in brick enclosures furnished with a closed shed in which they can spend the night.

Other areas where complex housing is provided are the West Indies – to prevent theft and savaging by dogs (Hupp and Deller 1983) – and Java where the animals are kept at night in bamboo pens or sheds with woven or slatted bamboo floors (Mason 1978).

Fig. 2.3 Night housing for agro-pastoral sheep in the Mossi area of Upper Volta (Photo R. T. Wilson)

Ewes and lambs

Almost invariably some extra attention is given to ewes at lambing and to ewes with young lambs. This, especially in extensive systems, may be limited to ensuring that the lamb finds its own mother and has a chance to suckle. Lambs born while the flock are out grazing during the day are usually carried by the shepherd to the camp or village. Usually in these systems lambs are penned in small shaded enclosures or tied in the shade of trees for the first few days of life. In slightly more sophisticated systems, and especially where flocks are large and births grouped in a relatively short period, lambs (or ewes with lambs) may be run as separate flocks and given preferential treatment in respect of grazing and watering.

In less extensive systems, lambs are subjected to management similar to that received by ewes. In all systems, one of a pair of twin lambs may be fostered on to a ewe which has lost its own lamb. Fostering of lambs to goats, which often have a greater milking capacity than sheep, is not an uncommon practice.

Weaning and castration

In almost all situations, weaning is allowed to occur naturally. This happens, depending on the breed, the time of year and the length of the parturition interval, at anything from 60 to 150 days.

Where there is a definite policy of weaning, various methods are used. Although it is possible to separate the young from their mothers, this is seldom done as it leads to complications in the daily herding cycle. More common practices are blocking the ewe's teats with dung, tying off the teats, or enclosing the teats in a small pocket made of home-cured leather. Occasionally muzzles are placed on the lambs, although where the weaning practice is aimed at preventing the lamb from suckling, rather than the ewe from being suckled, it is more usual to place a twig transversely in the mouth of the lamb and attach this round the back of the head with bark string. This method is used in many countries, including Mauritania where it is known as the *chbâba* (Fig. 2.4). It prevents suckling, but does not hinder grazing.

Castration of sheep in traditional management systems is not common. This is in part owing to the early removal of ram lambs from the flocks, and may occasionally be because of religious reasons. Where castration is carried out, it is rarely done by the open method, probably to avoid infection.

The Masai of East Africa castrate sheep in order to encourage fat deposition for their particular dietary requirements. Attrition of the spermatic cord is achieved by beating it between two sticks, or the two testicles are crushed by beating with stones. Where wool or hair is an important product of the system (see Table 2.3), a number of males may

Management

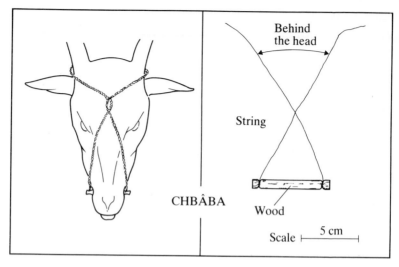

Fig. 2.4 The *chbâba* – a traditional Mauritanian practice for preventing suckling (From Baron 1955)

be castrated. The Moors castrate animals required for hair production within a few days of birth, while the Fulani castrate their Macina ram lambs required for wool production usually between 3 and 6 months of age (Wilson 1983). The Fulani methods of castration are similar to those of the Masai.

Sheep owners who are Muslim by religion are reluctant to castrate their sheep. This is because of the ritual value of entire rams for sacrifice at Idd el Adda. However, even orthodox Muslims will castrate for special purposes. The Bedouin of the Near East castrate sheep chosen as bell-wethers to lead the flock. The method they use is either to bite through the spermatic cord, or to tie a string tightly round the neck of the scrotum (Epstein 1982).

A method of castration used in modern intensive systems is elastration. A rubber ring is fitted over the scrotum within a few days of birth (making sure that both testes have descended). This tight ring causes the whole scrotum to wither and it eventually falls off.

Milk production

Keeping of sheep for milk is a specialised branch of husbandry. In the tropics and sub-tropics, it is found in North Africa and the Near East. In sub-Saharan Africa and India, sheep are milked only in the drier areas. The one specialised dairy breed in the tropics is the Awassi. The Chios, originally from the Greek island of that name, has been imported to Cyprus to improve the milk production of the indigenous sheep there

(Gall 1975; Lysandrides 1981). There are specialised large-scale commercial milk-producing systems in Israel based on the Awassi breed.

In many nomadic Bedouin flocks, the practice is not to start milking ewes for human use until the lambs are weaned at between 2 and 3 months. The breeding season in the Bedouin flocks is determined by the amount of grazing available. In Lebanon, Syria and Israel this means that ewes start to come on heat in June, with a peak number breeding in July and August, and only a few remaining in season in September. Lambs are thus born mainly in December and January when there is still some pasture left. The period for which the lambs are suckling depends very much on the individual shepherd's judgement of the state of grazing, the period of birth, the development of the lamb and his knowledge of the ewe's capabilities. Removal of milk for human consumption starts when a fair number of lambs can be weaned at one time. During the first 3 or 4 months ewes are milked twice a day, but during the last month of lactation ewes are milked only once. Milking is done in the open, and by the women of the family or camp (Epstein 1982). The average lactation period (including the suckling period) may be up to 250 days. In the 5 month milking period, about 100 kg milk is obtained (Bhattacharya and Harb 1973).

The milk production of Chios sheep under Cyprus conditions is typically 250 kg in a lactation period of 230 days. Most milk is produced in the first 8 weeks, and there is a marked decline in daily yield after the lamb is weaned (Louca 1972). These milk yields are considerably greater than the 10–25 kg of milk produced by the few Indian breeds that are milked (Acharya 1982).

Wool production

Relatively few breeds of sheep indigenous to the tropics are wooled. Fleece weights are low, the yearly total of greasy wool rarely exceeding 1 kg. The Macina of Mali produce about 0.7 kg (Wilson 1983), the Menz of Ethiopia about 0.5 kg (Galal 1983), while the majority of Indian breeds give in the region of 1–1.5 kg (Acharya 1982).

Almost all wool from tropical breeds of sheep is coarse and suitable for carpet manufacture. There is little specific management for wool production, although sometimes the soiled wool is removed (a process known as dagging) and the sheep may be washed before shearing. Washing does not normally reduce gross impurities such as twigs and small stones. Shearing is often done with a knife, which in the Near East is usually double-edged. Careful use of knives is needed otherwise numerous cuts in the sheep are made. In northern India shears are nowadays mostly used instead of knives (Taneja 1978). These are either in the form of scissors or else two blades held together by a handle which springs open after each cut to speed up the rate of shearing. Electric shears are rarely used in commercial wool production in the tropics and sub-tropics.

Sheep are shorn twice a year in Mali and Ethiopia, and in some parts of India shearing is done two or even three times a year. Frequent shearing leads to short staple lengths which may not be suitable for the international market, but which give a yarn strong enough for the primitive looms used in local industries.

Relationship with goats

Single-species flocks are unusual in the tropics and sub-tropics. Almost everywhere sheep are kept together with goats, although occasionally they are herded with cattle. However, the relationships of goat-owning to sheep-owning and goat numbers to sheep numbers vary considerably, as can be seen from Tables 2.4 and 2.5 for parts of Brazil and Africa respectively.

Goats are often considered to be more intelligent and tractable than sheep and might therefore be kept as leaders in mixed flocks. However, behavioural studies have shown that when grass is freely available, sheep behaviour is the dominant factor in the feeding patterns observed, whereas when browse is the main food supply, goats tend to lead the mixed flock.

Table 2.4 *Characteristics of small ruminant populations on sample farms in Northeast Brazil*

County	Average small ruminant population per farm (No.)			Producers with only goats (%)	Producers with only sheep (%)	Producers with sheep and goats
	Goats	Sheep	Sheep and goats*			
Granja	163	149	271	13	13	74
Sobral	60	115	145	0	56	44
Crateus	49	172	194	0	54	46
Independencia	213†	173	233	0	71	29
Taua	262	218	448	0	13	87
Parambu	90	49	118	0	14	79
Quixada	94	165	209	0	47	53
Quixeramobim	57	133	161	0	44	56
Morada Nova	58	108	121	6	40	54
Average	116	142	211	3	39	58

* The average sheep and goat population does not equal the sum of the two preceding columns since not all farms had both species.
† A very small number of farms had goats.
Source: From De Boer 1983a.

Table 2.5 *Ownership patterns of sheep and goats in an agro-pastoral area in Central Mali*

	Irrigated rice sub-system		Rainfed millet sub-system	
	Goats	Sheep	Goats	Sheep
Number of owners studied	27		16	
Number owning sheep or goats	26	15	16	9
Number owning goats but not sheep		12		7
Number owning sheep but not goats		1		0
Mean flock size (±sd)*	9.0(±6.03)	6.4(±13.51)	38.2(±27.75)	7.1(±14.81)
Mean flock size (±sd)†	9.3(±5.87)	11.5(±17.00)	38.2(±27.75)	12.6(±18.27)
Range of flock sizes	0–23	0–64	2–91	0–58

* Of all owners, i.e. irrespective of whether the holding of one species of stock is nil.
† Of only those flocks in which animals are held, i.e. nil holdings excluded.
Source: From Wilson *et al.* (1983).

There are, of course, advantages other than better year-round use of resources accruing from keeping mixed flocks of sheep and goats. In the Sahel these include less severe seasonal fluctuations in milk supply and meat availability due to the generally different main reproductive seasons of the two species (Wilson *et al.* 1983). Animals for possible sale are also available at a more constant rate over the whole year if the two species are kept together.

Identification

Traditional owners and shepherds usually have good knowledge of their animals and are generally able to recognise them individually. In addition they often have considerable powers of recall in respect of the reproductive histories of ewes and the relationships of all the animals in the flock to each other. The identification of an individual sheep is often related to its history. As well as colour and physical characteristics, past events, such as droughts, floods, and family and political mileposts are used to separate individuals. A combination of these factors may be reflected in the specific names given to each animal. Where control of breeding is practised this pool of knowledge is put to good use to prevent inbreeding, or to breed for perceived values, whether these be economic or fancier's traits.

Many owners put distinctive marks, such as ear notches or brands on their animals, to avoid controversy and possible litigation. Unlike in modern systems of production where each animal is given a unique mark, the same mark is applied to every member of the flock. Brands are usually placed on the face or muzzle, but may also be made on the ears.

Breeding policy

Although there are apparently few breeding practices aimed at genetic improvement of stock, some owners do follow a breeding policy. There are risks of inbreeding in small flocks, but these risks are reduced where communal herding or grazing is the normal practice. Similarly, communal herding reduces the risk of an owner attaining a low lambing rate through not having a breeding ram. In parts of India farmers act positively to prevent the risks of inbreeding even when they herd flocks individually; they either purchase rams from, or exchange rams with, other flock owners (Acharya 1982).

Governments or development authorities have on numerous occasions attempted to improve either breeds or reproductive performance in traditional systems. These attempts may take the form of the distribution of rams, or the use of artificial insemination (AI). Both the Indian government (Acharya 1982) and the Ivory Coast government have adopted AI as a strategy for improvement. In the Ivory Coast, attempts are also being made to unbalance the sex ratio in favour of females by administering vitamin D to the ewe at the appropriate point in the oestrous cycle (Moussa Touré, personal communication).

In some areas control of breeding is practised – either seasonally or by preventing individuals from service. Arab and northern African groups influenced by Islam usually use the *kunan* (Fig. 2.5): this consists of a cord tied round the neck of the scrotum and looped over the prepuce to prevent extrusion of the penis. An analogous method is used in parts of India, particularly in Rajasthan and Gujarat, in which the prepuce is tied up with cotton tape (Acharya 1982). Using these devices a relatively small number of rams can be used on a large number of ewes by the owner restricting the number of services. Similarly, inbreeding can be prevented if the owner allows the rams to serve only those ewes to whom they are not closely related. Another form of contraceptive, used particularly by the Masai of East Africa is the apron (Fig. 2.6), traditionally of leather, but often nowadays part of a motor-car inner tube or a piece of plastic attached to the thorax of the ram and hanging down in front of the penis.

Inadvertent contraception may occur in some special cases. In particular, the fat tail of the Awassi ewe causes problems for rams of other breeds which do not have the behavioural adaptations of Awassi rams. Such exotic rams normally have to be assisted by the herder for mating to be successful.

Control of breeding by separation of the sexes into separate flocks is seldom done as this would lead to an unacceptable increase in the amount of labour required to control the animals. In particular, in agropastoral systems this could lead to a serious labour constraint at the time of land preparation for sowing or harvesting of crops.

Fig. 2.5 A Desert Sudanese ram with a *kunan* to prevent penetration (Photo: R. T. Wilson)

Intensification and stratification

The production goals of individual owners are rarely those for which the central government aims. The government wishes to provide cheap meat for the expanding urban population, and to export live animals or meat to improve the balance of foreign exchange. Generally government objectives are attained by the policy of designating certain zones as breeding areas, intermediate zones as growing-out areas, and zones better endowed either climatically or with feed as finishing areas. Therefore, any national or international aid and development devoted to the sheep industry has often stressed intensification and stratification.

Few schemes for intensification or stratification have ever got beyond the planning stage. As can be deduced from Table 2.3 and other information provided earlier in this chapter there is, particularly in traditional flocks, a rapid turnover of animals, especially males not required for breeding. This turnover is often related to local customs or beliefs. For example, throughout the Muslim world offtake is geared to the annual Idd el Adda, in Ethiopia to the main Orthodox Christian festivals, and in Nigeria to the times of year when migrant workers or civil servants return to their home areas *en masse*.

Most attempts at intensification have suffered from lack of adequate

Fig. 2.6 A contraceptive apron on a Masai ram (Photo: R. T. Wilson)

management ability, failure to assure a constant supply of food and inability to overcome disease problems associated with large numbers of sheep in confined spaces. Stratification for specific products has proved successful in some areas (e.g. wool from Merinos in Kenya, pelts from Karakuls in Namibia and South Africa), but has been a complete failure in others. Stratification along modern lines has failed even where different functions for different classes of stock exist traditionally. For example, in Mali the Fulani stratify their Macina wool sheep (Table 2.6) according to their function in the flock. Fifty years of attempts by the colonial powers to produce wool in the same area failed completely (Wilson 1981).

Such failures may have had some influence on the attitudes of post-colonial governments to the development of the small ruminant subsector, although obvious and continued failures in developing the cattle industry have not prevented the proliferation of schemes for the stratification and intensification of cattle production. The economics of stratification, while perhaps being attractive to individual producers keeping one or two animals, also need to be carefully considered. It has

Table 2.6 *Stratification of Macina flocks with demographic characteristics and management objectives of each*

Name	Group size	Use	Composition			Notes
			General	Males (%)	Females (%)	
Beydi	Generally small	Nurse flock	Newly lambed females, advanced pregnancy, weak and aged animals	26	74	Kept in village: herded by infants. Regular movements of animals into and out of group
Tarancaradji	Medium	Sale/slaughter	Largely male, generally young, with some older females	60	40	Kept in village, generally not herded
Njarniri	Small	Slaughter	Overwhelmingly male	95	5	Individually tied and zero grazed. Responsibility of women
Bucal	Medium	Milk	Predominantly female	25	75	Individual ownership but commonly grazed on reserved pastures by paid herder or by family labour in rotation. Household milk supply. Most of village goats are in this group
Bendi						Similar to Bucal. Term used mainly by hair sheep owners
Horey	Large	Wool/meat	Predominantly female	24	76	Main flocks which transhume. Reserve for constitution of other groups as required. Milked by herders as required

Source: From Wilson 1983.

recently been shown that even where there is an enormous but rigidly confined spatial and temporal demand for meat or live animals (e.g. at Idd el Adda), the financial risks are very high (De Boer 1983a,b).

Current management strategies are almost all geared to the early offtake of males not required for breeding. Most of these are slaughtered at or near the place of herding or sale. With existing technologies and veterinary services, removal of animals at young ages while still in the relatively fast-growing stage is sound policy. If death-rates, particularly pre-weaning (which typically may be 30%) can be reduced and feed supplies assured, then animals surplus to traditional market requirements might become available for intensification and stratification schemes on a larger scale.

Assuming that these conditions can be met, and that there is suitable land, and a ready market for the products of stratification – whether these be milk, meat or fibre – a general form of stratification as shown in Fig. 2.7 can be postulated. The scheme is fully described by Morag (1972). Briefly, it would consist of crossing hardy and fecund ewes of desert breeds with either mutton- or dairy-type rams. All lambs from the mutton cross, and male lambs from the dairy cross would be transferred to a fattening unit. Female lambs of the dairy cross would be used for milk production. Their own offspring, resulting from back-crossing to a mutton-type ram, would be transferred to a fattening unit, whatever their sex. At some stage it might be envisaged that the prolificacy of the foundation breed be increased by introducing genes from highly prolific breeds.

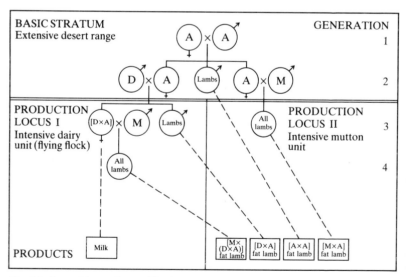

Fig. 2.7 General form of a stratification model from a basic sheep breed (A) crossed with dairy (D) and with mutton (M) type rams (After Morag 1972)

The stratification would depend on an adequate water supply at the fattening and milking units as well as readily and cheaply available quantities of concentrates and roughages. It is suggested that the system would be equally viable either as a smallholder enterprise or as an industrial concern.

Chapter 3

Health

Poor health is one factor limiting the productivity of sheep in the tropics. Animals are always subjected to disease challenges, and are never in a state of perfect health. It is important that poor health does not seriously limit the animals' chance of surviving and producing. Disease problems become of increasing economic importance as other problems (of nutrition, heat stress, poor management, etc.) are solved. Conversely, other problems become more important as disease problems are solved.

An economic assessment of the importance of diseases of sheep in the tropics is not the same as their clinical importance. Economic assessment may be based either on the losses in direct and indirect production that result, on the cost of veterinary and other measures taken to control the disease or treat infected animals, or on the cost of the damage that the disease would cause if it were not prevented or controlled. It is extremely difficult to get quantitiative estimates of the economics of diseases of sheep in the tropics according to any of these definitions. The few estimates that have been made (e.g. Seneviratna 1967) suggest that the economic losses due to disease are substantial.

Some diseases are of much more economic importance than others. A survey instigated by the author among veterinarians working in many tropical countries showed that they felt that the diseases causing the greatest economic losses to sheep production included helminthiasis, foot-rot, orf, external parasites especially mange, enterotoxaemia, pneumonia, pregnancy toxaemia, bluetongue, streptothricosis, PPR and caseous lymphadenitis. These eleven diseases are briefly described. Numerous other diseases occur and also cause economic losses. Details of diseases are given in textbooks such as Robertson (1976), Hall (1977), Blood *et al.* (1979) and Soulsby (1982).

1. Helminthiasis

Gastrointestinal roundworms

Roundworms (also known as nematodes) living in the digestive tract of sheep may be an important cause of mortality, as reported in India by

Minett (1950), but more often they depress productivity (e.g. Hunter and Heath 1984). Several genera are important, including *Haemonchus, Trichostrongylus, Oesophagostomum, Bunostomum* and *Gaigeria*. In the hot tropics *Haemonchus* spp, are the most serious parasites of sheep; other genera are more important in the sub-tropics and highlands.

All these worms have a life-cycle in which eggs are excreted from the sheep in faeces. The eggs hatch and develop through three larval stages on the pasture. Sheep eat the larvae together with vegetation. The larvae develop into egg-laying adults in the digestive tract of the sheep, and this takes 2–7 weeks depending on the worm species. The time between the deposition of eggs on the pasture and the presence of infective larvae is more variable. It depends on the temperature of the environment, the type of vegetation and the species of larvae. In favourable conditions the period on the vegetation may be only a few days, but it may be much longer.

Roundworms reduce the voluntary food intake of sheep and reduce their food-conversion efficiency. Internal parasites seriously affect post-absorptive metabolism, but surprisingly malabsorption is usually not important as there are compensatory increases lower down the digestive tract (Anderson 1982). Most sheep carry worms, but worms cause economic losses only when the infestation is severe. For instance, Darmono (1982) reported that 85% of sheep slaughtered at Bogor slaughter-house in Indonesia were infected with *Haemonchus contortus*. The concentration of infective larvae on the pasture is a major factor affecting the worm burden, and this depends on the number of eggs deposited in the pasture and on their survival. In addition, the susceptibility of the sheep affects the production losses. Young sheep are in general more susceptible than older animals which develop immunity. Some breeds are more susceptible than others. For example, Owen (1982) reported that in Papua New Guinea, Priangans are more resistant than Corriedales.

Roundworms can be controlled by grazing management to ensure that susceptible animals are not grazed on pasture containing high levels of infective larvae, and by anthelmintic drugs. Keeping sheep in stalls or tethered on bare ground (with faeces removed frequently) and carrying food to them ensures that they are subjected to few worms. Under some circumstances, nomadism and transhumance reduce the worm burden as the sheep always graze fresh pasture, but this will not be beneficial unless the period between the first and last flocks moving over a particular area is substantially longer than the survival of infective larvae.

In ranching systems roundworms are a serious problem particularly if the stocking rate is high. It is possible to keep the parasite burden low by grazing the area in rotation, allowing the larvae to develop on the ungrazed areas and their numbers to decline before that area is grazed again. By incorporating forage conservation into the grazing pattern, a clean break is introduced. The regrowth of pasture after cutting is practically free from larvae and is suitable for the most susceptible sheep, i.e. lambs which have recently been weaned. Similarly, crop stubbles provide clean grazing. Surprisingly few studies have been conducted on the longevity of eggs and larvae on pastures in different tropical climates, or on the effect of grazing management on helminthiasis in sheep in the tropics.

By grazing different species of animal in rotation or even together, the worm burden is reduced because worms are relatively species-specific. For instance few worms which infest cattle also affect sheep. However, sheep and goats share most species of worm.

Where sheep are grazed continuously, particularly on communal land, there is little chance of controlling the infection of the pasture. Apart from resorting to anthelmintics the only way to prevent worms becoming a serious problem in young sheep is to confine them where the eggs cannot develop into infective larvae, and feed them with clean vegetation. In the dry tropics pastures are sterilised during the long dry season and the worm burden is reduced, but in the humid tropics, conditions throughout the year are favourable for the development of larvae on pasture.

A list of anthelmintic componds which control roundworms is given in Table 3.1. Many anthelmintics are expensive and not available to most livestock owners in the tropics. Therefore old-fashioned anthelmintics such as carbon tetrachloride are more widely used than modern anthelmintics. Resistance of some species to certain modern anthelmintics has developed in Australia, but the importance of this resistance in most countries in the tropics and sub-tropics is probably minimal. At the beginning of the rainy season, inhibited larvae in the wall of the abomasum are an important source of infection. Many of the newer anthelmintics are able to prevent these inhibited larvae from maturing, and thus prevent the rise in infection which otherwise occurs soon after the onset of the rains.

Treatment with anthelmintics should preferably be carried out in conjunction with control of grazing, otherwise treatment must be repeated every few weeks. In a controlled system, for instance, lambs are dosed at weaning then moved on to clean pasture. Where sheep are stall-fed, disophenol injected every 3 months can keep the numbers of *H. contortus* in the faeces down to low levels (Soetedjo *et al.* 1980). Substantial gains in productivity result from the treatment of sheep with anthelmintics, but the cost of treatment is very high. The control of helminths largely by sensible grazing management is therefore preferable.

Lung worms

The most important roundworms living in the lungs of sheep are *Dictyocaulus filaria*. Like gastrointestinal worms, they have a life-cycle

Table 3.1 List of chemicals used as anthelmintics for sheep, and examples of products produced by international companies. The list of trade names is not complete, and omission of any particular name does not imply that the product is not recommended

Chemical name	Trade name	Company	Helminths controlled*
Benzimidazoles and pro-benzimidazoles			
Albendazole	Valbazen	SmithKline	GLTF
Febantel	Bayverm	Bayer	GTL
Febantel	Amatron	ICI	GTL
Fenbendazole	Panacur	Hoechst	GTL
Mebendazole	Ovitelmin	Crown Chemical Co.	GTL
Oxfendazole	Synanthic	Syntex Agribusiness	GTL
Oxfendazole	Systamex	Wellcome	GTL
Oxibendazole	Anthelworm	R. Young & CO.	G
Oxibendazole	Widespec	Rycovet	G
Oxibendazole + Oxyclozanide	Duospec	Rycovet	GF
Parbendazole	Helmatac	SmithKline	G
Parbendazole	Topclip wormer	Ciba-Geigy	G
Parbendazole	Triban	Crown Chemical Co.	G
Thiabendazole	Thibenzole	Merck, Sharp & Dohme	GL
Thiophanate	Day's worm drench	Day & Co.	G(L)
Thiophanate	Nemafax	May & Baker	G(L)
Thiophanate + Brotianide	Flukombin	Bayer	GF
Thiophanate + Brotianide	Vermadex	May & Baker	GF
Carbon tetrachloride	—	—	BF(H)
Diamphenethide	Coriban	Wellcome	F
Diethylcarbamazine	Caulicide	Bimeda	L
Diethylcarbamazine	Dicarocide Forte	Willows Francis	L
Diethylcarbamazine	Husk injection	Crown Chemical Co.	L
Diethylcarbamazine	Superil	Crown Chemical Co.	L
Haloxon	—	—	G
Ivermectin	Ivomec	Merck, Sharp & Dohme	GL
Levamisole	Cyverm	Cyanamid	GL
Levamisole	Nemicide	ICI	GL
Levamisole	Nilverm	ICI	GL
Levamisole	Ripercol	Crown Chemical Co.	GL
Levamisole + Clostridial vaccine	Nilvax	ICI	GL
Levamisole + Oxyclozanide	Nilzan	ICI	GFL
Morantel	Paratect	Pfizer	G
Niclosamide	Mansonil	Bayer	T
Nitroxynil	Trodax	May & Baker	FH
Oxyclozanide	Flukol	R. Young & Co.	F
Oxyclozanide	Zanil	ICI	F
Phenothiazine	—	—	G
Rafoxanide	Flukanide	Merck, Sharp & Dohme	FH
Rafoxanide + Thiabendazole	Ranizole	Merck, Sharp & Dohme	GFL

* Helminths controlled – G: gastrointestinal roundworms (H: haemonchus only, B: hookworm only); T: tapeworms; F: liver fluke; L: lungworms.

which takes place partly in the sheep and partly on pasture. The worms in the bronchi lay eggs which are swallowed; these hatch in the small intestine and the larvae are passed out in the faeces. In the pasture the larvae develop into infective forms within a few days under optimal conditions. When these larvae are ingested by a new host they penetrate the wall of the gut and migrate via the lymphatic circulation to the lungs where they move from the alveoli up into the bronchi.

The characteristic symptoms of dictyocauliasis are coughing and poor body condition. Dictyocauliasis is found in the cooler parts of the tropics, but is also common in the Near East. Sheep lose most of the worms within 2 months of infection, and then are resistant to reinfection. Therefore the most serious problem occurs in lambs.

Liver fluke

Fascioliasis is a disease caused by infection with *Fasciola gigantica* or, in South America, Australia, tropical highlands and temperate regions, *F. hepatica*. It is particularly important in the wetter areas of Africa and Asia, although it is also found in the dry tropics if suitable habitats are present. Fascioliasis affects all types of ruminant; acute fascioliasis results in death but chronic fascioliasis resulting in poor body condition is more common.

The life-cycle of the fluke involves a snail (*Lymnaea* spp.) as a secondary host. The snail host for *F. gigantica* requires well-oxygenated permanent water. Fascioliasis may therefore be controlled by drainage. Under field conditions molluscicides rarely kill all the snails (Robertson 1976). Biological control by fish, ducks and other species is being investigated. Modern anthelmintics are effective against most stages of fluke in sheep, but are very expensive. Carbon tetrachloride is reasonably effective against adult fluke and is still widely used.

Dicrocoelium spp. are another type of fluke which infest the livers of sheep. They are found in dry climates, having a land snail and an ant as intermediate hosts. Most infections are sub-clinical and the largest numbers of flukes are found in old animals. Control is difficult but *Dicrocoelium* spp. can be treated successfully with anthelmintics.

Tapeworms

Sheep are hosts to several species of tapeworm. They acquire infection in a variety of ways. Adult worms may grow to a few metres long in the intestines and segments of worm can be seen in the sheep's faeces. The effect of most tapeworms on productivity is surprisingly small considering their dramatic appearance. One exception is *Taenia multiceps* whose larvae migrate to the brain and spinal cord of sheep and cause nervous dysfunction known as sturdy or gid. Dogs are the final hosts of *T. multiceps* and become infected by eating the worm cysts in dead sheep.

The total losses from all types of helminth are very large. Herlich (1978) estimated that the world-wide annual mortality from helminthiasis is about 30 million sheep and goats. The number of sheep and goats suffering morbidity as a result of helminth infestations must be substantially greater. Losses are probably proportionately worse in the tropics where nutrition is poor. When sheep are severely undernourished after a long dry season, helminthiasis causes high mortality rates at the beginning of the rains (Vassiliades 1981). A high plane of nutrition greatly reduces the mortality from helminthiasis (Sharma and Kidwai 1971). Good nutrition does not necessarily prevent the establishment of the infestation, but may mitigate its effects (Gordon 1960).

2. Foot-rot

Foot-rot is caused by a mixed infection of *Bacteroides nodosus* (otherwise known as *Fusiformis nodosus* or *Sphaerophorus nodosus*) and *Spirochaeta pernotha*. *Fusiformis necrophorus* has also been implicated, and it is possible that other organisms may also be involved in ovine foot-rot in some parts of the world. The infective organisms live in the foot under the hoof. They cause decay which has a characteristic smell, and the hoof becomes overgrown as the sheep does not put its full weight on the foot.

The symptom of foot-rot is lameness and the sheep may even graze in a kneeling position. Badly affected sheep lose body condition. Foot-rot is worst in damp areas (Fig. 3.1), but occurs throughout the world. Attempts to evaluate the production losses resulting from foot-rot have been made in temperate countries (e.g. Littlejohn 1964) but not in the tropics. In the hills of Nepal foot-rot was first observed in about 1976 (LAC 1978), and is thought to have been introduced with the importation of exotic sheep. Since its introduction into Nepal, foot-rot has resulted in serious losses in transhumant flocks as the sheep become unable to graze properly, lose body condition and either become unproductive or die.

Foot-rot is treated by cutting away the overgrown hoof and sole and applying an antibiotic. Badly infected animals should be isolated and treated regularly before they are allowed to rejoin the flock. Sheep in an infected flock which are not lame can be protected from foot-rot by slowly walking through a foot-bath containing a solution of either formalin or copper sulphate and then putting them on new pasture. Although some resistant forms of the causal micro-organism may remain in the soil for several years, removing sheep from a pasture for only 2 weeks removes most of the micro-organisms. A vaccine against *B.*

Fig. 3.1 Sheep on wet ground conducive to the spread of foot-rot

nodosus has been developed, but the duration of resistance to infection that it gives may be only a few weeks (Skerman *et al.* 1982).

3. Orf

Orf, or contagious ecthyma, is caused by a virus and is seen as scabby lesions on the lips, muzzle and legs of lambs. Lesions are also found on the udders of lactating ewes, and these may result in mastitis, with the effect that the ewe may be culled. Orf occurs world-wide in both sheep and goats, rarely resulting in mortality, but reducing the productivity of individuals. As orf is endemic to most flocks it is difficult to prevent. Live vaccines are available.

4. External parasites

External parasites of sheep include ticks, mites, lice and flies. They cause loss of productivity in many ways: by interfering with feeding through annoyance, etc.; by causing wounds which allow diseases such as streptothricosis to enter; by damaging the skin so that its value after slaughter is low; and by acting as disease vectors. External parasites cause little loss in small flocks of sheep which are handled regularly because the shepherd removes the parasites. They are a much more serious problem in extensive systems and large intensive units. The survival of parasites is affected by humidity, and they are more numerous in the wet season than in the dry season.

Mange is caused by mites, and may be either sarcoptic, psoroptic, chorioptic or demodectic according to the species of mite. All stages of the life-cycle of the mite (egg, larva, nymph and adult) remain on the sheep, feeding on the skin surface or in some cases burrowing beneath the epidermis to the hair follicles. The life-cycle is only 10-15 days so that large populations build up rapidly. Mites spread by direct contact and do not survive long away from the host. Head mange is a common problem in arid areas. In general, mange is more of a problem in goats and sheep, but Okoh and Gadzama (1982) reported that the morbitity rate of sheep in an outbreak of sarcoptic mange in Plateau state, Nigeria, was 60%. Mange is controlled by dipping. Psoroptic mange is known as sheep scab, and infected sheep are seen to scratch and bite their skin. Cutaneous lesions ooze serum which binds the fibres of the coat together.

5. Enterotoxaemia

Enterotoxaemia is the name given to diseases in which there is rapid multiplication within the gut of *Clostridium perfringens* (of differing types). All are characterised by rapid death. It includes pulpy kidney, struck and lamb dysentery. Vaccines are available.

Pulpy kidney occurs world-wide in sheep of all ages, but most commonly in lambs aged 3-12 weeks. It is especially common where lambs have been moved from poor pasture to a good diet. It gets its name from the decomposition of the kidneys which takes place. Struck or haemorrhagic enterotoxaemia is a similar disease in adult sheep found mainly in the USA and the UK.

Lamb dysentery affects lambs in the first few weeks of life, most commonly strong lambs 2-5 days old. Lambs are seen in acute pain, have diarrhoea and soon die. Lamb dysentery occurs mainly in Europe and South Africa, but is likely to become more important as intensive systems are introduced into less developed countries.

6. Pneumonia

Pneumonia is a term which refers to inflammation of the lungs, and may be caused by bacteria, viruses, mechanical factors or a combination of these. Enzootic pneumonia (also known as acute exudative pneumonia, shipping fever or pasteurellosis) is caused by a massive infection of the lungs by *Pasteurella haemolytica* or *P. multocida*. It occurs throughout the world and in sheep of all ages, but particularly in lambs a few months old. The first indication of enzootic pneumonia in a flock may be the death of some individuals and the heavy breathing of others. *Pasteurella haemolytica* is present in healthy sheep, and causes disease only when the animal's resistance is weakened by stress such as transportation or wind and rain coupled with undernutrition. Vaccines against *P. haemolytica* are available, but their value is not yet fully established (Barlow 1982).

7. Pregnancy toxaemia

Pregnancy toxaemia is also known as twin lamb disease or ketosis. It is a disease which occurs in ewes in late pregnancy and results from metabolic stress. The early signs of pregnancy toxaemia are an aimless attitude and loss of appetite. Later the ewe collapses, goes into a coma and dies. Body temperature is never raised. If the ewe aborts she may survive.

Treatment is often difficult. The ewe should be given access to good-quality food and shelter. In the early stages she may respond favourably to dosing with glycerol, glucose or another form of sugar. Some authors advocate the removal of lambs by Caesarean section, but this is unlikely to be justified in the tropics.

Good nutrition is the key to the prevention of pregnancy toxaemia. A good plane of nutrition is necessary in late pregnancy but overfat ewes also suffer from pregnancy toxaemia because fatness reduces food intake. The importance of pregnancy toxaemia in traditional systems is not known, but it is likely to increase with intensification.

8. Bluetongue

Bluetongue is a disease found only in sheep (although it is widespread as a silent infection in cattle), and is usually limited to exotic and crossbred animals. It is common in Africa and the Near East, but outbreaks have been reported in other parts of the world (e.g. Lonkar *et al.* 1983). Bluetongue is caused by a virus transmitted by midges and so appears soon after the beginning of the rainy season. Infected animals suffer fever, nasal discharge, mouth lesions and lameness. Animals may die from the virus itself or from secondary infections, particularly pneumonia. Annual vaccination against the appropriate group of strains before the rains begin is satisfactory in controlling bluetongue. However, ewes should not be vaccinated in the 3 weeks before mating or in early pregnancy.

9. Streptothricosis

Streptothricosis or dermatophilus infection causes skin lesions mostly along the spine, and can result in 'lumpy wool disease' in wool sheep. It is caused by the bacterium *Dermatophilus congolensis,* although the skin must be damaged before streptothricosis develops. Streptothricosis affects many species of animal, and is most important in West Africa, but it is also found in other parts of Africa and in Australia, South America and Asia (Lloyd and Sellers 1976). The highest incidence occurs in the rainy season. Streptothricosis can be treated by removal of the scabs and application of an antibiotic, or alternatively by injection of antibiotic. Many other treatments such as dipping the whole lamb in zinc sulphate, have been used, but their efficacy has not been evaluated scientifically (Hyslop 1980).

Streptothricosis is considered a serious disease because of its detrimental effect on skin quality. When infected skins are tanned they show a strongly pitted grain and so have a low value. In a survey in East Africa, Bwangamoi and DeMartini (1970) found that 2.2% of sheepskins were affected by streptothricosis. In addition, streptothricosis lesions increase the susceptibility of wooled sheep to blowfly strike (Gherardi et al. 1981).

10. PPR

The disease PPR (peste des petits ruminants) or kata is endemic in West Africa. It is found only in goats and sheep, and is similar to rinderpest in cattle (ILCA 1979b). Infected animals suffer fever, mucosal erosions and diarrhoea, and often die. Animals that survive usually develop an exudative labial dermatitis and then scabs form. Usually, PPR occurs in cycles with major outbreaks every 1–3 years (Mack 1982), which are set off by newly purchased animals. Rinderpest vaccine protects small ruminants against PPR.

11. Caseous lymphadenitis

Caseous lymphadenitis is a chronic disease of sheep in which abscesses are formed in the lymph nodes by *Corynebacterium pseudotuberculosis.* The abscesses are filled with greenish-yellow pus which in the early stages is soft, but later becomes dry. The general health of the sheep is not affected unless the disease becomes generalised, spreading to the lungs and other internal organs. However, carcasses may be condemned in slaughter-houses because of the presence of the abscesses. The organism is susceptible to penicillin, but treatment is rarely attempted. There is no satisfactory vaccine.

12. Other diseases

There are many other diseases which may be of importance – specifically pox, brucellosis, heartwater and coccidiosis. Infertility, abortion and neonatal deaths caused by a variety of factors are discussed in Chapter 6. Malnutrition and bad management are often serious contributory factors if not the main causes of poor health.

Mortality and morbidity rates

Although some mortality and morbidity rates are published in the literature (and summarised in *Animal Disease Occurrence* published by the Commonwealth Agricultural Bureaux) few have been obtained in the field. For the humid zone of West Africa, ILCA (1979b) cite estimates of the mortality of adult sheep of between 5 and 35%. The mortality rate of lambs is usually higher than that of adults. The morbidity rate (the proportion of animals which are diseased) is probably more important in determining the economic losses, but is less often recorded. Mortality and morbidity rates vary substantially from year to year and from season to season. The survival of disease microorganisms is closely correlated with ambient humidity, so that most diseases are more prevalent in the wet season that in the dry season (e.g. Smith and Olubunmi 1983).

Veterinary education and services

Effective veterinary education and services are essential if optimal sheep health is to be attained. Veterinary education ranges from university or college training for veterinarians, through short courses for veterinary assistants to extension sessions for shepherds. It is important that each person receives training appropriate for his level of understanding and future work. Present veterinary courses in universities are criticised for being too narrow in outlook, and should contain aspects of sociology and economics as well as animal health and production (Moor 1975; Campbell 1976). In addition to being suitably trained, personnel must be motivated (financially or otherwise) to put into practice what they have learned. Too often veterinarians are relegated to desk work and are reluctant to undertake field posts because these are not looked upon favourably with respect to promotion in the future (Harding 1981). Several countries have trained veterinary assistants ('barefoot vets') to undertake routine field-work and relieve the shortage of qualified veterinarians.

In most less developed countries, veterinary services are run and subsidised by the state. Ellis (1974) discusses the structure of comprehensive veterinary services. A national service should allow cooperation between veterinarians in different parts of the country in controlling or eradicating disease. In some cases international cooperation to combat disease and share technology is desirable (FAO 1966).

In all parts of the tropics, sheep are also treated by 'non-qualified' people using traditional practices. Nwude and Ibrahim (1980) listed over forty herbal remedies used for sheep in Nigeria. Examples include *Coleus dazo* leaves pounded and applied to an inflamed eye, *Erythrina senegalensis* bark boiled in water and used as a drench against fascioliasis, and *Hibiscus cannabicus* flowers fed to ewes suffering dystocia or a retained placenta. The concepts of animal disease and treatment traditionally held by the herdsman are often startlingly close to the scientific view (Ibrahim *et al.* 1983). These methods are mostly not documented in veterinary literature, although some are related in more general literature (e.g. ICAR 1964). Such practices are often regarded by scientists as mere superstition, but it is likely that there is a scientific reason why some traditional medicines are successful, and a scientific investigation may provide veterinarians with alternative (and cheaper) methods of treatment.

Effect of sheep-production systems on health

Sheep-production systems can be classified into migratory, smallholder mixed farming and large-scale systems. The health problems of each type of system are very different, as are the veterinary requirements. The following discussion was derived largely from a document produced by ILCA (1979c) which concentrated mainly on cattle.

1. In dry areas with nomadic and other migratory flocks the transmission of contagious diseases such as pox is favoured by the flocks moving over large areas and mixing with other flocks at watering-places and on dry-season grazings. These diseases occur in epizootics and may cause high mortality. Diseases which impair the productivity of individuals are of less importance. On the whole, disease is of minor importance compared with other constraints such as scarcity of food or water.

    Veterinary services are normally at a very low density in dry areas. They are organised as mobile teams and most of their work consists of mass vaccination to reduce the risk of high mortality caused by disease. Sandford (1983) discusses the problems of providing veterinary services for pastoralists. Veterinary inputs into migratory systems on marginal land have resulted in the rapid expansion of flocks until the pasture is said to be overstocked and overgrazed by livestock and overpopulated by people, so that in a drought famine is severe and the ability of the pasture to recover is impaired.

2. In smallholder systems sheep are usually only a small part of a mixed farming system. There are large numbers of small isolated flocks, and infectious diseases tend to smoulder continuously through the population rather than cause epidemics.

The veterinary input into these systems tends to be restricted, consisting mainly of mass treatments (e.g. vaccination campaigns) and veterinary posts carrying out extension activities, selling drugs and treating individual animals. Work in smallholder communities tends to be time consuming because of the need to deal with large numbers of owners. The low response of sheep production to veterinary services is limited by the low level of other inputs into the system. Also the actual benefits to individual sheep owners depend to a large extent on the actions taken by the surrounding sheep owners. Sheep have a low priority for inputs in mixed smallholder systems and all resources are so limited that villagers often fail to take even the simplest 'rational' preventive veterinary measures (McCorkle and Jimenez-Zamalloa 1982).

3. Large-scale ranches and feedlots are relatively independent of other flocks in the surrounding area so that contagious disease is a less serious problem than in other livestock systems, although high stocking rates can favour the incidence of internal parasites. Sheep entering these largescale units can introduce many diseases as they probably come from several different areas and are then trekked over long distances.

A wide range of veterinary activities is normally carried out on ranches. Most activities are performed by the ranch staff themselves. On feedlots the veterinary service provided is usually kept to a minimum by treating store lambs prophylactically before they join the main flock of sheep on the feedlot. Under normal circumstances there is little need for treatment during the short finishing period.

Environmental and other stresses are important in initiating disease and reducing production (Jericho *et al.* 1974). Intensification changes the types of disease and unless management is good, intensification increases the health problems.

Economics of animal health programmes

An economic assessment of animal health programmes is outlined by Ellis and James (1979). It comprises estimation of the costs and benefits of disease control, taking into account social aspects such as public health as well as direct financial aspects. There have been few assessments of the economics of animal disease, and none concentrating on sheep in the tropics. Morris and Meek (1980) discussed the measurement and evaluation of the economic effects of parasitic diseases on sheep in Australia. From the few relevant studies, some general principles emerge.

Maximum productivity in a given system of production emerges when disease control is optimal. However, the best financial returns result when disease control is less than this optimal value, at a level where the cost of the last veterinary input equals the marginal improvement in productivity that it gives (Morris 1969). In practice in less developed countries, the veterinary resources available are limited, and the question becomes one of getting the best from these limited resources, i.e. in deciding which approach will give the highest marginal returns. Information is needed on the response of sheep production to different treatments or courses of action.

Some approaches to disease control which are relatively inexpensive give good prevention. For instance, international quarantine regulations for islands minimise the risk of introducing serious epizootic diseases (Sudiana *et al.* 1981). If a good vaccine is available, prophylactic measures are more effective than treatment of diseased animals. Genetic resistance to disease is widespread and found for virtually all diseases in which it has been sought (Spooner 1982; Gavora and Spencer 1983). For instance, some sheep are more resistant to helminthiasis than others (Whitlock 1955; Altaif and Dargie 1978, Assoku 1981), Blackhead Persian and Masai sheep are more tolerant of trypanosomiasis than Merino sheep (Griffin and Allonby 1979), and fleece rot in Merino sheep can be reduced by selection (Atkins *et al.* 1980). Genetic resistance to disease is particularly important where there is no vaccine against a disease, or where treatment is expensive, difficult or becoming unsuccessful because of drug resistance.

Chapter 4

Nutrition

Digestive physiology

Sheep, in common with other ruminants (cattle, goats, buffaloes and camels) have four stomachs. Food ingested passes with saliva down the oesophagus into the first stomach known as the rumen, where it is broken down by micro-organisms. After a period of eating, a sheep ruminates; food from the rumen returns to the mouth, is chewed again, and is then reswallowed. This remasticated food is further digested in the rumen and reticulum, and some absorption takes place. When food passes into the third stomach (omasum), much water is removed from it. In the fourth stomach (the abomasum), the food is acidified and further broken down by enzymatic secretions. After leaving the stomachs, food passes through the small intestine, where more nutrients are absorbed, through the large intestine, and is finally voided as faeces.

Because of the nature of its digestive system, a sheep can digest cellulose, and thus can live on a diet of low-quality roughage which could not support a monogastric animal. Digestion in the sheep is essentially a two-stage process: firstly, food is digested by micro-organisms in the rumen, and secondly, there is further digestion and absorption in the abomasum. Each digestion process is an inefficient conversion of substrate into product so that the overall efficiency of food conversion is usually lower for a ruminant than for a monogastric animal.

The type of micro-organisms in the rumen and reticulum of a sheep depend on its diet. Micro-organisms are unable to cope with large quantities of unfamiliar foods, so that if the diet of a ruminant is changed, it must be done slowly to prevent digestive upset.

Young lambs, like the new-born of other ruminants, have a digestive system in which the rumen does not function. The physiology and function of the digestive tract of lambs is similar to that of calves which is reviewed by Roy (1980). In the very young lamb, liquid food moves directly from the oesophagus to the abomasum along a tube formed by the closure of the oesophageal groove, and thus bypasses the rumen. Therefore the young lamb needs a liquid feed digestible by the abomasum and intestines, i.e. milk proteins, fat and the sugars lactose and glucose. Digestion of sucrose and starch in the young lamb is limited.

When a lamb nibbles at solid foods these pass into the rumen where micro-organisms become established and rumen fermentation begins. Gradually the lamb obtains more of its nutrient requirements from the products of rumination, but it will continue to obtain milk from its dam for several months unless it is weaned.

Dentition

An adult sheep with a complete set of teeth has, on its lower jaw, three pairs of incisor teeth and one pair of canine teeth. Usually all these four pairs are referred to as incisors. There are no incisor teeth on the upper jaw, but there is a horny 'dental pad' against which the lower incisors bite. The adult sheep also has, on both its upper and lower jaws, three pairs of premolars and three pairs of molars.

A new-born lamb has few, if any, teeth. Its teeth erupt at intervals in the first weeks of life and comprise four pairs of incisors on the lower jaw, and three pairs of premolars on both jaws. This first set of teeth are known as milk teeth. As the lamb matures, its milk teeth are replaced by larger, permanent teeth. The first (permanent) molar teeth erupt on the lower jaw at an age of about 3 months. These are followed by the first molars on the upper jaw, and the second and third sets of molars, with all twelve molars erupted by the age of about 2 years. The permanent premolars begin to erupt at about 18 months, and are complete by 2 years.

The incisor teeth are easy to see, and the age at which they are replaced is reasonably constant for any breed of sheep in any environment. The number of permanent incisor teeth is therefore used to give a rough estimate of the age of a sheep. In general the first pair of incisors erupts at about 14 months of age, and the final (fourth) pair at about 3 years. The mean ages at eruption recorded for Blackhead Persians and Dorpers in Rhodesia, for Sahel sheep in Mali and for sheep in the UK are shown in Table 4.1. These values show that the age of eruption depends on the breed of sheep. The teeth of the Blackhead Persians, indigenous to Africa, appeared earlier than those of the crossbred Dorpers. The values obtained in the UK were similar to those of the African sheep. However, Sisson (1953) writing in the USA reported older ages of eruption (3 years 6 months to 4 years 0 month for the fourth pair of incisors). Wiener and Purser (1957) and Arrowsmith *et al.* (1974) demonstrated that feeding a better-quality diet reduces the age at eruption of incisor teeth. Similarly, Wilson and Durkin (1984) found that the teeth of lambs in agro-pastoral rice systems erupted earlier than those of lambs in millet systems in which the nutrition was poorer.

Table 4.1 Mean ages at which permanent incisor teeth erupt (years:months)

		Blackhead* Persian	Dorper*	Sahel†	UK‡
Incisors	1	1:2	1:3	1:3	1:0–1:3
	2	1:9	2:0	1:10	1:9
	3	2:3	2:7	2:4	2:3
	4	2:8	3:2	3:2	2:9–3:0
Molars	1	—	—	—	0:3
	2	—	—	—	0:9
	3	—	—	—	1:6
Premolars	1	—	—	—	1:9
	2	—	—	—	1:9
	3	—	—	—	2:0

Sources: *From Arrowsmith *et al.* 1974; †From Wilson and Durkin 1984; ‡From Miller and Robertson 1945.

Sheep older than about 4 years begin to lose their teeth, and are said to become broken-mouthed. Although the rate at which teeth are worn down depends on the diet, there appears to be little relationship between diet or breed and the age at which teeth are lost. If a sheep is supplied with an adequate diet and does not need to bite coarse vegetation, the loss of teeth is of little consequence. On the other hand, if a sheep is given poor-quality grazing only, the loss of teeth means that it is unable to satisfy its food requirements, its body condition declines and (assuming it is a ewe) it fails to conceive.

Nutritive requirements

The nutritive requirements of sheep can be conveniently considered as requirements for the essential components of food (energy, protein, minerals, vitamins, free fatty acids and water) within the limits of food intake. The ways in which these requirements can be met are discussed in Chapter 5.

Food intake

Most sheep are fed *ad libitum* so that the intake of nutrients is limited by voluntary food intake. The voluntary intakes of foods commonly used for sheep in the UK have been collated by the Agricultural Research Council (ARC 1980) and values for a 20 and a 40 kg sheep are given in Table 4.2. Intake increases with animal size (and weight), but intake per unit weight decreases as size increases. It is therefore convenient to express intake as dry matter (DM) consumed divided by metabolic weight, i.e. g $kg^{-0.75}$, a measure which is less dependent on animal weight. Typically the intake of a coarse diet with an energy density (M/D) of

Table 4.2 Dry-matter intake of growing sheep, in $g\ kg^{-0.75}\ d^{-1}$

Diet	M/D (MJ kg^{-1} DM)	20 kg sheep	40 kg sheep
Coarse	7.0	31	37
	8.0	37	43
	9.0	42	49
	10.0	48	54
	11.0	54	60
	12.0	59	66
	13.0	65	72
Fine	9.0	103	96
	10.0	99	91
	11.0	95	87
	12.0	91	83
	13.0	87	79

9 MJ kg^{-1} DM is 45 g kg$^{-0.75}$, and the intake of a fine diet with an M/D of 9 MJ kg^{-1} DM is 100 g kg$^{-0.75}$.

The intake of fine foods such as concentrates or ground roughages is considerably greater than that of coarse foods with or without concentrates. According to the data given by the ARC (1980), the type of diet has a considerably greater effect on the intake of small sheep (20 kg) than larger sheep (40 kg). However, in the UK data, size is confounded with age and it seems likely that age as well as size affects intake. Older sheep are better able to cope with coarse diets than lambs.

The M/D of foods affects their voluntary intake, but the direction of this response depends on the type of food. For coarse foods an increase in M/D is associated with an increase in intake. However, for fine foods an increased M/D gives a decreased food intake.

Figure 4.1 shows graphically how the intake of metabolisable energy by a 20 kg sheep eating a coarse diet changes with the M/D of the diet. The relationship was calculated from the data given in Table 4.2. The dotted line in Fig. 4.1 also shows the maintenance requirement for energy (calculated according to eqn [4.1] (p. 61)). For simplicity, the maintenance requirement is assumed to be constant, though in practice it declines slightly with increasing M/D. The two lines meet when M/D is about 9.2 MJ kg^{-1}, and at this point the ME intake is just sufficient to satisfy the sheep's maintenance requirements for energy. For diets with higher M/D values there is surplus energy which may be used for production: the amount of surplus energy corresponds to the shaded area on the graph. The ME available for production increases with increasing energy density until, when M/D is 13 MJ kg^{-1}, it is approximately equal to that used for maintenance. The intake of diets with energy densities lower than 9.2 MJ kg^{-1} is insufficient to meet even the maintenance requirements for energy. This deficit must be made up by utilizing body tissues. For a 40 kg sheep consuming a coarse diet a

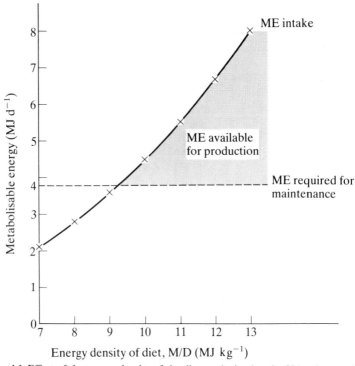

Fig. 4.1 Effect of the energy density of the diet on the intake of a 20 kg sheep eating a coarse diet, and its relationship with the metabolisable energy requirement for maintenance

similar picture is seen, but the critical M/D value for maintenance is somewhat lower at about 8.2 MJ kg^{-1}.

Values of food intake cited in Table 4.2 vary from 31 g kg$^{-0.75}$ d^{-1} for a 20 kg sheep consuming a coarse diet with a low M/D to 103 g kg$^{-0.75}$ d^{-1} when eating a fine diet with an M/D of 9 MJ kg^{-1}. These figures indicate that the voluntary food intake of a 20 kg sheep is between 290 and 970 g d^{-1}. The upper end of this range is consistent with the daily intake of 4–5% of liveweight which would be expected for a sheep on a good-quality diet, and on a moderate-quality diet, intake is about 3.5% of body weight (e.g. Maheswari and Talapatra 1967). The data of Brinckman (1981) for concentrate feeding of sheep in Nigeria show similar intakes (between 89 and 104 g kg$^{-0.75}$ d^{-1}). Those of Combellas and González (1972) for sheep eating buffel grass in Venezuela are between 70 and 80 g kg$^{-0.75}$ d^{-1}. The intake of Merino rams eating berseem in Egypt was 64 g kg$^{-0.75}$ d^{-1}, while that for Ossimi and Rahmani rams was about 54 g kg$^{-0.75}$ d^{-1} (Khalil and Morad 1977). For Dorper sheep in Zambia eating poor-quality grass, intake was 35 g kg$^{-0.75}$ d^{-1} (Gihad 1976b). Intakes as low as 31 g kg$^{-0.75}$ d^{-1} rarely occur in practice because extremely low intakes of low-quality forage are inadequate for maintenance.

The food intake of a sheep depends on its productive state. The intake of ewes is considerably greater when lactating than when dry (Owen *et al.* 1980). The ARC (1980) suggest that lactating ewes eat 20–70% more than comparable dry ewes, the difference being larger with good diets and for ewes nursing twins. There is some evidence that the intake of ewes is reduced towards the end of pregnancy because of the physical limitation imposed by the foetus (Oyenuga and Akinsoyinu 1977). High food intakes are also associated with compensatory growth following a period of poor nutrition. Shearing can increase the intake of wool sheep (Minson and Ternouth 1971): this may be because shearing alleviates chronic heat stress which reduces appetite. Forbes (1982) reported that under experimental conditions food intake and growth were substantially greater when daylength was long. This effect was more marked with unimproved (temperate) sheep than lowland breeds, but the provision of artificial light in commercial systems has not shown any consistent benefit.

The voluntary intake of grass depends on the species, month of cutting and level of fertiliser nitrogen (Minson 1973). These differences arise because of differences in retention time (the period which the food remains in the rumen) and the density to which the food is packed in the rumen (Thornton and Minson 1973). Similarly, the higher intake of leaves than stems can be attributed to a shorter retention time (Hendricksen *et al.* 1981). Retention time depends on the digestibility of the food and on the size of the particles, so that physical treatments such as grinding and pelleting of low-quality coarse foods increase food intake (Laredo and Minson 1975). Because supplements such as cottonseed (Haggar 1970) and urea (Tudor and Morris 1971) stimulate micro-organism function in the rumen, they reduce retention time and thus increase the intake of poor-quality foods.

The ARC (1980) note that the intake of grass silages by sheep is always considerably less than the intake of other roughages with similar nutrient density. Their data show that the intake of silage is about 20–30% lower than that of grass with a similar energy density. Legume and maize silages give higher intakes than grass silages.

There has been a lot of interest in the relative nutrient requirements of sheep and goats. Comparing the intakes of *Hyparrhenia* spp. by the two species in Zambia, Gihad (1976b) found that the dry matter intake of goats was greater than that of sheep. Other workers have found no significant differences between the two species except for low-energy diets when goats appear to have higher intakes than sheep. The results of comparative experiments appear to depend heavily on the breed and age of the animals studied. Devendra and McLeroy (1982) state that tropical breeds of goats have intakes up to about 3% of body weight, considerably lower than temperate breeds.

Energy

Energy is derived from organic constituents of food and used as fuel for body functions and production. The unit of energy is the joule and is given the symbol J. One calorie is equivalent to 4.184 J. In ruminant nutrition the kilojoule (1 KJ = 10^3 J) and the megajoule (1 MJ = 10^6 J) are frequently used. Many methods have been used to calculate the energy requirements of sheep: that described here is the metabolisable energy system devised in the UK by Sir K. L. Blaxter, and based on an understanding of energy use in the animal.

The partition of food energy within a ruminant is shown in Fig. 4.2, and Table 4.3 shows the energy values of some foods for sheep. The gross energy (GE) is the total energy of a food that would be obtained on complete oxidation. The gross energy of starch is 17.7 MJ kg^{-1}, while that for proteins and fats is higher (24.5 MJ kg^{-1} for casein and 39.0 MJ kg^{-1} for vegetable oils). For diets commonly fed to sheep a GE value of 18.4 MJ kg^{-1} is often assumed. The digestible energy (DE) of the food is the GE less the energy lost in the faeces.

The metabolisable energy (ME) is the DE minus the energy lost in methane and urine. On average ME = 0.82 DE. Not all the ME can be utilised usefully by the animal: heat is produced as a by-product of metabolism, by the micro-organisms in the rumen, and in the muscular and chemical work of digestion. The energy remaining is known as the net energy (NE) which may be used either for maintenance or for

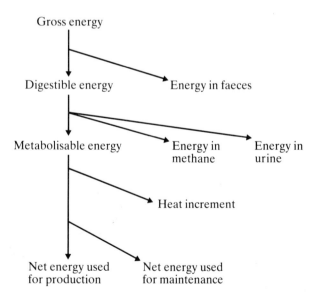

Fig. 4.2 The partition of food energy in a ruminant

production. The efficiency of utilisation of ME, i.e. NE/ME, known as k decreases with decreasing ME value of the food, and with increasing feeding level.

Assessment of energy requirements

The energy requirements of sheep can be calculated from a series of equations based on data from experiments. Several publications recommend equations which give differing energy requirements: the information given here comes largely from the MAFF (1975). Only the simplest forms of the equations will be given because it is felt that refinements are not justified when neither the energy requirements for tropical sheep nor the energy values of the foods are well known. The total energy requirements of a sheep are the sum of its requirements for maintenance and production (growth, lactation and pregnancy). The recommendations given include a 5% safety margin.

Maintenance

The basic NE requirement for maintenance is assumed to equal the fasting metabolism which is the total heat produced by a fasting, resting animal. In practice an activity allowance is added to supply energy for movement. Thus the ME requirement for maintenance (ME_m in MJ d^{-1}) of sheep kept outdoors is related to its weight (W in kg) by

$$ME_m = 1.8 + 0.1\ W \qquad [4.1]$$

for pregnant and lactating ewes, and

$$ME_m = 1.4 + 0.15\ W \qquad [4.2]$$

for growing and fattening sheep. Where sheep have to walk long distances or are under other stresses, these recommendations may have to be increased.

Growth

The ME requirement for growth (ME_g in MJ d^{-1}) is calculated from the NE requirement (NE_g in MJ d^{-1}) for liveweight gain and the efficiency of utilisation of ME for growth (k_g):

$$ME_g = NE_g/k_g \qquad [4.3]$$

k_g is calculated from the concentration of ME in the dry matter (M/D in MJ kg^{-1}) according to

$$k_g = 0.0435\ M/D \qquad [4.4]$$

Table 4.3 *Energy values of foods for sheep in MJ/kg^{-1} DM*

Food	Gross Energy	Energy lost in			Metabolisable energy
		Faeces	Methane	Urine	
Barley grain	18.5	3.0	2.0	0.6	12.9
Young dried ryegrass	19.5	3.4	1.6	1.5	13.0
Mature dried ryegrass	19.0	7.1	1.4	0.6	9.9
Mature grass hay	17.9	7.6	1.4	0.5	8.4

Source: After McDonald *et al.* 1981.

which indicates that diets with a high concentration of ME are more efficiently utilised than diets with a low M/D. Here, NE_g depends on liveweight gain (LWG in g d^{-1}) and body weight (W) such that

$$\log_{10} NE_g = 1.11 \log_{10} LWG + 0.004W - 2.10 \qquad [4.5]$$

The energy cost of deposition of fat is considerably greater than that of muscle, so that NE_g increases with weight of lamb. For lambs of a given weight, NE_g is higher for small breeds than for large breeds, and is slightly greater for female lambs than male lambs, but these effects will be ignored in the present calculations.

Lactation

The ME requirement for lactation (ME_l in MJ d^{-1}) is calculated from the product of daily milk yield (MY in kg d^{-1}) and the energy value of milk (NE_l in MJ kg^{-1}), divided by the efficiency of utilisation of ME for lactation (k_l). For ewes NE_l may be assumed to be 4.6 MJ kg^{-1}, a value greater than that for cows because of the greater concentration of solids in ewe's milk. Here, k_l is about 0.62, so that

$$ME_l = 7.4 \text{ MY} \qquad [4.6]$$

Pregnancy

The ME requirements for pregnant ewes are related to the stage of pregnancy, body weight (W in kg) and the number of lambs carried:

Single lambs: $ME_m = (1.2 + 0.05W) \exp 0.0072t$ \qquad [4.7]

Twin lambs: $ME_m = (0.8 + 0.04W) \exp 0.0105t$ \qquad [4.8]

where t is the number of days pregnant (and exp is the exponent of the base of natural logarithms). Thus for a 40 kg ewe the ME requirement rises to 9.2 MJ d^{-1} at the end of pregnancy if she is carrying a single lamb, and to 11.2 MJ d^{-1} if she has twins.

Energy requirements of tropical sheep

Compared with the values calculated by these equations, the energy requirements of tropical sheep may be slightly different. The resting metabolism of indigenous Indian lambs has been reported to be 10–20% less than that of temperate sheep (Pandey et al. 1972a). On the other hand, Butterworth (1966) reported that the maintenance requirements of adult sheep in Trinidad and Venezuela were not significantly different from the requirements of sheep in temperate areas. The energy requirements for maintenance of pen-fed sheep in Malaysia reported by Devendra (1981) are similar to those derived using eqn [4.2]. The requirements for temperate sheep were derived from experiments conducted in thermoneutral environments. Thus, any differences between the results of temperate and tropical experiments cannot be attributable to the effects of cold stress in increasing metabolic rate, although in the tropical experiments food intake may be reduced by chronic heat stress.

Tropical breeds are generally smaller than temperate breeds, so that a tropical lamb will be more mature than a temperate lamb of the same weight, and on a given diet will tend to deposit more fat. Thus its energy requirement for liveweight gain will be greater. This hypothesis is supported by the observation of Gómez and Hernández (1980) that the ME requirements for liveweight gain of Pelibuey lambs in Mexico were up to 25% greater at high liveweight gains than those predicted by UK feeding standards.

Example of calculation

A 20 kg lamb is given a diet of 11.0 MJ ME kg^{-1} DM. Assuming it eats *ad libitum*, and that energy is the limiting nutrient, how fast will it grow?

DM intake (from Table 4.2) = 54 g $kg^{-0.75}$ d^{-1} = 511 g d^{-1}

ME intake = 11.0 × 0.511 = 5.62 MJ d^{-1}

ME_m = 1.4 + 0.15W = 4.4 MJ d^{-1}

ME available for growth = ME intake – ME_m = 5.62 – 4.4 = 1.22 MJ d^{-1}

k_g = 0.0435 × M/D = 0.0435 × 11.0 = 0.48

NE_g = ME_g × k_g = 1.22 × 0.48 = 0.58 MJ d^{-1}

From eqn [4.5]

$\log_{10}$ 0.58 = 1.11 $\log_{10}$ LWG(g d^{-1}) + (0.004 × 20) — 2.10

64 *Nutrition*

$$\log_{10} \text{LWG}(\text{g d}^{-1}) = (2.10-0.08-0.24)/1.11 = 1.60$$

$$\text{LWG} = 40 \text{ g d}^{-1}$$

Protein

The system described here for calculating the protein requirements of sheep is that proposed by the ARC (1980) to replace the previous system based on digestible crude protein. Although the requirements of the animal are for nitrogen and specific amino acids, these are traditionally expressed as requirements for protein, where crude protein = nitrogen × 6.25.

The digestion and metabolism of nitrogenous compounds in the rumen is shown in Fig. 4.3. Some of the protein entering the rumen is degraded by the rumen micro-organisms. That protein which is degraded, together with non-protein nitrogen such as amino acids and urea, is known as rumen degradable protein (RDP), and that which is not degraded is called undegradable protein (UDP). Thus the protein entering the small intestine is the sum of the UDP and the microbial protein derived from the RDP.

The protein content and degradability of some foods are shown in Table 4.4. Crude protein contents range from about 3% for straw and cassava roots to 65% for fish meal. Whole cereals contain between 9.5 and 13.5% crude protein, grass between 8 and 20% and legumes typically 20%. Approximate values of the degradability of the proteins in these foods are also shown in Table 4.4.

Calculation of protein requirements

The calculation of protein requirements takes into account the nitrogen requirements of the micro-organisms and the requirement of the sheep for undegraded protein. The amount of RDP (in g d^{-1}) required depends on the energy available for microbial growth in the rumen and is related to ME intake by the equation

$$\text{RDP} = 7.8 \text{ ME} \qquad [4.9]$$

where ME is the intake of metabolisable energy in MJ d^{-1}. The RDP in the diet, calculated as the product of crude protein content, food intake and degradability, must satisfy this requirement for RDP.

The amount of amino acid supplied to the tissues by the rumen micro-organisms (TMP in g d^{-1}) depends on the RDP, and so is also related to ME intake:

$$\text{TMP} = 0.42 \text{ RDP} = 3.3 \text{ ME} \qquad [4.10]$$

The total tissue requirement of sheep for protein (TP) is the sum of its requirements for endogenous urinary nitrogen, dermal and fleece

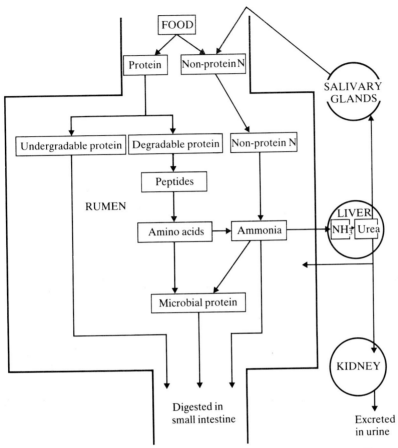

Fig. 4.3 The digestion and metabolism of nitrogenous compounds in the rumen (From McDonald *et al.* 1981)

protein, protein in growth, protein associated with pregnancy and protein in milk. The TP requirements are shown in Table 4.5.

If TMP is greater than or equal to TP, then the sheep's protein requirements will be met totally by the RDP supply. If TMP is less than TP, the deficit must be supplied in the form of UDP. The UDP requirement is calculated from

$$UDP = 1.9\ TP - 6.3\ ME \qquad [4.11]$$

Protein requirements of tropical sheep

Although there have been several studies demonstrating the response of tropical sheep to protein supplementation, few of these give enough quantitative details to allow calculation of daily protein requirements in

Table 4.4 *The crude protein content of foods and its degradability*

Food	Crude protein* (%)	Degradability of crude protein†
White-fish meal	65	0.4
Meat and bone meal	45–55	0.4
Oil-seed meals	20–50	0.5
Dried grass	18.7	0.7
Lucerne, early flowering	17.1	0.7
Grass silage, moderate digestibility	16.0	0.7
Ryegrass, ear emergence	13.8	0.7
Millet	12.1	0.7
Maize silage, whole crop	11.0	0.7
Barley grain	10.8	0.8
Sorghum	10.8	0.7
Maize grain	9.8	0.6
Grass hay, moderate digestibility	8.5	0.7
Rice, polished	7.7	0.7
Cassava roots	3.5	0.8
Wheat straw	3.4	0.6

Sources: *From McDonald *et al.* 1981; †Based on data by ARC 1980.

Table 4.5 *Protein requirements (TP) of sheep expressed as g protein d^{-1}*

P_m, Endogenous urinary nitrogen	$0.1465W + 3.375$
P_g, Liveweight gain of males and castrates	$LWG \times 10^{-3} \times (160.4 - 1.22W + 0.0105W^2)$
P_g, Liveweight gain of females	$LWG \times 10^{-3} \times (156.1 - 1.94W + 0.0173W^2)$
P_w, Wool growth	$3 + (0.1 \times P_g)$
P_l, Milk production	$48 \times MY$
where: W is body weight (kg)	
LWG is rate of liveweight gain (g d^{-1}) MY is milk yield (kg d^{-1})	

Source: After ARC 1980.

a form comparable to those given by the ARC. Farid *et al.* (1983) found that the protein requirement for maintenance of Barki rams was 350 mg $kg^{-0.73}$ d^{-1} of apparently digestible nitrogen. This is equivalent to a requirement of 614 mg $kg^{-0.73}$ d^{-1} or, for a 50 kg sheep, 67 g d^{-1} of crude protein. Calculation using eqns [4.2] and [4.9] gives a similar RDP requirement (69 g d^{-1}) and subsequent calculations show that in this case the TP requirement for maintenance is satisfied by the TMP. Native sheep in both Rhodesia (Topps 1963) and India (Gill and Negi 1971) were observed to have lower protein requirements for maintenance than would be predicted from the ARC equations. Topps (1963) suggested that this low maintenance requirement was an adaptation to the low protein content of vegetation in the dry season.

Example of calculation

A female lamb weighing 20 kg is fed on grass (ME = 11.0 MJ kg^{-1}, CP = 8.0% and degradability of CP = 0.70). Will the protein in the grass be adequate? If not, how much urea and protein should be given as a supplement?

DM intake = 54 g kg$^{-0.75}$ d^{-1} = 511 g d^{-1}

ME intake = 11.0 × 0.511 = 5.62 MJ d^{-1}

which provides enough energy for a liveweight gain of 40 g d^{-1}.

RDP required = 7.8 ME = 7.8 × 5.62 = 43.8 g d^{-1}

RDP supplied by diet = 0.08 × 511 × 0.7 = 28.6 g d^{-1}

which is less than that required. This deficit of RDP can be rectified by adding non-protein nitrogen to the diet:

RDP deficit = 43.8 — 28.6 = 15.2 g d^{-1}

Nitrogen deficit = 15.2/6.25 = 2.43 g d^{-1}

Urea contains 46% nitrogen, but the efficiency with which it is used to provide extra RDP is only about 80%, so that the amount of urea required is

2.43/(0.46×0.8) = 6.6 g d^{-1}

The protein supply to the tissues from the micro-organisms (TMP) is

TMP = 3.3 ME = 18.6 g d^{-1}

The sheep's total requirement for protein (TP) is the sum of its requirements for maintenance, growth and wool/hair production (Table 4.5):

P_m = 0.14675 W + 3.375 = 6.3 g d^{-1}

P_g = LWG(156.1−1.94 W + 0.0173 W^2) = 5.0 g d^{-1}

P_w = 3 + 0.1 P_g = 3.5 g d^{-1}

TP = 6.3 + 5.0 + 3.5 = 14.8 g d^{-1}

which is less than TMP so that there is adequate UDP in the diet, and no need for a protein supplement if 6.6 g d^{-1} urea is given.

Minerals

In addition to nitrogen (and carbon, oxygen and hydrogen), seven elements are essential in relatively large quantities: calcium, phosphorus, potassium, sodium, chlorine, sulphur and magnesium. Several other elements are required in very small quantities, and these are known as trace elements. Many of these trace elements are required in such minute quantities that deficiencies under practical conditions are extremely rare. Deficiency symptoms can arise if one or more of these essential minerals is deficient in the diet, or toxicity symptoms can result from excesses in the diet.

Mineral deficiencies and toxicities occur in many parts of the world. For instance, in the llanos of Colombia, Miles and McDowell (1983) reported that pasture is deficient in calcium, phosphorus, sodium, zinc, copper, cobalt, sulphur and selenium, but iron and manganese levels are too high. However, in general few problems are found under low production, extensive systems, and mineral imbalances become more serious under improved and more intensive husbandry when other nutritional problems have been eased. Some areas of the tropics are known to be deficient in certain minerals, and the seasonal movement of sheep under nomadism or transhumance minimises the effects of local deficiencies. The eating of soil by domestic animals is observed on soils reputed to have a high mineral content (French 1945), but there has been little scientific study of this phenomenon known as geophagia. There is evidence from New Zealand that the ingestion of soil together with herbage enhances the absorption and retention of several minerals (Healy 1972).

Underwood (1981) reviewed the mineral nutrition of livestock throughout the world. A mineral imbalance arises in an animal because the mineral content of its food is either deficient or in excess. Mineral imbalances in vegetation arise, in turn, because of mineral imbalances in the soil: coarse, sandy soils are most frequently deficient in essential minerals. The pH of the soil affects the availability of minerals to the plant, and some species of plant are accumulators of certain minerals. The mineral content of individual plants falls as they mature (Perdomo *et al.* 1977), and agronomy practices to give higher herbage yields usually result in reduced concentrations of micro-nutrients within the herbage.

The mineral requirements of sheep depend on their age, productivity and adaptation to the area. Because some minerals can substitute for others and for other nutrients, and for some minerals it is the ratio between minerals which is important, the mineral requirements of ruminant animals are difficult to define. In addition, if the minerals in the food are in a chemical form in which they are not available to the sheep, then chemical analysis of the food will not give a true picture of the mineral intake of the animal.

Although acute symptoms of mineral imbalances arise under

experimental conditions, generally chronic symptoms only are seen. These chronic symptoms include low growth-rate, poor fertility, loss of appetite or depraved appetite, loss of hair and diarrhoea. These nonspecific symptoms also arise from inadequate energy and protein intakes, and as a result of helminths and other diseases, so that it is extremely difficult to diagnose mineral imbalances by clinical observation of sheep.

It is possible to identify a deficiency of some minerals by analysis of the soil and herbage, but there are so many complicating factors that analysis of animal tissues and fluids is generally more satisfactory. Analysis of blood, milk, urine, hair and saliva can be done without damaging the sheep, and the concentration of minerals in the blood is widely used. For some minerals (cobalt, copper, manganese and selenium) a better estimate of mineral status can be obtained by analysis of tissue from the liver. The ranges of micro-mineral concentrations found in forage are shown in Table 4.6.

Mineral deficiencies may be rectified either indirectly by adding fertilisers or directly by supplementation of the animal. Fertilisation of the pasture is used for the trace elements copper and selenium, but for most minerals direct supplementation is more satisfactory. Table 4.7 gives the recommended daily intakes of phosphorus and magnesium. The choice of method of direct supplementation depends on the practicalities of management. If the sheep are kept under an intensive system, addition of minerals to the diet or to the water supply is possible, but under grazing conditions other methods are more appropriate. Mineral licks are based on salt so that intake is restricted, but there is no guarantee that all the animals receive adequate supplementation. Oral dosing results in correct dosage, but the labour cost may be prohibitive and unless the animals are dosed at frequent intervals oral dosing is suitable only for those minerals which are stored in the body. For

Table 4.6 *Micro-minerals: concentrations in serum, estimated requirements and toxic levels in ppm*

Mineral	Range in forage	Concentration in serum	Estimated requirement	Toxic level
Cobalt	0.03–0.3	0.005	0.05–1.0	>50
Copper	5–8	1.0	5–10	100
Iodine	0–10	—	0.2	50–100
Iron	60–300	1.5	35	500
Manganese	500–900	0.02–0.1	20	50–900
Molybdenum	0.2–10	0.06–0.60	0.01–1.0	5–10
Nickel	0.9–1.0	0–0.3	—	>100
Selenium	0–50	0.06–1.1	0.08–0.15	8–10
Zinc	15–40	0.8	25–40	1000

Source: After Hansard 1983.

Table 4.7 *Requirements of sheep for phosphorus and magnesium in g d^{-1}*

	Phosphorus	Magnesium
20 kg sheep, LWG 0.1 kg d^{-1}	1.1	0.35
40 kg sheep, zero LWG	0.8	0.45
40 kg sheep, end of pregnancy	1.9	0.77
40 kg sheep, milk yield 1 kg d^{-1}	3.1	0.99

Source: After ARC 1980.

minerals such as cobalt of which a continuous supply is necessary, slowly dissolving bullets in the rumen can be used. Some minerals may be injected intramuscularly as inorganic complexes which are slowly absorbed. Intra-ruminal devices which can release any one or several minerals at a controlled rate are now being developed (Siebert and Hunter 1982), and these should revolutionise the treatment of mineral deficiencies.

Calcium

Calcium is rarely the primary limiting factor in the diets of sheep, yet responses have been reported to dietary supplementation with calcium (Singh *et al.* 1968) and to the application of calcium fertiliser (Rees and Minson 1976).

Phosphorus

Phosphorus, together with calcium and vitamin D, is required for bone metabolism, but a deficiency of phosphorus is usually seen as anoestrus and poor reproductive performance. Phosphorus deficiency is common in cattle in the tropics, but less so in sheep (Cohen 1980). It arises when animals are fed on mature non-leguminous herbage, including straw, particularly when grown on acid soils. Supplementation of the diet with phosphate is the easiest way to correct the deficiency, but fertiliser grade phosphate contains fluoride and so should not be used unless it is first detoxified (Agarwala *et al.* 1971).

Potassium

The potassium content of plants is high compared with the potassium requirements of sheep, so that potassium deficiency is rare.

Sodium and chlorine

Deficiencies of chlorine are rare, but sodium deficiency occurs in tropical Africa and the inland parts of Australia (Underwood 1981). Sodium deficiency is most likely to occur in sheep when their salt losses are high as a result of heat stress or lactation. Supplementation of sodium using salt licks is the most common method because ruminants ingest the right quantity to satisfy their own requirements, although following a period of deficiency, salt should be offered gradually. Excess salt reduces food intake and food conversion efficiency. The tolerance level is thought to be between 0.7 and 5% salt in the dry matter (ARC 1980), depending on the availability of water, breed of sheep, type of diet and the concentration of other ions.

Sulphur

Sulphur is found in the amino acids cystine, cysteine and methionine, and in the vitamins biotin and thiamin. Wool contains a high proportion of cystine and thus contains about 4% sulphur. The quantity of sulphur required by the rumen micro-organisms depends on their rate of protein metabolism, and so is directly related to their nitrogen requirements. The ARC (1980) suggests that for every gram of nitrogen required (as RDP), 0.07 g of rumen degradable sulphur is required. Traditionally, adequate sulphur has been supplied in the form of protein, but with the increasing use of non-protein nitrogen there may be benefit in adding sulphur (as sulphate) to the diet (McDonald *et al.* 1981).

Magnesium

Although magnesium is abundant in most common foods relative to the requirements of animals (Underwood 1981), in adult sheep hypomagnesaemic tetany is associated with low blood levels of magnesium, and can result in death. The exact cause of hypomagnesaemia is uncertain, although the primary factor appears to be inadequate absorption of magnesium from the digestive tract. The severity of hypomagnesaemia can be reduced by increasing the magnesium intake of sheep by feeding a mineral mixture or by application of fertiliser.

Iodine

The only function of iodine is for the synthesis of thyroid hormones. Iodine deficiency can result in anoestrus in ewes, poor libido in rams and goitre in lambs. Certain areas of the world are known to be deficient in iodine: these areas are generally away from the sea as maritime rain has a high iodine concentration. The incidence of goitre is also affected by the presence of goitrogens which interfere with the synthesis of thyroid hormones. Goitrogens are found in most *Brassica* spp. and in the grasses *Cynodon nlemfuensis, Panicum coloratum* and *Paspalum dilatatum* (Rodel 1972).

Copper

Copper deficiency is widespread in the tropics and sub-tropics in Australia (Harvey 1952b), India (Kapoor *et al.* 1972) and South America (Mitidieri *et al.* 1959). The availability of copper in the diet depends on the concentration of other minerals. In particular high levels of molybdenum and sulphate decrease the availability of copper. The symptoms of copper deficiency may be non-specific although swayback is a characteristic symptom in lambs, and in wool sheep fleece quality may be low. Copper toxicity, too, can be a problem. Sheep are particularly intolerant of excess copper, and care must be taken to ensure that when they are fed on concentrates these do not contain high levels of copper.

Cobalt

Cobalt is required for the synthesis of vitamin B_{12} in the rumen. Cobalt deficiency generally results in unthriftiness known by a variety of names including pining. In Australia, New Zealand and South Africa sheep grazing pastures containing *Phalaris tuberosa* have been reported to suffer either an acute poisoning or a chronic nervous disorder known as phalaris staggers (Underwood 1977). Both these diseases can be prevented by cobalt supplementation. Sheep require a continuous supply of cobalt: it may be added to the diet, applied to the soil as cobalt sulphate or most commonly as cobalt bullets. These are dense pellets of cobalt oxide and iron which are put into the reticulo-rumen where they should yield a steady supply of cobalt. A coating of calcium oxide develops on the pellet and two pellets may be administered to each sheep so that abrasion between them keeps their surfaces clean.

Selenium

Selenium is an essential mineral, but selenium toxicity is more common than selenium deficiency. Seleniferous soils have been identified in Africa, Australia, the USA and Israel. All the vegetation growing on these areas contains a high concentration of selenium, and certain plants accumulate selenium. Native plants may be selenium accumulators, but accumulation and toxicity are more common when forage plants are introduced into the area (McCray and Hurwood 1963).

Fluorine

Mild fluorosis affects the teeth and bones of sheep (Harvey 1952a). The ARC (1980) recommends that fluorine levels should not be more than 60 ppm in the diets of ewes and 150 ppm for finishing lambs. Fluorosis is never caused by fluorine in the herbage because plants do not absorb fluorine, but it can result from the continuous consumption of rations containing a high fluorine concentration or from the ingestion of herbage contaminated by airborne fluorine from industrial sites.

Vitamins

At least fifteen vitamins are essential, but not all of them are of practical importance in the diets of sheep. The ten most important vitamins are listed in Table 4.8. Vitamins B and C are water-soluble, whereas vitamins A, D, E and K are fat-soluble. Although deficiencies of vitamins do occur in sheep in the tropics, in general vitamin deficiencies are of minor importance compared with deficiencies of other nutrients.

Table 4.8 *Vitamins important in animal nutrition*

Vitamin	Chemical names
A	Retinol
B_1	Thiamin
B_2	Riboflavin, nicotinamide
B_6	Pyridoxine, pantothenic acid, biotin, folacin, choline
B_{12}	Cyanocobalamin
C	Ascorbic acid
D_2	Ergocalciferol
D_3	Cholecalciferol
E	Tocopherols
K	Phylloquinone (and other naphthoquinone derivatives)

Source: After McDonald *et al.* 1981.

Vitamin A

Vitamin A or retinol is necessary for vision and many other body processes. Retinol is manufactured by the sheep from provitamins which are widely distributed in plants. The most important provitamin A is β-carotene. As herbage matures, the carotene it contains becomes less utilisable by sheep (ARC 1980). In a study in Egypt, Ghanem and Farid (1982) found that a diet of wheat straw was deficient in vitamin A. Unless supplementary vitamin A was given, Barki lambs suffered from blindness followed by a fall in body weight and incoordination and finally death. In Australia, the natural vegetation provides adequate vitamin A for adult Merinos (Gartner and Anson 1966; Gartner and Johnston 1969), but clinical deficiency symptoms in young sheep can be expected during long drought periods.

The requirements of vitamin A are difficult to define. The recommended daily intake for UK sheep is 10 μg kg^{-1} body weight of retinol, or 60 μg kg^{-1} of β-carotene, for maintenance and growth (ARC 1980). During pregnancy it is recommended that this level is increased by 100%, and that it is increased by 50% during lactation. Vitamin A is stored in the body, so that supplementation need not be given daily, but care must be taken because an overdose can be toxic.

Vitamin B

Many of the B vitamins are synthesised by micro-organisms in the digestive tract of ruminants, but pre-ruminant lambs depend on a dietary supply. Several B vitamins are essential for lambs: thiamin, riboflavin, vitamin B_6, pantothenic acid, nicotinic acid, vitamin B_{12}, folic acid, choline and probably biotin. Lambs which suckle or receive a substitute based on milk receive adequate quantities of B vitamins, but if the diet is based largely on soya bean or fish then it may be deficient in B vitamins (ARC 1980).

Vitamin C
Sheep are able to synthesise vitamin C so that deficiencies normally do not arise.

Vitamin D
Vitamin D together with calcium and phosphorus is necessary for bone metabolism. Experimental deficiencies cause rickets in growing lambs, but deficiencies are rarely found in practice. Ruminants are able to manufacture vitamin D if they are exposed to sunlight.

Vitamin E
Vitamin E is a non-specific biological antioxidant and has a specific action, associated with selenium, in which it protects phospholipids from peroxidative damage. Young herbage and cereal grains contain adequate quantities of vitamin E, but mature, leached herbage and straw contain little. Deficiency symptoms may be seen in lambs fed on poor-quality roughage and supplements containing little vitamin E. The acute form of the deficiency is known as 'stiff lamb disease'.

Vitamin K
Vitamin K is one of the more recently discovered vitamins, and is rarely a limiting factor for sheep. Bacterial synthesis in the digestive tract provides an adequate supply.

Fatty acids
Linoleic acid or its derivatives are essential in the diets of ruminants, but there is no authenticated case of essential fatty acid deficiency in a sheep with a functioning rumen (ARC 1980). Deficiencies are rare too in pre-ruminant lambs, although if lambs are reared on an artificial diet at least 1% of the energy content must be in the form of linoleate.

Water
Water deprivation in a hot climate results in a loss of body weight. The physiological effects of water deprivation are discussed by King (1983). Lambs in Rajasthan, India, given water only once every 2 or 3 days from 3 months of age weighed an average of only 19 kg compared with 24 kg for a control group given water every day (More *et al.* 1976). Part of this loss is the direct consequence of the lower water content of the body and part is the effect of inadequate water on voluntary food intake. Some reports (e.g. Osman and Fadlalla 1974) state that restriction of water has little effect on digestibility, while more recently, Bohra and Ghosh (1983) have found that water restriction reduces food intake and increases digestibility.

In practice it is the frequency of watering that limits water intake rather than the quantity of water available each day. In the dry season in semi-arid areas, water is available only at wells or other watering-places. The vegetation around these is overgrazed so that the shepherds must balance the advantages of frequent watering against obtaining superior grazing some distance away from the water supply.

Taneja (1965) reported that when Marwari wethers in India were watered every 2, 3 or 4 days for 12 days, their weight losses were 6, 9 or 12% of body weight respectively. He concluded that it is safer to water sheep every third day rather than every fourth day as this allows them to maintain respiration rate and body temperature at approximately normal levels. In an experiment with Yankasa sheep in northern Nigeria, Umunna et al. (1981) found that there was a linear decrease in water intake as watering frequency was increased from 1 to 4 days. Food intake declined only slightly but there was a marked decrease in liveweight gain, so that food conversion efficiency was low at the infrequent watering intervals.

The water requirement of sheep is closely related to their food intake, and non-lactating, temperate sheep in a thermoneutral environment require between 1 and 2 kg of water for every kg of dry matter consumed (ARC 1980). For tropical breeds this requirement may be lower in a thermoneutral environment, but it is elevated because of heat stress. In the experiments conducted by Umunna et al. (1981) the total intake of water by Yankasa sheep was 1.5 kg kg^{-1} food in the rainy season when maximum air temperature was 28 °C, and 2.5 kg kg^{-1} food in the dry season when temperatures were as high as 35 °C. Similarly, Singh (1980) recommended that sheep are given water weighing 8% of their body weight per day in the winter and rainy seasons and 13% of their body weight per day in the summer.

The water requirements of a ewe are increased when she is pregnant or lactating. In the last month of pregnancy the water requirement rises to about 140% of the pre-pregnant value for a ewe carrying a single foetus, and to 200% for a ewe carrying twins (ARC 1980). For non-dairy ewes the water requirement during lactation is increased by about 50% in early lactation and 25% in late lactation. No figures are available for dairy ewes, but by analogy with dairy cows, the extra water required for lactation is equivalent to the amount of milk secreted.

Slagsvold (1970) studied the water content of faeces of five breeds of sheep in Kenya. A higher water content was found in faeces of Romney Marsh than Somali, Merino or Karakul sheep on the same pasture. These breed differences probably parallel breed differences in tolerance to water deprivation: breeds which have evolved in dry areas are less affected by water shortage. Within breeds there is some evidence to suggest that sheep with high concentrations of potassium in the blood are better adapted to water deprivation than are those with low potassium concentrations (Purohit et al. 1973; More et al. 1981).

In Zambia, Gihad (1976b) found that goats required less water than sheep. Sheep and goats are less dependent on water than cattle (Maloiy 1973), but have higher water turnovers than camels and oryx (MacFarlane et al. 1971) and kangaroos (Dawson et al. 1975).

Sheep obtain water from three sources: by drinking, as water contained in food, and by oxidation of food and body tissue. Oxidation of 1 kg carbohydrate yields about 500 g water, a substantial amount compared with that coming from the other two sources. Water obtained by oxidation together with water contained in food may completely satisfy the water requirements. For instance, sheep can survive without losing weight if they eat the pods of *Prosopis spicigera* which contain nearly 70% water (Taneja 1973).

A high concentration of salt in the diet or in the drinking-water increases the requirement for water. Typical recommendations indicate that sheep can tolerate a 1.0% solution of sodium chloride and the presence of other salts in the drinking-water. Magra and Marwari ewes in India drink saline water containing up to 3500 mg litre^{-1} total soluble salts without ill-effects (Mittal and Ghosh 1983). In New Zealand, ewe lambs given water that was contaminated with either iron or soil were healthy and had similar growth-rates to lambs given pure water (Bircham and Crouchley 1976).

Evaluation of foods for sheep

The nutritive value of foods depends on their chemical composition. The gross energy of most foods is reasonably constant – a value of 18.4 MJ kg^{-1} is often assumed. Also the ME content is approximately proportional to the digestible energy (ME = 0.82 DE) so that the difference in energy value between foods is largely due to their differing digestibilities. The magnitude of the digestibility of energy is similar to the digestibility of organic matter, or of dry matter.

The digestibility of foods can be measured *in vivo* by a feeding trial, and is calculated as

$$\text{Digestibility} = 100 \times (DM_{food} - DM_{faeces}) / DM_{food} \quad [4.12]$$

where DM_{food} is the dry matter consumed and DM_{faeces} is the dry matter excreted in the faeces. To obtain reliable results from a feeding trial several animals are needed with collections made over at least 2 weeks following an adaptation period of 1 week.

A method which has been developed more recently is the determination of digestibility by putting samples of food in small nylon bags in the rumen (through a fistula), and recording the digestion after 48 hours. This technique is described by Minson *et al.* (1976) and Ørskov *et al.* (1980). Alternatively, an indigestible marker, such as chromic oxide ($Cr_2 O_3$) is fed to sheep, and the concentration of marker in the faeces is used to calculate digestibility (McDonald *et al.* 1981). Acid-insoluble ash is a naturally occurring marker which can be used to determine the digestibility of roughage diets (El-Hag and El-Hag 1983).

Digestibility can be measured by *in vitro* methods. Formulae have been devised which enable calculation of digestibility from the chemical composition of a food (see for example Crowder and Chheda 1982; Martin 1982), but for tropical foods neither the equations nor the chemical composition are known accurately. Digestibility can also be determined by incubating a food sample firstly in rumen fluid and then in pepsin (Cowlishaw and Unsworth 1976). This method is simple and many samples can be analysed simultaneously. High accuracy can be obtained by including one or two standards which have been measured *in vivo*.

The crude protein (CP) content of foods is calculated from their nitrogen content (CP = N × 6.25) as determined by kjeldahl or similar method (AOAC 1975).

Chapter 5
Feeding

Food is one of the major inputs into a sheep-production system and inadequate feeding is usually regarded as the major factor limiting sheep production particularly in the dry tropics. The nutrient requirements of sheep were discussed in Chapter 4. Sheep use food for maintenance and for production (increase in body weight, growth of wool or hair, sometimes for milk production and sometimes for foetal growth). The partition of nutrients between these uses is difficult to predict; it depends on the sheep's genotype and physiological status. For instance, a dairy ewe may respond to increased feeding with a substantial increase in milk production, whereas a lactating ewe of a non-dairy breed will give little extra milk but will gain weight. Milk production has a high priority for nutrients compared with wool production so that lactation can retard the wool production of ewes of both dairy (Goot 1972a) and wool breeds (Heydenrych 1977), although even under very severe undernutrition, wool continues to grow slowly.

Feeding involves making decisions about which animals should receive which food, how much and when. If food is limited, how can this best be used? In a stress period, such as during a drought, a decision must be taken whether to give the available food to those animals whose individual production is suffering most (pregnant ewes and young lambs) or whether it is in the interests of the flock for these animals to perish. Is it worth while to grow food specially for use by sheep, and can such a practice be economic?

Natural pastures

Natural pastures form the entire diet of most sheep in the tropics. It is convenient to regard pastures as systems taking in energy from sunlight and water and minerals from the soil to produce food for sheep. Different species of plant produce different quantities and qualities of food and the production system may be modified by man.

Most tropical grasses are known as C_4 grasses, a term used to describe their biochemical pathways and which indicates that they are better able to use high levels of solar radiation than temperate or C_3 species (Bogdan 1977). The quantity of dry matter produced by a pasture is proportional to the amount of sunlight intercepted, provided that water or other nutrients are not limiting. Thus maximum productivity occurs when the grass canopy fully covers the ground and when there is adequate moisture in the soil; dry matter yield is impaired if the pasture is severely overgrazed or if the grass suffers water stress. There are species differences between grasses in productivity.

Unesco (1979) adopt a practical classification for natural grassland consisting of:
1. *Savannah* – grasses plus shrubs or trees. Sub-types of savannah are categorised according to the number and type of shrubs or trees. Most savannahs have an annual dormant period during the dry season. Apart from certain soil conditions, the climate of savannah regions is normally that where forest could occur and savannahs constitute a non-natural state maintained by fire.
2. *Steppe* – grasses, sometimes mixed with woody plants. Steppes are found in areas with an annual rainfall of less than 500–600 mm, and with a dry season of 8–9 months. Much of Rajasthan (India) and Pakistan, the *Acacia* grasslands of East Africa, and the *Eucalyptus* grasslands of Australia are categorised as steppe.
3. *Grasslands* – there are relatively few true grasslands in the tropics. Aquatic grassland is found where there is seasonal flooding as in the black cotton soils of East Africa, around Lake Chad, in southern Sudan and in the llanos of Venezuela and Colombia. Grasslands also occur at high altitudes in the tropics and sub-tropics.

The potential stocking rate or carrying capacity of a pasture is determined principally by the quantity of vegetation, whereas the productivity of individual sheep on the pasture is a function of the quality or nutritive value of the vegetation. Both the quantity and quality of the vegetation produced are affected by a number of factors, and some of these are given under the headings below.

Climate and season

The vegetation in arid areas is largely annual grasses or perennial succulents, whereas perennial grasses dominate in savannahs. As rainfall decreases it becomes a more important determinant of range productivity (Jahnke 1982) so that carrying capacity is directly related to rainfall (Table 5.1).

In the humid tropics where there is little annual variation in climate, there is little annual variation in the quality and quantity of vegetation. In the sub-humid and drier areas, however, there is little pasture growth in the dry season. The duration of the dry season is a measure of the aridity of the climate, and in the true desert (e.g. in the Sahara or in the Australian Desert) several years may pass without any rainfall. As the dry season proceeds the nutritive value of the vegetation falls; its fibre

Table 5.1 Fodder production and carrying capacity of rangelands in lowland tropical Africa. Consumable fodder is calculated as 30% of aerial growth

Annual rainfall (mm)	Annual production of consumable fodder (kg DM ha^{-1})	Carrying capacity (sheep ha^{-1})
200	150	0.7
400	300	1.3
600	450	2.0
800	530	2.3
1000	620	2.7
1400	780	3.4
1800	960	4.2

Source: After Jahnke 1982.

content increases and its digestibility, ME value, protein content and mineral content fall. Soon after the onset of rains there is a flush in growth with a high water content and a high content of sugars. Seasonal changes in the nutritive value of natural pasture in semi-arid India are shown in Fig. 5.1.

Burning

Burning of grazing land is common in the tropics. Either fires are started accidentally, or they may be part of a planned programme of pasture management. Burning rangeland is a controversial topic, and in Botswana, for instance, burning of communal pastures is prohibited by law (Botswana, Animal Production Research Unit 1981).

The benefits of burning on pasture management are several: burning controls the growth of undesirable bushes, it removes old fibrous vegetation which has been rejected by livestock and which otherwise hampers the growth of desirable species, and it can be used to encourage livestock to graze areas they would otherwise avoid. It is usually accepted that burning during the dry season stimulates the growth of grasses – in particular it stimulates tillering and may break the dormancy of seeds with hard coats. Burning therefore generates high-quality herbage even when the nutritive value of most pastures is extremely low. Fire may also be used to remove bush as part of a programme to suppress the population of tsetse flies, and may be used to remove vegetation before reseeding pasture.

On the other hand, frequent removal of the aerial parts of pasture plants and stimulation of out-of-season growth decreases plant reserves and has a deleterious effect on the long-term productivity of fire-susceptible species. Many improved species of grass are reasonably tolerant of fire, but most pasture legumes are very susceptible. Regular burning always decreases the number of shrubs and trees in rangeland, and if these are an important source of fodder in the dry season, burning

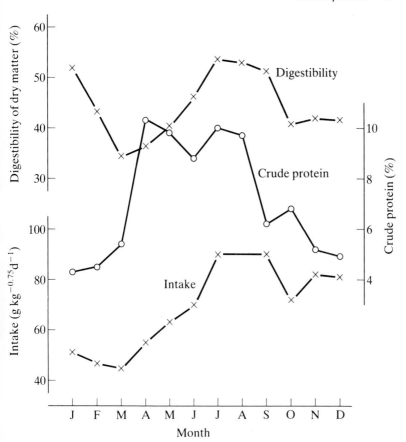

Fig. 5.1 Seasonal changes in the crude protein content and digestibility of pasture in semi-arid Rajasthan, and intake of pasture by Malpura rams (Data from Bhatia *et al.* 1973)

is undesirable. Another deleterious effect of burning is the removal of vegetation so that, particularly in the semi-arid tropics, soil erosion is aggravated.

Thus burning alters the productivity and botanical composition of the sward. Whether or not these changes are beneficial depends on many factors. It is usually recommended that for burning to have a beneficial effect, pastures should be burnt not more frequently than every 3 years. To obtain a satisfactory burn which removes undesirable woody species there must be adequate dry vegetation so that the stock should be removed from the pasture for a considerable period before burning to allow vegetation to accumulate. It is also recommended that stock are kept off the pasture for a few weeks after burning.

Legume content

Legumes have a beneficial effect in pasture. The rhizobia on their roots fix nitrogen so that legumes contribute to the nitrogen economy of the pasture, and usually have a better nutritive value than grass (particularly a higher protein content and digestibility). Their role in natural pastures is very variable. In the wetter African savannahs there are few legumes, whereas in semi-arid East Africa legumes are frequent and may even predominate (Unesco 1979). However, these legumes are mostly unpalatable species and so are usually of little value as food. In contrast, in America, particularly in the llanos of Venezuela and Colombia, *Stylosanthes* spp. are important species for livestock.

Browse

Sheep browse on palatable trees and bushes. Trees have deeper roots than grasses and so remain green in the dry season when other vegetation has senesced. Similarly, trees and bushes are better able to survive overgrazing. Another advantage of browsing is that it gives a lower parasite burden than grazing. The importance of trees and shrubs as fodder increases as rainfall decreases. Ibrahim (1981) reported that in arid areas the dry matter yield of trees and shrubs may be as much as 2 t ha^{-1} y^{-1}, four times greater than the yield of natural grasses. Sheep can eat vegetation up to about 1 or 2 m high depending on their size and ability to stand on their hind legs. Higher vegetation may be cut by the shepherd using a long-handled sickle (Fig. 5.2) or by climbing the tree. The management of trees for fodder is critical, because too frequent and severe pruning has an adverse effect on productivity, and may even cause the tree to die. Sheep eat not only leaves but also twigs, seeds, pods and even bark. Many palatable trees and shrubs are legumes and contain over 20% protein (Wilson 1969). Important natural species are *Acacia, Atriplex, Ficus, Prosopis* and *Zizyphus*.

Grazing behaviour

Sheep have a cleft upper lip which, though not prehensile permits very close grazing (Hulet *et al.* 1975). Sheep discriminate against plants contaminated with sheep urine or faeces, and can select desirable species and parts of plants. The diet selected by sheep may be different from that selected by goats, cattle and wild animals such as kangaroos (Griffiths and Barker 1966) so that the competition for food between species is less than that between animals of the same species. However, Tetteh (1974) found that in Ghana all domestic ruminants preferred *Andropogon gayanus, Panicum maximum, Setaria anceps* and *Digitaria decumbens* to other species of grass. Sheep are able to select vegetation of high nutritive value; Topps (1967) reported that sheep in Rhodesia maintained on herbage with an average crude protein content of 2.2%,

Fig. 5.2 Cutting browse to feed sheep and goats, Rajasthan, India

ingested herbage with a crude protein content of 10.7%. After rains sheep will consume succulent grasses, but during a dry season are forced to eat less palatable species, pods and stems (Weston and Moir 1969). Young cattle exhibit a greater degree of selection than older animals (Zoby and Holmes 1983), but no comparable observation has been made for young sheep. Zeeman et al. (1983) reported that during the dry season, Dorper and Merino sheep selected vegetation with higher digestibility than goats and cattle.

The ability to select a good diet is obviously an adaptive advantage to sheep but it does cause problems in feeding trials! The voluntary intake and nutritive value of a food fed to sheep in pens depends on the form in which it is fed and the quantity fed. If excess of a heterogeneous food is given the sheep will eat the better fraction of it and thus the food will appear to be of a better quality than if only a limited quantity of food is provided (Moore and Dolling 1961; Zemmelink 1980). This problem can be overcome by pelleting the diet so that selection cannot take place or by taking account of selection by analysing refusals.

Sheep which are free to determine their own grazing pattern generally begin grazing at sunrise. The grazing pattern of the whole flock tends to be synchronised and they graze for periods interspersed with rumination, idling and sitting. Grazing may cease for a period of a few hours at midday either if the animals are distressed by the heat and seek shade or if they are on good-quality herbage and have temporarily satisfied their appetite. Grazing usually stops at sunset (Hughes and Reid 1952), although under some circumstances animals may graze for part of the night (e.g. Chakravarty and Subbayyan 1969). The duration of grazing varies inversely with the quality of diet. Many sward characteristics affect the herbage intake by sheep (Hodgson 1982). The rate of intake depends on the height of the herbage (rather than the quantity present) up to 7 cm; sheep can partially compensate for unaccessibility of herbage by increasing grazing time (Allden and Whittaker 1970). Ewes with high nutritional requirements graze for longer than those with lower requirements; Tribe (1950) found that in late pregnancy ewes carrying twins grazed for 1 hour longer than those carrying single foetuses.

For sheep which are herded, the grazing time is limited by the shepherd's desire to have breakfast after sunrise and his wanting to come home at night. The shepherd, too, selects what areas his flock grazes and the speed at which they move, so that the day-to-day management of herded flocks is a primary factor affecting their productivity (Wilson 1982). Even when the sheep are herded for only 8 hours a day, they may spend more than half of this in activities other than grazing and browsing (Swain 1982). Tethered sheep are able to exercise little preference over where or when they graze. For both tethered and stalled animals the contamination of food with excreta and soil can result in rejection and waste, so that feeding arrangements are of prime importance.

For sheep on ranches, certain areas are often favoured for resting and idling. These are generally close to rubbing-posts or in the shade, and are known as camping sites. When feed is abundant the vegetation near the camping sites is more heavily grazed than in more distant parts. This results in an uneven defoliation of the pasture and redistribution of nutrients from the grazed areas to the camping sites (Arnold 1981). Similarly, the location of water has a dominant effect on the distribution of grazing in a semi-arid area.

There are reputed to be breed differences in the degree of dispersion of flocks while grazing. Flocks which stay close together make shepherding easier but allow adequate intake only if there is a lot of vegetation or if the flock is roaming. Arnold *et al.* (1981) reported the dispersion during grazing of Dorset Horns, Merinos and Southdowns, and found that in contrast to the other two breeds, the Merinos dispersed into sub-groups only under extreme food shortage.

The number of sheep that can be shepherded together is limited by the degree of dispersion of the flock and the visibility through the undergrowth. For sheep under kola and oil-palms, Asiedu (1978) recommended that one man could satisfactorily look after not more than 100 animals. In mixed flocks of sheep and goats, the goats are usually regarded as the leaders which set the pace of the group while the sheep follow. Even where sheep are more profitable than goats, a few goats may be kept as leaders.

It is often said that sheep graze while goats browse. This is a misleading generalisation, for both species will graze and browse depending on the vegetation. Carew *et al.* (1980) found that in the humid zone of Nigeria the proportion of feeding time spent grazing was 98.7% for goats and 92.6% for sheep. In contrast, in semi-arid areas in Kenya goats spent 56% of feeding time browsing but sheep did not browse (Wilson 1982).

Poisons

The ingestion of certain plants and other foods is associated with detrimental effects which may include death, chronic illness, photosensitisation, haematuria, abortion and birth defects. Several publications discuss the poisoning of ruminants; three useful reviews are by Keeler *et al.* (1978), McDonald (1981) and Hegarty (1982).

The losses which result from the ingestion of poisonous plants are very difficult to quantify. They are most serious in intensive systems or when vegetation is scarce so that the sheep are compelled to eat species they would normally avoid (Merrill and Schuster 1978). The number of known and suspected toxic plants in the tropics is very large. The toxins in these plants can be grouped into organic toxins (alkaloids, cyanogenetic glycosides, goitrogenic substances, amino acids, oxalates and isoflavins) and inorganic substances (fluorine, copper, molybdenum, selenium and nitrate). Toxins may be produced by a fungus (e.g. aflatoxins in mouldy foods, particularly in groundnuts) or by the deposition of a toxin (e.g. lead or fluorine) on vegetation. Probably no plants are entirely free from potentially harmful compounds! Many of the most important pasture plants are demonstrably harmful under certain conditions (McDonald 1981), but deleterious effects are seen only when the sheep's detoxification system is incapable of dealing with the toxins.

There have been a number of successful plant breeding programmes for both grasses and legumes which have resulted in the production of cultivars which do not have a toxic effect (Hegarty 1982). Alternatively, the deleterious effects of poisonous plants can be minimised by sensible management. For instance, sheep should not be moved to an area with which they are not familiar when they are hungry (Krueger and Sharp 1978). There appear to be big differences between individual sheep in their ability to select against or tolerate poisonous plants.

Problems of Grass Production and Utilisation

Three major problems of grass production and utilisation are: (1) the poor quality of vegetation available for sheep; (2) the seasonality of production; and (3) the unreliability of production, particularly in droughts. There are many technical solutions to these problems, but substantial improvement of pasture is rarely economic. Approaches to the solution of these problems include improving pasture management (e.g. correct stocking rates and keeping a grazing reserve), feeding supplements, feeding during droughts, improving pastures, conserving fodder, growing fodder crops and integrating sheep production with arable systems.

In semi-arid parts of the tropics and sub-tropics, inadequate food is the major constraint to sheep production. This problem is becoming worse as the human population in dry areas is increasing, more marginal land is being cultivated, and the numbers of domestic animals are increasing. As a consequence of these pressures, pasture is often overgrazed. Although the technical solution to overgrazing is simply to regulate grazing, it is extremely difficult to implement this.

Supplementation of roughages

Supplements added to low-quality roughage diets (grass or straw) generally consist of one or more of concentrates, molasses, non-protein nitrogen (NPN) and minerals. Concentrates can be those which supply energy but little protein (e.g. cereals) and those which supply significant amounts of protein as well as energy (e.g. soya-bean meal). The effect of supplements can be predicted by calculation if there is adequate knowledge of the nutritive value of the supplements and the roughage in the diet. Frequently however, the analysis of feedstuffs is not known and a feeding trial may be carried out to find the most appropriate supplement and the level at which it should be fed.

As low-quality roughages contain only low concentrations of ME, CP and minerals, there is generally a beneficial response to any form of supplementation. The response to supplementation by molasses, NPN or minerals alone is, however, likely to be small because each of these three supplements satisfies only one deficiency and the diet remains deficient in another essential constituent. Supplementation with molasses must therefore be accompanied by nitrogen supplementation and, if necessary by mineral supplementation. A further complication arises because urea, the most common form of NPN supplement, is unpalatable to sheep and must be disguised in molasses or concentrates. Nitrogen from urea gives a smaller response than nitrogen from soya-bean meal, even when the diets have the same energy content if the diet is deficient in undegradable protein.

Supplements result in improved animal performance in several ways. Firstly, by providing essential nutrients for the rumen micro-organisms they speed up fermentation of food in the rumen so that the passage of food residues through the gut is faster and more roughage can be consumed. Secondly, the enhanced micro-organism activity in the rumen gives better digestion of the roughage (although this may be offset by the increased rate of passage through the rumen). Thirdly, the supplements provide nutrients for the sheep.

As an example of the magnitude of the effect of supplements, consider an experiment conducted with Chios sheep in Cyprus (Economides *et al.* 1981). Non-lactating ewes were fed long barley straw *ad libitum* and a mineral–vitamin mixture. Three groups of the ewes were given supplements of either (1) soya-bean meal and barley, (2) urea and barley or (3) barley alone. The quantities of supplements and the resulting composition of the diets are shown in Table 5.2. All three forms of supplementation substantially increased the M/D and crude protein contents of the diet from the basal levels of 5.5 MJ kg^{-1} and 3.6% respectively. Diets (1) and (2) had similar energy and crude protein contents (8.5 MJ kg^{-1} and 7.6%). However, in diet (1) the protein was supplied by soya-bean meal and in diet (2) the protein was largely in the

Table 5.2 *The effect of suplements on the overall composition of diets, digestibility of organic matter (DOMD), their intake by ewes and the cost. B is barley and SBM is soya-bean meal*

Diet	Straw	1 Straw + 80 g SBM + 170 g B	2 Straw + 10 g urea + 240 g B	3 Straw + 250 g B
M/D (MJ kg^{-1})	5.5	8.5	8.5	7.6
CP (%)	3.6	7.6	7.7	5.2
DOMD (%)	42	57	58	53
DM intake (g d^{-1})	680	1120	1080	1050
ME intake (MJ d^{-1})	3.7	9.5	9.2	7.9
Cost of supplement (C£ ewe^{-1} d^{-1})	0.000	0.025	0.018	0.016
Extra ME intake per unit cost of supplement (MJ C£$^{-1}$)	—	232	306	263

Source: After Economides *et al.* 1981.

form of urea. The food intake and ME intake were slightly greater for the ewes given the soya-bean supplement than the urea, but compared with the control diet both these treatments increased the digestibility of the organic matter in the dry matter by about 15 percentage units (an increase of 36%) and the ewes ate about 170 g d^{-1} more straw and increased their intake of metabolisable energy by a factor of 2.5. Supplementation with 250 g barley grain (treatment 3) increased the energy density of the diet to 7.6 MJ kg^{-1}, and the crude protein content to 5.2%. These increases were associated with a 50% increase in dry matter intake and a 110% increase in ME intake. Although the addition of barley to the straw had a significant benefit, the crude protein in the barley grain (10.4%) was inadequate to meet the protein needs of the sheep.

The cost of the supplements is also shown in Table 5.2. The urea and barley supplement gave the greatest increase in ME intake per unit cost of supplement. Although the soya-bean meal and barley supplement resulted in the greatest intake of ME, its cost was also the highest so that of all the three supplements its benefit per unit cost was the least. The economics of supplementation in other countries can be evaluated in a similar way. In general the prices of soya-bean meal and barley are very high compared with urea, so that the use of urea as a source of nitrogen is recommended from the economic viewpoint. The optimum level of supplementation depends on the relative value of supplements and the enhanced productivity. In general, supplements give greater economic benefits in more intensive systems and those in which the value of animal products is high.

Supplements may be fed separately to sheep or may be incorporated into a complete diet. Micro-elements such as cobalt or selenium are usually administered individually to animals. In Australia, non-protein nitrogen and macro-minerals such as sulphur, phosphorus or sodium are usually supplied as licks (Siebert and Hunter 1982). Energy and protein supplements may also be incorporated into licks which are then known as feed blocks. For feeding supplements in other forms, such as a dry diet or liquid supplement, specialised feeders may be needed if the supplement is to be fairly distributed to the sheep. Feeding supplements in a complete diet gives a more uniform intake per animal and may result in better performance, but the roughage must be ground, mixed with the supplements and possibly pelleted. These processes are expensive.

Drought feeding

In arid and semi-arid areas droughts are said to occur when the rains fail or when there are seasons with below-average rainfall. Droughts can have a disastrous effect on sheep production; for instance, during the Australian drought of 1892–1902 the sheep population of Australia was halved (Franklin 1951).

Problems of grass production and utilisation 89

If no control measures are taken to help sheep survive in drought years there is bound to be a cyclic variation in sheep population. During a succession of years with good rainfall, sheep numbers will increase. In subsequent drought years there will usually be a decline in numbers. These changes result from the effect of nutrition on reproductive rate as well as the direct effect of nutrition on the mortality of adult sheep. Individual sheep owners and governments must decide what decline in sheep numbers is acceptable during droughts, and what measures should be taken to reduce losses.

The main problem in a drought is the poor quality and inadequate supply of vegetation, but coupled with this is an inadequate supply of water. The importance of watering frequency in determining food intake is discussed in Chapter 4. In practice a limited number of watering-points results in an uneven distribution of grazing pressure, aggravating the problem of food shortage in the watered areas. Overstocking is a contributory factor to drought losses; a large number of animals per unit area and little vegetation in reserve both add to the problem.

In a drought sheep are forced to eat whatever vegetation is available. This means that they will have to eat those plants and parts of plants which are less palatable and which they normally reject. They may dig down to expose the sprouting stumps of shrubs (Figs 5.3 and 5.4). In a

Fig. 5.3 Scraping behaviour of sheep in the dry season (Photo: Central Sheep and Wool Research Institute, India)

Fig. 5.4 Newly sprouted *Crotolaria burhia* shrub in the dry season (Photo: Central Sheep and Wool Research Institute, India)

drought, the diet is of low quality; the response of sheep to undernutrition depends on their physiological status. Ewes lose weight and cease to display oestrus. Young animals stop growing, but even 2-month-old lambs can survive prolonged nutritional restriction such that their weight remains constant, with no serious long-term effects (Allden 1968). After a period of restricted feeding, sheep exhibit compensatory growth. However, if the drought is prolonged, the low quality of the diet and its low intake become inadequate to keep the sheep alive. The most susceptible animals are lambs (Egan and Doyle 1982) and exotic breeds.

Giving supplementary feed during droughts can greatly reduce deaths and losses resulting from low productivity, but the practicalities and economics of drought feeding must be considered. In Queensland, Australia, feeding of lambs with hay and maize during droughts may give substantial economic benefit (Le Gros 1963). However in less developed parts of the tropics the availability of food reserves and their transport to drought-stricken areas cause large problems. A unit of metabolisable energy is cheaper to transport as concentrate than as roughage, but the cost of concentrates is high and it is difficult to ensure that limited quantities of concentrates are fairly distributed among the flock; it is easy for the largest and greediest ewes to receive a lot more than the weaker ewes. For this reason, and to save labour costs, weekly feeding of a limited quantity of food so that there is a temporary abundance is more satisfactory than daily feeding (Franklin 1951).

For sheep kept on extensive ranches, a major cost of drought feeding is the labour to gather the ewes to the feeding area or to distribute the food over a large area. For easy feeding it may be better to confine the sheep during the drought, even though this means they are unable to forage.

Because of the uncertainty of how long the drought will last it is difficult to devise feeding strategies. Belschner (1959) recommended starting feeding early in order to maintain the body condition of the sheep. If the drought is long, exporting the sheep to another area may be a cheaper solution than importing food. In Australia sheep are moved on trucks; in traditional systems in semi-arid areas transhumance and nomadism have evolved to give maximum productivity in dry seasons and drought periods.

If it is decided to give supplementary food how much should be given? At the beginning of a drought, Moule (1951) recommended that Merino sheep are given 57 g d^{-1} of a protein-rich supplement to supplement the poor-quality vegetation. This should be increased to satisfy 30–50% of the maintenance requirement as the drought progresses, and finally the complete requirement of 340 g d^{-1} may have to be fed. It is difficult to calculate the maintenance requirements of sheep in a drought from their initial body weights; for cattle the optimum economic level of feeding causes them to lose weight so that their maintenance requirements are lowered (Elliott *et al.* 1966), and the same must be true for sheep.

Assuming that the sheep reaches an equilibrium weight of 20 kg then its maintenance requirement for metabolisable energy will be 4.4 MJ d^{-1} (calculated from eqn [4.2]). This ME can be supplied by 370 g concentrate with a M/D of 12 MJ kg^{-1}. The sheep's requirement for protein is 34 g RDP d^{-1} (eqn [4.9]), and 6.3 g UDP d^{-1} (eqn [4.11]). If the degradability of the protein in the concentrate is 0.7, the protein needs of the sheep will be satisfied only if the concentrate contains at least 13% crude protein. Thus it will be necessary to feed urea in addition to a concentrate with a low protein content.

Improved pastures

The productivity of grassland may be improved by changing its botanical composition and by improving the management. In dry parts of the tropics, economic factors limit the improvement of pastures to simple measures such as grazing control or the sowing of new species by broadcasting. McIvor *et al.* (1982) recently reported on a programme to select pasture species for semi-arid tropical Australia. There is more potential for improving pastures in the sub-humid and humid parts of the tropics (Lane 1981). The introduction of species of grass which have a better nutritive value and dry matter yield results in there being a larger quantity of better-quality vegetation available for the sheep. For instance, the average dry matter yield in both the wet and dry seasons of para grass (*Brachiaria mutica*) after 59 days' growth in Venezuela was 4.8 t ha^{-1} (Combellas and González 1973). The estimated annual yield would be about 30 t ha^{-1}. An annual yield of 130 t ha^{-1} has been reported for *Pennisetum purpureum* in Colombia (Crowder and Chheda 1982). More realistically, average yields of dry matter of 4–6 t ha^{-1} y^{-1} can be expected in the humid tropics (ILCA 1979b), and of the forage produced about only 30% is consumed by livestock.

However, the improved productivity is unlikely to be sustained unless pasture management is also improved; fertiliser application and grazing control are required if improved pastures are not to revert to their original botanical composition. Mixtures of improved grass and legumes may not need such high levels of fertilisation as pure swards, but care must be taken to ensure that the grass and legumes are well matched. An aggressively growing grass will swamp a less competitive legume, and vice versa. Thus before a pasture can be considered suitable it must be tested under realistic management. Desirable attributes of pasture plants include the ability to grow on a low-fertility soil, to withstand desiccation in the dry season and waterlogging in the wet season, to withstand severe grazing pressure and to spread over an easily erodable area or on a steep slope where sowing is difficult. Table 5.3 gives a list of useful species of tropical grasses and legumes. The establishment and maintenance of pastures is a complex subject and is discussed in specialised texts (e.g. Whiteman 1980; Crowder and Chheda 1982).

Fertiliser increases the yield of improved species of grasses and improves their quality so that digestibility and intake are improved (Minson 1973). Nitrogen (N) is generally the most limiting mineral for pasture, but the response to phosphorus (P), potassium (K) and other minerals may also be substantial. Legumes are more responsive to phosphorus and potassium than to nitrogen, so that the legume content of a pasture is enhanced by high levels of P and K, but low levels of N. Under humid or irrigated conditions the highest yield of grass may be obtained when the nitrogen application is over 1000 kg ha^{-1}y^{-1} (Crowder and Chheda 1982), but the economic optimum depends on the relative values of fertiliser and pasture and is likely to be at a much lower level of application. The response of tropical grasses to moderate levels of nitrogen fertiliser is usually in the range 20–50 kg DM kg^{-1} N (Humphreys 1978), substantially more than that of temperate grasses.

The response of pasture to light is asymptotic so that at high intensities of radiation there is little benefit of more intense radiation. Solar radiation has a heating effect and enhances evaporation from the sward so that high levels of radiation cause high rates of evaporation and may result in moisture stress. Different species have different optimal light intensities; many legumes (e.g. *Calopogonium mucunoides, Centrosema pubescens, Desmodium heterophyllum* and *Pueraria phaseoloides*) and some grasses (e.g. *Brachiaria mutica* and *B. humidicola*) are shade tolerant and are thus suitable for pastures under trees (Whiteman 1980).

Table 5.3 *Tropical grasses and legumes*

Latin name	Common name
Grasses	
Andropogon gayanus	Gamba grass
Axonopus compressus	Carpet grass
Bothriochloa insculpta	Sweet pitted grass
Brachiaria brizantha	Palisade grass, signal grass
B. decumbens	Surinam grass
B. humidicola	Koronivia grass, creeping signal grass
B. mutica	Para grass
B. radicans	Tanner grass
Bromus unioliodes	Rescue grass
Cenchrus ciliaris	Buffel grass, African foxtail
Chloris gayana	Rhodes grass
Cynodon dactylon	Star grass, Bermuda grass
C. nlemfuensis	Star grass
C. plectostachyus	Giant star grass
Dichanthium caricosum	Nadi blue grass
Digitaria decumbens	Pangola grass
Eragrostis curvula	Weeping love grass
Heteropogun contortus	Spear grass
Hyparrhenia rufa	Thatching grass, jaragua grass
Imperata cylindrica	Cogon grass
Ischaemum indicum	Batiki blue grass
Melinis minutiflora	Molasses grass
Panicum maximum	Guinea grass
Paspalum dilatatum	Dallis grass
Pennisetum clandestinum	Kikuyu grass
P. purpureum	Elephant grass, napier grass
Setaria anceps (or *S. sphacelata*)	Setaria grass
S. splendida	Giant setaria
Themeda australis	Kangaroo grass
T. triandra	Red oat grass
Tripsacum dactyloides	Gama grass
Pasture legumes	
Cajanus cajan	Pigeon pea
Calopogonium mucunoides	Calopo
Centrosema pubescens	Centro
Clitoria ternatea	Cordofan pea
Desmodium heterophyllum	Hetero
D. intortum	Green leaf desmodium
D. uncinatum	Silver leaf desmodium
Glycine wightii	Glycine
Lablab purpureus	Lablab bean
Leucaena leucocephala	Leucaena, ipil-ipil
Lotononis bainesii	Miles lotononis
Macroptilium atropurpureum	Siratro
M. lathyroides	Phasey bean
Medicago sativa	Lucerne, alfalfa
Phaseolus atropurpureus – see *Macroptilium atropurpureum*	
Pueraria phaseoloides	Puero
Stylosanthes guianensis	Stylo, finestem stylo

Table 5.3 *(Cont)*

Latin name	Common name
S. humilis	Townsville stylo
Trifolium alexandrinum	Berseem, Egyptian clover
Vigna unguiculata	Cowpea
V. vexillata	—

Some values of the digestibility and crude protein content of tropical grasses and legumes are given in Table 5.4. Further values for grasses reported in the literature are listed by Butterworth (1967) and values for Kenyan feedstuffs are summarised by Abate *et al.* (1984). For grasses the digestibility of dry matter is typically between 45 and 65%, and the crude protein content between 4 and 15%. For legumes, the digestibility values are similar to those of grasses, but the crude protein contents are typically between 10 and 20%, somewhat higher than the values for grasses. Although there are some species differences in nutritive value, these are masked by the large differences within species. The major factor affecting the nutritive value is the stage of maturity. Figure 5.5 shows how the digestibility of *Panicum maximum* declines with increasing maturity, and this decline is accompanied by a decline in intake by sheep. Similarly, the crude protein content decreases as grass becomes more mature (Fig. 5.6).

Tropical grasses are said to be less digestible than temperate species (Whiteman 1980), but this difference is small if they are compared in a tropical environment. There are large differences in pattern of digestibility decline between species; Minson (1972), comparing the growth of grasses in Queensland, found that the quality of some species (*Chloris gayana, P. maximum* and *Setaria splendida*) declined rapidly with advancing age, whereas others (*Digitaria decumbens, Paspalum dilatatum* and *Pennisetum clandestinum*) declined only slowly. The ability of grasses to maintain their nutritive value into the dry season has received relatively little attention, yet is one of the most important factors affecting sheep performance in the dry season.

Although the quality of grass decreases as it becomes more mature, its growth-rate increases as the cutting interval is lengthened (e.g. Combellas and González 1972). The aim of grazing or cutting management is therefore to arrange that vegetation is used at an appropriate time to strike a balance between the quantity and quality of vegetation on offer to the animals. If pastures are fenced the stocking rate can be controlled. During the growing season grazing can be confined to one part of the pasture, the other being reserved for fodder conservation, and in the dry season the flock can be allowed to graze the whole area. Although fertilising the pasture at the end of the wet season may stimulate some growth during the dry season, continuous fodder production is possible only if the pasture is irrigated. The nutritional

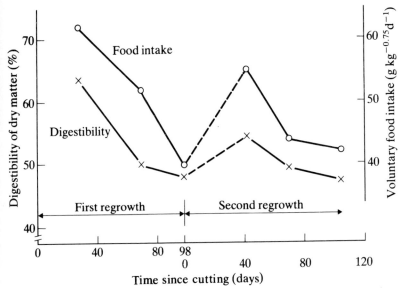

Fig. 5.5 Effect of maturity of *Panicum maximum* on its digestibility and intake by Merino wethers (Data from Minson 1972)

needs of a ewe and her lambs are not constant through the year; the ewe's needs are highest in late pregnancy and early lactation, and lambs need good-quality pasture when they are weaned. By seasonal breeding the nutritive requirements of the flock can be matched to pasture growth.

The successful utilisation of improved pasture depends on grazing control, which is aided by fencing. Most forms of fence are expensive and used only in ranching systems. The more expensive forms of sheep fence include walls, wooden fences, high-tensile wire and wire netting. Barbed wire and electric fences (Bishop 1979) are usually cheaper. In particular, electric fences which can have wide spacing of wires and posts are likely to become more popular in less developed countries now that good solar-powered energisers have been developed. Hedges traditionally comprise non-palatable shrubs (Singh *et al.* 1982), but more recently the use of the fodder plants gliricidia (Chadhokar 1982) and leucaena (Sumberg 1983) have attracted scientific interest. Figure 5.7 shows gliricidia used as a corner post for an electric fence.

Fodder conservation

Fodder can be conserved as hay, silage, haylage and dried grass. In theory, grass and other vegetation can be dried in the sun and stored as hay until it is needed. Once dry there is little fall in the feeding value of grass. Figure 5.8 shows haymaking in Ethiopia. However, haymaking is

Table 5.4 Digestibility of dry matter (Dig DM) and crude protein content (CP) of tropical grasses and legumes

Species	Dig DM (%)	CP (%)	Location	Reference
Grasses				
Brachiaria mutica	55–61	6.5–13.2	Venezuela	Combellas and González 1973
Cenchrus ciliaris	60–77	8.6–17.3	Venezuela	Combellas and González 1972
C. ciliaris	44	9.7	India	Singh 1975
Chloris gayana	62–67	—	Australia	McLeod and Minson 1974
C. gayana	48–63	4.9–13.1	Australia	Minson 1972
C. gayana	—	8.2–17.0	Saudi Arabia	Ruxton 1975
Cynodon dactylon	47–53	9.1–11.7	Venezuela	Combellas et al. 1972b
C. dactylon	—	8.4–9.6	Saudi Arabia	Ruxton 1975
Digitaria decumbens	39–63	—	Cuba	Funes 1975
D. decumbens	66	7.2	India	Kulshrestha et al. 1972
D. decumbens	49–67	6.0–14.9	Australia	Minson 1972
Hyparrhenia spp.	54	6.5	Zambia	Gihad 1976b
Panicum maximum	47–64	6.4–16.0	Australia	Minson 1972
Paspalum dilatatum	43–59	5.9–13.3	Australia	Minson 1972
Pennisetum clandestinum	47–62	5.9–17.3	Australia	Minson 1972
P. purpureum	61	4.5	Brazil	Lima et al. 1972a
Setaria anceps	60	10.5	India	Sharma et al. 1972
S. splendida	63–66	—	Australia	McLeod and Minson 1974
S. splendida	51–67	5.7–13.6	Australia	Minson 1972

Legumes			
Desmodium uncinatum	47–54	10.9–18.2	Australia	Milford 1967
Lablab purpureus (leaf)	56	14.4	Australia	Hendricksen et al. 1981
L. purpureus (stem)	49	10.0	Australia	Hendricksen et al. 1981
Lotononis bainesii	60	20.0	Australia	Milford 1967
Macroptilium atropurpureum	42–71	13.8–18.0	Brazil	Lima et al. 1972b
M. atropurpureum	50	16.8	Australia	Milford 1967
M. atropurpureum	55	16.7–23.0	India	Saxena et al. 1971
M. lathyroides	42–62	7.6–19.2	Australia	Milford 1967
Stylosantheses guianensis	57–64	12.4–13.7	Nigeria	Ademosun 1970
S. guianensis	48	11.8	Australia	Milford 1967
Trifolium alexandrinum	68	20.6	India	Sharma and Murdia 1974
Vigna vexillata	58–69	16.2–20.3	Australia	Milford 1967

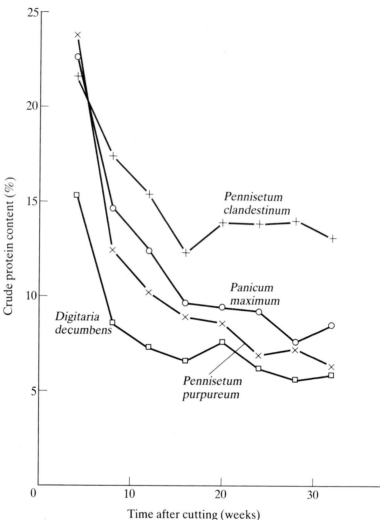

Fig. 5.6 Effect of maturity on the crude protein content of four tropical grasses in Brazil (From Gomide *et al.* 1969)

not popular in the tropics. It is difficult to make hay of a satisfactory quality. After the first few weeks of the dry season the only vegetation available is of a poor quality, and in the wet season there is a high risk of rain wetting the grass during haymaking and causing serious losses of nutrients. The end of the wet season and the beginning of the dry season would appear to be the optimum for haymaking, but these are times of peak labour demand for arable crops, and a farmer must devote his time to cultivating and harvesting rather than making hay. An important

Problems of grass production and utilisation 99

Fig. 5.7 Gliricidia used as a corner post for an electric fence (Blenheim Sheep Project, Tobago)

Fig. 5.8 Piles of hay, Ethiopia

advantage of hay compared with silage is that it can easily be sold and transported.

Silage-making is a method of preserving grass by excluding air and lowering the pH so that there is little bacterial decomposition (Fig. 5.9). To make good-quality silage, good-quality grass must be used. It must be harvested at an early stage of growth before its nutritive value has fallen too far. On the other hand very early harvests give low yields and high water contents. Losses occur in silage-making if the silage clamp is not airtight and aerobic decomposition of the grass takes place, or if the silage produces effluent (Wilkinson 1983). To reduce the losses the grass may be wilted for a short period to lower its water content before ensiling and the clamp consolidated during filling.

The availability of strong plastic sheeting and large plastic bags at relatively low prices greatly aids silage-making. It is possible to make silage in small quantities in plastic bags so that air is not allowed to enter into the whole clamp as the silage is removed. It is necessary to protect the plastic from physical damage (Fig. 5.10) and from vermin. Acid added to the grass at cutting or ensiling rapidly lowers its pH thus encouraging the growth of desirable micro-organisms. Tropical grasses

Fig. 5.9 Small clamp of silage, Nepal

Fig. 5.10 Silage clamp covered with a plastic sheet held down by rubber tyres. The tear in the sheet allows air to enter and reduces the nutritive value of the silage.

contain only low levels of sugars (Catchpole and Henzell 1971) so that the addition of sugar (as molasses) has a beneficial effect in lowering pH and directly contributes to the energy value of the silage. Silages are generally less palatable to sheep than fresh vegetation and good-quality hay so that voluntary food intake is less than that predicted by the data in Table 4.2 (ARC 1980).

Haylage is a form of silage in which the forage is wilted to give a high dry matter content and is chopped into short lengths. The loss of nutrients in the effluent from haylage is less than from conventional silage and it is popular for highly mechanised systems in developed countries.

The nutrient losses incurred in artificial drying are low compared with other conservation methods. However, at present, methods of grass drying necessitate the use of large quantities of fossil fuel and so are unsuited to most less developed countries. The development of appropriate technology to harness solar energy for drying would seem to have potential in the tropics.

Fodder crops

Crops grown specifically for animal food are known as fodder crops. Figure 5.11 Shows a crop of oats grown as animal food in a semi-arid area of India. Fodder crops usually require substantial inputs in terms of land, labour and fertiliser which must be justified in the context of these resources being in demand for producing crops for human consumption. Fodder crops may be justified if they are catch crops (i.e. grow in a period when the land would otherwise not be used), break crops (which improve the yield of subsequent crops by reducing pests and diseases or improving soil fertility) or intercrops (which grow between rows of another crop and shade it or, give protection from wind and rain, and prevent soil erosion). Fodder crops must give substantially better yields or quality food than pastures. For instance, legume crops such as leucaena, berseem clover, gliricidia and lucerne are grown to provide a high-protein supplement, and high-quality food in the dry season. Care must be taken to ensure that when sheep graze legume crops they do not suffer bloat, the accumulation of gases in the rumen.

Some fodder crops such as leucaena (Fig. 5.12) are receiving substantial scientific attention at present. The dry matter yield of leucaena is between 2 and 20 t ha^{-1} y^{-1} (NAS 1977). Leucaena contains mimosine which can cause goitre if animals are fed large quantities of leucaena over long periods. Sheep are said to be more susceptible to mimosine than cattle, but in many parts of the world sheep eat leucaena with no apparent ill-effects.

Fig. 5.11 Forage oats, Rajasthan, India

Fig. 5.12 Leucaena at the Sugar-cane Feeds Centre, Trinidad

There are many other fodder species which can be of use as food for sheep in the tropics. Details of species and methods of cultivation and use can be obtained from texts such as Bogdan (1977) and NAS (1979).

Integration of sheep with arable systems

In all but extensive pastoral or range systems, sheep are integrated into agricultural systems which also produce crops and other livestock. Sheep complement the other enterprises. They are able to utilise the vegetation growing on waste land and field boundaries, they consume by-products such as straw, and they graze on arable fields once the crop is harvested and during the non-growing season. In return, sheep produce manure which benefits the crops. Sheep may be grazed in rotation with crops, on grass leys or on fodder crops. Rotation benefits crop production by improving soil fertility, reducing weeds and reducing crop pests and diseases. For the sheep, rotation with arable crops can reduce the prevalence of parasitic diseases.

An example of the integration of sheep with an arable system is seen in northern Syria (ICARDA 1982). Here, sheep production and barley production are symbiotic. Much of the barley grain is used for intensive finishing of sheep. Immature barley in the wetter areas is grazed by sheep, and sheep eat failed crops which are not worth harvesting. After mechanical harvesting of the barley grain, up to 80 kg ha^{-1} remains in the field, and this is either eaten by sheep grazing on the stubble, or it germinates and is eaten by sheep in the subsequent fallow.

Agricultural by-products used as food for sheep

There are many agricultural by-products which can be used as sheep food. Information of the nutritive value of many tropical by-products is summarised by Göhl (1981). As well as problems of formulating appropriate diets, unless sheep are integrated into arable farms it is necessary to collect, store and distribute the by-products. For instance on the reclaimed delta land in Egypt (Naga and El-Shazly 1983) there are many by-products but few animals so that by-products are processed commercially and transported to Alexandria where they are sold.

Straw

Straw is a by-product of cereal production, and is of little use except as a roughage feed for ruminants. Straw from rice (*Oryza sativa*) is probably the most common in the tropics (Figs 5.13 and 5.14), but in some areas maize (*Zea mays*), millet, sorghum (*Sorghum bicolor*), wheat (*Triticum aestivum*), oats (*Avena sativa*), barley (*Hordeum vulgare*), teff (*Eragrostis tef*) and rye (*Secale cereale*) straw are available. Several by-products from other crops, such as sugar-cane, gram and cocoa, share many chemical properties with straw and the methods for their utilisation are similar.

The nutritive value of straw is generally low (Table 5.5). The energy content ranges from 5.5 to 9.6 MJ ME kg^{-1} DM. Within each cereal category the ME value depends on the variety and the growing conditions. The lignin content of straw is high and its digestibility correspondingly low. The minimum crude protein requirement to break down the lignocellulose in roughages is claimed to be 3.8–5% (Göhl 1981), and Table 5.5 shows that few straws have protein contents above this range. In addition the mineral content (especially phosphorus) of straw is low. All these factors limit the activity of micro-organisms in the

Fig. 5.13 Small stack of rice straw, Nepal

rumen, and, together with the physical form of straw they contribute to a low rate of passage through the digestive system and thus a low voluntary intake of straw.

Straw can usefully be incorporated into the diets of sheep and other ruminants by adding supplements of energy, protein (generally as NPN) and minerals, to give an adequate nutrient supply to the rumen microorganisms and the sheep itself. In addition to supplementation, the nutritive value of straw itself can be improved in three ways:

1. By physically grinding or chopping the straw, or heating it with steam to increase the availability of nutrients. In this way the voluntary intake of straw will be increased but the cost of these treatments is generally too high to make the small increase in nutritive value worth while.
2. By treatment with chemicals, particularly alkalis and nitrogenous compounds, to increase the digestibility and nitrogen content. Practical problems such as the availability of chemicals and the difficulty of getting the techniques right, prevent this practice being widespread at present. However, there is considerable interest in chemical treatment of roughages, and this approach seems to have

106 *Feeding*

Fig. 5.14 Rice straw forming a well-insulated roof of an animal house, Andhra Pradesh, India

potential provided that practicable and economic methods can be developed.
3. By biological processing or semi-solid fermentation. This is a new approach and deserves attention.

Physical treatment

The grinding and pelleting of straw increases consumption and weight gain, and the greatest effects are seen when roughages of low digestibility are fed alone. However, the maximum improvement in intake of digestible organic matter as a result of grinding is about 30% (Greenhalgh and Wainman 1972), and the benefit of grinding is considerably lower when straw comprises less than half of the diet. Treatment of grass straw with steam at high pressure can increase digestibility by 50% (Guggolz *et al.* 1971).

Chemical treatment

There has been a lot of research into chemicals which improve the digestibility of straw, and this has recently been reviewed by Greenhalgh (1980). Treatment with alkalis results in hydrolysis of the link between lignin and hemicellulose, saponification of ester linkages and swelling of the lignin framework, thus exposing the digestible components to bacterial attack. The chemicals which have been used are sodium

Table 5.5 Nutritive value of straw: energy density (M/D), crude protein content (CP), digestibility of dry matter (Dig DM), digestibility of organic matter (DOM) and phosphorus content (P)

	M/D (MJ kg⁻¹ DM)	CP (%)	Dig DM (%)	DOM (%)	P (%)	Source
Barley	5.5	3.6	41	42	—	Economides et al. 1981
Barley	7.3	3.8	—	49	—	MAFF 1975
Buckwheat	6.6	5.7	—	45	—	MAFF 1975
Gram	—	4.0	43	—	0.14	Agarwala et al. 1970
Maize stover	7.3	5.9	—	51	—	MAFF 1975
Oat	6.7	3.4	—	46	—	MAFF 1975
Rice	—	4.2	—	—	0.14	Devendra 1975
Rice	—	3.1	—	41	—	Dolberg et al. 1981
Rye	6.3	3.6	—	43	—	MAFF 1975
Wheat	7.5	3.3	56	60	0.08	Bhargava and Ranjhan 1976
Wheat	9.6	2.9	—	—	—	Göhl 1981
Wheat	—	4.8	43	—	—	Ratan et al. 1979
Wheat	5.7	2.4	—	39	—	MAFF 1975

hydroxide (NaOH), potassium hydroxide (KOH), calcium hydroxide (Ca(OH)$_2$), sodium carbonate (Na$_2$CO$_3$), ammonia (NH$_3$) and ammonium hydroxide (NH$_4$OH). Urea and urine have also been investigated.

The use of sodium hydroxide has been extensively investigated, and it has the greatest effect on digestibility. At its best, alkali treatment will improve the organic matter digestibility of straw by about 20 percentage units, from say 44 to 65%; but because alkali raises the ash content of straw the improvement in the digestible organic matter in the dry matter (DOMD) will be less, say from 42 to 56% (Greenhalgh 1980). Alkali treatment also increases the intake of straw, so that digestible energy intake and ME intake are increased considerably through the dual effects on digestibility and intake. The actual improvements in productivity are difficult to predict, partly because straw is rarely fed alone. In some instances alkali treatment has failed to give significant improvements in animal performance, and the reason for lack of success is not always clear (Jackson 1978).

Several methods can be used to distribute the alkali over the straw. Originally straw was soaked in a dilute solution of sodium hydroxide for a period of 4–24 hours, and then washed. This method, known as the Beckmann method, is laborious, wasteful of alkali and water and gives a product which is difficult to handle and store. On small farms in Norway a process has developed in which bales of straw are immersed in a 1.5% NaOH solution for up to 1 hour, are removed and allowed to drip for 2 hours, and are ready for feeding after a period of a few days.

Alkali can be sprayed on to the straw using a concentrated solution (at least 10%). The method of spraying can vary from a sophisticated purpose-built machine, possibly on a forage harvester, to a simple watering-can. Ensiling the straw is achieved by mixing it with alkali (approximately 5% solution) and putting it in an airtight and watertight silo.

Other alkalis have been investigated. Potassium hydroxide is more expensive than NaOH and has a more variable effect, but is said to give a more palatable product. Calcium hydroxide and calcium oxide are less effective and slower than NaOH; higher concentrations are necessary, but they are cheaper and safer to handle. Straws treated by Ca(OH)$_2$ appear to be more susceptible to mould growth. Mixtures of chemicals may prove to be more effective than single alkalis.

Ammonia treatment has the considerable advantage over alkali treatment that nitrogen is bound on to the straw and is available to the rumen micro-organisms. The method of treating stacks of bales with either anhydrous ammonia or aqueous ammonia is described by Sundstøl *et al.* (1978). It is essential that the straw is enclosed by airtight material (usually plastic sheeting). The quantity of ammonia, the dry matter content of the straw and the temperature all affect the process (e.g. Cloete *et al.* 1983). The recommended duration of treatment depends on temperature. In an environment with a mean temperature of 10 °C 4 weeks is needed, but this decreases with increasing temperature so that at 20°C 18 days are needed, and at 30 °C only about 13 days.

A faster method of ammonia treatment, suitable for commercial use, is possible if the temperature of the straw can be increased; forced circulation of heated ammoniated vapour through a stack at about 75 °C shortens the treatment time to 24 hours. During treatment the straw becomes browner and softer, and its palatability improves. Problems of using ammonia are its non-availability and the difficulty of handling it. Ammonia is an unpleasant gas, and air containing a high proportion of ammonia is explosive on ignition.

In an experiment in Guatamala in which lambs were fed a diet of 75% wheat straw and 25% maize silage with 340 g d^{-1} of concentrate, treating the straw with ammonia increased the liveweight gain of the lambs from 56 g d^{-1} to 130 g d^{-1} (Tejada *et al.* 1979). Ammonia treatment of the straw resulted in a 13% increase in intake and an increase in food conversion efficiency from 0.09 to 0.19. Similarly, for maize stover, Oji *et al.* (1977) found that ammonia treatment increased the digestibility from 52 to 60%, and the crude protein content from 9 to over 17%, and the intake of ammonia-treated stover by wethers was over 950 g d^{-1} compared with 660 g d^{-1} for the animals on untreated stover.

Urea is cheaper than ammonia, and generates ammonia in the presence of urease found in most plant materials. Dolberg *et al.* (1981) treated rice straw in Bangladesh with a concentrated urea solution and sealed it in pits or bamboo baskets. After 3 weeks the digestibility had increased from 41 to 54% and the intake by sheep increased from 52 to 69 g kg$^{-0.75}$ d^{-1}. The crude protein content of the treated straw was 6.7% compared with 3.1% for the untreated straw.

Urine is a readily available and cheap source of nitrogen. In Bangladesh, Saadullah *et al.* (1980) treated rice straw with sheep urine and fed it to sheep. They added the urine to the straw (1 litre kg^{-1}), stacked and covered it with bamboo mats and plastered it with cowdung and mud. After 20 days the stack was opened. The digestibility of the straw had increased from 38 to 51% and its crude protein content from 3.3 to 5.6%. These authors report that urine-treated straw is palatable, but further work needs to be done on its utilisation and on the implications for animal health.

Biological processing.

Fungi can be used to decompose the lignin in straw and thus enhance its digestibility (Stapleton 1981). Research into fungal improvement of straw is still in the early stages of development. The problem is to find a fungus which will decompose the lignin in straw, while not breaking down the cellulose, and which will thrive under conditions (of pH and temperature) at which pathogens are inhibited. Several fungi which occur naturally in straw, including *Humicola lanuginosa* and *Caprinus cinereus*, are being investigated. Alternatively, fungi found in ruminant dung may be suitable. Fungi from wood readily decompose lignin, but flourish only in slightly acid conditions which also favour the growth of pathogens. There seems to be ample scope for research into this field; modern techniques such as genetic manipulation may help to give fungal strains which thrive in the right environment and which produce the required enzymes. An alternative approach is to manufacture specific enzymes and treat the straw directly with these. Enzyme treatment is much quicker than treatment with fungi, but the processes are likely to be complex, so it will be more suited to a factory rather than a farm.

Practicalities of improving straw for sheep

Any method used in the field or as an industrial process must give economic benefits; i.e. the enhanced animal productivity resulting from the improvement in straw quality must be reliable and must justify the cost of treatment.

The benefits gained from treatment are typically up to a 30% increase in ME intake for physical treatment and up to 100% for chemical treatment. The cost of physical treatment and the equipment it requires have prevented its widespread use. Chemical treatment using sodium hydroxide or ammonia is expensive and the chemicals are not generally available in less developed countries. The cheaper chemicals such as calcium hydroxide, urea and urine give less reliable results than the more expensive ones such as sodium hydroxide and ammonia. Insufficient research has been conducted on biological methods to assess their cost and practicability.

In conclusion, the economics of these methods prevent their widespread use, but the development of more appropriate chemical or biological methods could result in substantial benefits.

Sugar-cane (*Saccharum* spp.)

Molasses is produced mainly as a by-product of cane sugar, but is also obtained from citrus, beet, dates and wood. All types of molasses have a high concentration of sugars, giving an ME value of about 10 MJ kg^{-1} DM, and a low CP concentration. Molasses is generally included in the diets of sheep at a level of up to 10%, although in areas of the world where sugar-cane is grown, higher levels may be given. Unless water is available *ad libitum*, molasses toxicity can occur. The palatability of diets for sheep is enhanced by molasses, and this is particularly beneficial for diets based on cottonseed hulls and bagasse (Ghauri *et al.* 1964), cereal straw (Bhargava and Ranjhan 1974) and hay (Chicco *et al.* 1972; Ramadan and Robinson 1973). Molasses is also used to bind dry diets and reduce dust. It may be used as a carrier for urea, which, together with minerals may be fed directly to sheep grazing low-quality pastures (Entwistle and Knights 1974).

The residue of sugar-cane from which the sugar has been extracted is known as bagasse. Both fibrous bagasse and pithy bagasse contain about 20% lignin and have a low feeding value. Bagasse is not suitable for young ruminants, but may be fed to older sheep (Ghauri et al. 1964). To form a satisfactory food it must be supplemented with energy (e.g. molasses) as well as nitrogen (e.g. urea) and minerals. The digestibility of bagasse itself can be increased a little by grinding or chemical treatment (Al-Tawash and Alwash 1983), but these processes are too expensive to be used commercially. Torres et al. (1982) found that a mixture of urea and poultry litter can give a substantial increase in the digestibility of bagasse.

Low market prices for sugar have created interest in feeding whole cane to ruminants. Whole cane contains about 42% sugar in the dry matter (Gooding 1982), but its protein content is only 1–3%, so that supplementation with protein or non-protein nitrogen is necessary. Cane harvesting is normally done in the dry season. For a continuous supply of animal food, cane must either be conserved or harvested throughout the year. In the rainy season cane becomes contaminated with soil and it is difficult to take harvesting machinery on to the land. Attempts have been made to conserve sugar-cane as silage (SFC 1983), but have not been entirely satisfactory. The high levels of water-soluble carbohydrates ferment to give alcohols. In laboratory studies these alcohols were contained in the silage, but under farm conditions the clamp was inadequately sealed to prevent gases escaping. In some countries such as Costa Rica, sugar-cane for animal feeding is stored simply in a stack for feeding during the following months.

Cocoa (*Theobroma cacao*)

Cocoa husks or pods are produced as a by-product of the cocoa industry, particularly in West Africa. They are used as a mulch, and attempts have been made to use husks as food for sheep (Adeyanju et al. 1975; Otchere et al. 1983). To give a reasonable intake the husks must be dried and ground. Their feeding value is low; they have a high fibre content, low crude protein and an ME value of only 7 MJ kg^{-1} DM. The nutritive potential of cocoa husks is therefore similar to that of low-quality straw, and they are utilised in a similar way.

Coffee (*Coffea arabica*)

Coffee by-products include pulp, hulls, molasses and wastes. Dried coffee pulp contains about 10% crude protein, 21% fibre and 2.5% fat (Hutagalung 1981) and its ME value is 4–8 MJ kg^{-1} DM. Feeding more than 20% in the diets of ruminants is not recommended because it depresses performance. Coffee hulls can be used as a carrier for molasses and urea, but are of little value as a nutrient source.

Cotton (*Gossypium* spp.)

The by-products of cotton are used in the diets of adult sheep, but not for young lambs which are more susceptible to the toxic effects of gossypol contained in the kernel. Cottonseed meal or cake is a valuable protein supplement for all types of livestock, but the economics of feeding it to adult sheep are viable only if it is cheap (McIntosh et al. 1976). Cottonseed hulls are widely used as a roughage food for sheep; although the crude protein content of hulls may be about 14%, their digestibility is extremely low, and their ME content is about 7 MJ kg^{-1} DM only. The nutritive value of cotton-wood (i.e. stems, branches and leaves) is very low.

Sisal (*Agave sisalana*)

Sisal pulp may be fed to sheep. Bores et al. (1983) reported the feeding of sisal pulp supplemented with only molasses and poultry manure to sheep in Mexico.

Cassava (*Manihot esculenta*)

Cassava is grown in many parts of the tropics for its roots which are used for human food. It is also known as manioc, tapioca, Brazilian arrowroot and yuca. The roots, either dried or fresh, have been satisfactorily used as the basis of diets to fatten lambs (Chicco et al. 1971). There are, however, three limitations to the use of roots. Firstly, they contain the compounds limarin and lotaustralin which when digested produce hydrocyanic acid (McDonald et al. 1981). Limarin is found mainly in the peel of roots and in bitter varieties. It can be eliminated by drying or cooking. Secondly, the crude protein content of roots is very low (about 2.5% only). Thirdly, cassava roots are in demand as food for humans, so that the peelings only may be fed to sheep.

Cassava leaves and stems have a very high crude protein content (about 25%) and are a useful cheap source of UDP. Cassava silage can be made from the whole plant or from the waste from processing factories. Adebowale (1981) fed cassava silage to West African Dwarf lambs and found that although the silage gave lower growth-rates than maize, its low cost made it an attractive food.

Soya bean (*Glycine max*)

Soya beans are largely grown in the USA and China. After the oil has been extracted, the residual meal contains about 1% oil (McDonald et al. 1981), about 50% crude protein and its ME content is about 12 MJ kg^{-1}. Because it is such a desirable food for non-ruminants, soya-bean meal is very expensive.

Groundnut (*Arachis hypogaea*)

Many parts of the groundnut plant are fed to sheep. Groundnut haulm (the aerial part of the plant after the nuts are harvested) is an excellent food provided it is harvested immediately the pods are ripe, and it can be used as hay (Combellas *et al.* 1972a). Groundnut cake is produced when the oil is extracted from the nuts. Undecorticated cake which includes the hull has a high fibre content, but decorticated cake is a desirable food for all classes of livestock, with a crude protein content of 50–60% and an ME value of about 14 MJ kg^{-1} DM. A toxic factor (known as aflatoxin) is sometimes found in groundnut cake as a result of fungal growth, but has little deleterious effect on sheep (McDonald *et al* 1981). Groundnut hulls have a crude fibre content of 32% and only 5.5% crude protein (El-Hag and George 1983). They may be used as poultry litter and subsequently fed to sheep.

Oil-palm (*Elaeis guineensis*)

Oil-palm products used as sheep foods include palm kernel meal or cake, palm press fibre and oil-palm sludge. Palm kernel meal is a good source of energy and protein and is exported from producing countries (e.g. Malaysia) for use in concentrates in many parts of the world. Palm press fibre is the waste separated from the nut in the depericarper. It has a high fibre content and low palatability and is traditionally used as fuel. However, with appropriate supplementation it can be utilised by ruminants (Hutagalung 1981). Oil-palm sludge is the effluent from the processing of oil-palms. When dried it can be stored and transported but drying is expensive. Fresh sludge is used close to the processing plants for ruminant feeding together with other palm by-products.

Date (*Phoenix dactylifera*)

Date stones are a good source of energy but have a low protein content. When crushed they can form up to 50–70% of ruminant rations (Alwash and DePeters 1982).

Coconut (*Cocus nucifera*)

The main by-product of coconuts which is fed to sheep is oil-cake. This is the residue after the oil has been extracted from the copra, and depending on the extraction process, it contains between 1 and 20% oil so that it may rapidly become rancid. It has a high crude protein content (over 20%) and a high fibre content.

Citrus (*Citrus* spp.)

Oranges, grapefruit and other citrus fruits are fed fresh (preferably sliced) to sheep when there is a glut. More usual is the feeding of the residue of the fruits after canning or juice extraction. Because of the high water content of pulp, and its perishable nature, fresh pulp is used only in sheep units close to processing factories. It has an ME value of about

13 MJ kg^{-1} DM and is added to silage to lower the pH and encourage a desirable fermentation. If citrus pulp is dried, lime must first be added (otherwise the hydrophilic pectin absorbs water and it becomes slimy). The dried pulp is known as citrus meal and is used to replace cereal in concentrates for sheep; Devendra (1973) found the optimum level of inclusion of citrus meal was 20%. Molasses made from citrus is used in a similar way to sugar molasses.

Banana (*Musa* spp.)

Bananas or plantains rejected for human consumption (especially when destined for export) are available for livestock food. They seem to be less palatable for sheep than for cattle and pigs. Bananas have low protein, mineral and fibre contents, so they are fed together with roughage and supplements of minerals and protein. The leaves of banana trees have a low digestibility (García *et al.* 1973) because of their high tannin content. Fruits, leaves and pseudo-stems may all be ensiled.

Macadamia nut (*Macadamia integrifolia*)

Macadamia nuts are grown in Australia, the Pacific islands and the USA. Feed-grade nuts have been fed to sheep, substituting for grain (Sherrod and Ishizaki 1966). However, because of their high fat content (about 70%) an inclusion rate of not more than 10% is recommended.

Guar (*Cyamopsis tetragonoloba*)

Guar or cluster bean is a drought-resistant herb grown for grain or as a vegetable. Gum is extracted from the grain and the residue, known as guar meal, has a very high crude protein content but is unpalatable to sheep and contains a toxic factor. Nevertheless, guar meal has been satisfactorily used to constitute up to 10% of the diet of fattening lambs (Anwar *et al.* 1965).

Bran

Bran is produced from most cereals, but particularly rice and wheat. It comprises the outer parts of the hulled grains, and for rice it amounts to about 10% of the total grain weight. Rice bran contains about 15% oil which results in rancidity during storage, and to avoid this the oil may be removed or the bran heated soon after milling to reduce the rate of increase in free fatty acids. Bran is reasonably palatable and is a valuable foodstuff (Chenost and Geoffroy 1975). Its crude protein content is about 12% and its ME content at least 12 MJ kg^{-1} DM.

Manure

Poultry manure is successfully used as a protein supplement in the diets of sheep (Gihad 1976a). Depending on the method of collection and the origin of the litter, its dry matter has a very high crude protein content of at least 15% and more often about 30%. A large proportion of the nitrogen (about 40%) is present as amino-acid nitrogen (Leibholz 1969). Because there is a lot of uric acid in poultry manure, the rumen microorganisms take a long time to adapt to it, and the manure should be gradually added to the diet over a period of about 3 weeks.

The mineral content of poultry manure is high but its energy content only moderate (about 10 MJ kg^{-1} DM). Fresh manure ferments rapidly and can give toxic products. It can be preserved by drying, ensiling (Hadjipanayiotou 1982) or chemical treatment to suppress harmful micro-organisms. Sheep will consume up to about 15% sawdust in the diet (Leibholz 1969), and Devendra (1976) recommended that poultry manure can be included at 20 or 30% in the diet of mature sheep. A ration containing 40% broiler litter, 30% cereal, 15% milling by-products and 15% molasses is recommended by Shah and Müller (1983). Figure 5.15 shows Pelibuey ewes eating a supplement of molasses and chicken manure. Kim (1981) recommended ensiling poultry manure with straw (previously treated with sodium hydroxide) and including this silage as 40% in the diets of growing ruminants. Other methods of preserving poultry manure and using it in diets for sheep are given by Müller (1982).

Pig manure may also be included in sheep diets, but care must be taken to ensure that it does not contain a high level of copper to which sheep are very sensitive. Human sewage can be fed to sheep in the form of activated sludge, up to about 8% of the diet (Göhl 1981). Cattle manure, together with wheat straw and poultry manure has been fed to mature ewes (Benjamin *et al.* 1982).

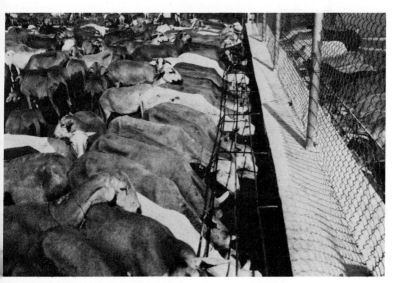

Fig. 5.15 Pelibuey ewes eating chicken manure and molasses, Yucatan, Mexico (Photo: S. M. Broom)

Integration of sheep with tree crops

The integration of ruminants with tree crops has focused mainly on cattle in coconut, oil-palm and rubber plantations (Thomas 1978). Sheep may also be grazed under these crops (Perera 1972; Devendra 1979; Salleh and Tan 1982; Vanselow 1982) and under kola, citrus (Asiedu and Appiah 1983), mango and cashew (Asiedu et al. 1978), coffee, kapok and timber. Figures 5.16 and 5.17 show sheep grazing under coconuts in Tobago and mangos in India. The commercial integration of sheep with plantations does occur, but the grazing of subsistence flocks in plantations is more widespread. The plantation owner (often the government) usually charges the sheep owner an annual fee for each sheep grazed in the plantation.

The primary income from a plantation comes from the tree crop, while that from the sheep is secondary. It is important therefore, that the sheep do not reduce the crop income by damaging trees. Sheep are much preferred to goats in this respect and have a shorter reach than cattle.

Fig. 5.16 West African ewe and lamb tethered under coconuts, Tobago

Sheep can improve crop productivity by grazing the vegetation beneath the trees, thus reducing the competition from weeds and, for coconuts, making nut gathering easier. The herbage growth under rubber and coconut plantations in Malaysia allows a carrying capacity of about four sheep per hectare (Devendra 1979), while *Brachiaria milliiformis* under coconuts in Sri Lanka is capable of carrying at least ten ewes per hectare (Perera 1972). Some species of grass (particularly *Brachiaria* spp.) and legumes (*Desmodium heterophyllum* and *Centrosema pubescens*) are well suited to the low intensities of light found in plantations. The methods of establishment and management of pastures under coconuts are discussed by Plucknett (1979) and Reynolds (1980). Cultivation costs must be kept to a minimum if the integration of ruminants is to give economic benefits. Careful management is needed to control helminth parasites which thrive in the shade beneath the trees (Vanselow 1982), but the trees also have a beneficial effect in providing shade and shelter for the sheep.

Fig. 5.17 Nellore flock herded under mangos, Andhra Pradesh, India

Food conversion efficiency

Food conversion efficiency (FCE) is defined as the quantity of product arising from unit input of food. The product is usually an increase in body weight, but it may also be milk or wool yield. The quantity of food consumed may be expressed on a fresh-weight or air-dry basis, or it may be the quantity of dry matter consumed. The FCE gives a measure of the efficiency of the sheep in converting food into product. It is a particularly useful measure for a system in which food is the primary input, and in which the output is easily quantifiable. Thus it is used to compare systems of finishing lambs, rather than to compare extensive systems which produce lambs, milk and less tangible products.

Typical values of FCE measured for finishing lambs in the tropics are between 0.07 and 0.15 kg liveweight gain per kg dry matter intake (Brinckman 1981; El-Hag and Mukhtar 1978). The highest values are recorded when sheep are eating a high-quality ration and are growing rapidly, and the lowest values when the diet quality is poor. An FCE as low as zero will be recorded if the liveweight gain is zero. The FCE for milk production is more difficult to measure because of problems of measuring milk yield accurately and because of the complication of changes in body weight in the lactating ewe. For ewes in Egypt, an experiment by Aboul-Naga *et al.* (1981a) over a 12-week period indicated that ewes produced milk with an FCE of 0.42 kg liquid milk per kg dry matter intake. The corresponding value of kg dry matter in milk per kg dry matter intake are 0.086 and 0.076, assuming that milk has a dry matter content of 19.3%. The efficiency of conversion of food into wool is much lower; the data of Dolling and Moore (1961) give an average value of 0.0081 kg clean wool per kg dry food consumed.

There has been much controversy regarding the relative efficiency of sheep and goats. There is some evidence that goats can digest low-quality roughages slightly more efficiently than goats (El-Hag 1976), but the overall food conversion of the two species depends on many other factors including diet selection and the relative production of meat, milk and wool. There is thus insufficient information to allow a true comparison of FCE under realistic management.

Factors affecting FCE

For a growing animal the FCE can be defined as the liveweight gain (LWG) divided by the dry matter intake (DMI):

$$FCE = LWG/DMI \qquad [5.1]$$

The DMI is related to the energy density of the diet (M/D), the net energy required for growth (NE_g), the efficiency of utilisation of ME for growth (k_g) and the metabolisable energy required for maintenance (ME_m) according to

$$DMI = (NE_g/k_g + ME_m)/(M/D) \qquad [5.2]$$

so that

$$FCE = LWG(M/D)/(NE_g/k_g + ME_m) \qquad [5.3]$$

This equation shows the factors which affect FCE. In practice these factors are interrelated; their individual effects are:
1. An increase in M/D results in an increase in FCE.
2. High maintenance requirements (of a large animal, or through work or other activity) give a low FCE for the same LWG.
3. A high efficiency of utilisation of ME for growth (k_g) favours a high FCE.
4. The effect of LWG on FCE is more complicated. Here, NE_g is related to LWG and more energy is needed to deposit unit weight of body tissue in a fat animal. Rams require less food than ewes and castrates for a given LWG; at a given body weight, rams contain more protein and less fat so that the energy cost of gain is less (South Africa, Department of Agricultural Technical Services 1981). However, if LWG is rapid, then the proportion of energy used for maintenance is small and FCE correspondingly large. Similarly, the efficiency of conversion of food into milk is greater when energy intake and milk yield are high (Dattilo and Congiu 1979).

Interrelationship between FCE and economics

The profitability of a finishing system depends on the relative prices of inputs and outputs, and on the efficiency of conversion of inputs into outputs. Assuming that the prices per unit liveweight of finished and store lambs are similar, then the economics of the finishing system can be described by

$$F_q \cdot F_p + I + P = O_q \cdot O_p \qquad [5.4]$$

where F_q is the quantity of food consumed, F_p is the unit price of food, I is the total cost of other inputs but excluding the cost of the store lamb, P is the profit, O_q is the increase in weight of lamb, and O_p is the unit price of finished lamb.

If I and P are expressed per unit output ($I=i \cdot O_q$; $P=p \cdot O_q$) then

$$F_q \cdot F_p = O_q (O_p - i - p) \qquad [5.5]$$

Food conversion efficiency is defined as the ratio of lamb output to food input

$$FCE = O_q/F_q = F_p/(O_p - i - p) \qquad [5.6]$$

so that the relationship between profit and FCE is

$$p = O_p - i - (F_p/\text{FCE}) \qquad [5.7]$$

Suppose that FCE = 0.1 kg liveweight kg^{-1} cereal; O_p = $2 kg^{-1} liveweight; i = $0.30 kg^{-1} liveweight; and F_p = $0.15 kg^{-1} cereal, then, according to eqn [5.7], p = 2−0.30−(0.15/0.1) = $0.2 kg^{-1} liveweight, i.e. the profit is 20 cents for every kg of meat produced. The finishing system can make a profit only if (O_p-i) is greater than F_p/FCE, or in this case if O_p is at least $1.8 kg^{-1} liveweight. Alternatively, assuming that the price obtained for finishing lambs is $1.50 kg^{-1} liveweight, the enterprise will make a profit only if the cereal price is less than $0.12 kg^{-1}.

This calculation could be further complicated because of a differential in price between store lambs and finished lambs. In a well-developed marketing system, such as in Europe, the price per kg of store lambs is greater than that of finished lambs so that technical factors must be even more favourable than the calculations above indicate if the system is to make a profit.

Chapter 6

Reproduction

Female reproduction

The oestrous cycle

In the ewe, ovulation occurs in close relation with a period of sexual receptivity known as oestrus or heat. This ensures that deposition of spermatozoa in the female tract is synchronised with ovulation. For ewes indigenous to temperate areas, oestrus occurs, during the breeding season, every 14–19 days with a mean of 17 days (Terrill 1974). Some reports show similar values for tropical ewes (Narayanaswamy and Balaine 1976; Gaillard 1979; Yenikoye *et al.* 1981). However, El-Wishy *et al.* (1971) reported mean cycle lengths of 28–30 days for Ossimi and Awassi ewes in Egypt. In their study approximately one-third of the ewes observed had lengths greater than 26 days, indicating either a failure to detect oestrus or showing that not all ewes cycled every 14–19 days. More surprising is the fact that 25% of Awassi ewes had cycles shorter than 14 days. The fertility of these ewes was high, with conception rates consistently over 90%. Cycle lengths of up to 70 days were recorded for Fulani ewes in Niger by Yenikoye *et al.* (1982) who observed that abnormally long cycles were generally limited to certain seasons of the year. It therefore seems that the duration of oestrous cycles of tropical ewes can be erratic, making prediction of oestrus difficult.

Oestrus is difficult to detect from the behaviour of ewes. Reports indicate that oestrus of tropical ewes lasts for between 12 and 72 hours (Taparia 1972; Castillo Rojas *et al.* 1977; Gaillard 1979; Yenikoye *et al.* 1981) with an average of about 45 hours. These values are not inconsistent with the 24–36 hours reported for temperate sheep (Hunter 1980), but suggest that some individuals have short oestrous periods which can make detection by the shepherd difficult, although short oestrus should not significantly reduce the chance of natural mating unless cues for the ram are weak or the ewe:ram ratio is too high.

The oestrous cycle is conveniently divided into the follicular and the luteal phases. Conventionally the oestrous cycle is said to start on the second day of the follicular phase which is when oestrus begins. The follicular phase lasts for about 3 days, and during this period follicles mature on the ovaries. In the luteal phase which starts at ovulation and lasts for about 14 days the evacuated follicles have been converted to corpora lutea. In the extremely long cycles recorded by Yenikoye *et al.* (1982) the luteal phase was extended, but cycles a few days longer than average resulted from a lengthened follicular phase.

Several factors are involved in the control of the oestrous cycle, and the most important of these are shown in Fig. 6.1. The output of

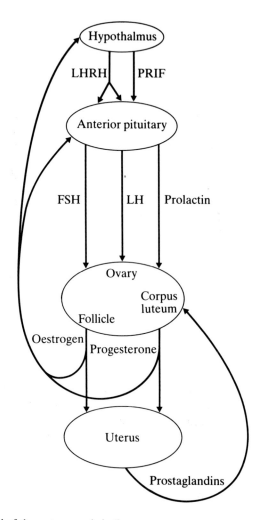

Fig. 6.1 Control of the oestrous cycle in the ewe

pituitary gonadotrophins (follicle-stimulating hormone known as FSH, luteinising hormone known as LH, and prolactin) is under the control of releasing hormones from the hypothalamus. Release of FSH and LH is stimulated by LHRH. Prolactin release, on the other hand, seems to be controlled largely by an inhibiting factor, PRIF. The production of LHRH is partly regulated by negative feedback of the gonadal steroids oestrogen and progesterone to the brain.

Follicles on the ovary develop in response to stimulation by FSH and pulses of LH. As these follicles mature they secrete oestrogens, particularly oestradiol. Oestrogens are an important factor in the initiation of sexual receptivity. Initially there is negative feedback of oestrogens on LH secretion, but later positive feedback allows a large burst of LH to be released and this initiates ovulation. The collapsed follicle develops rapidly to form a corpus luteum and produces progesterone which inhibits any major development of more follicles on the ovaries. Prolactin and LH are both necessary to maintain the corpus luteum, but in the normal oestrous cycle regression of the corpus luteum appears to result not from inadequate LH and prolactin levels, but from prostaglandins produced by the uterus.

The concentrations of progesterone, LH and oestradiol in the peripheral blood of a normal, non-pregnant, temperate ewe during the oestrous cycle are shown in Fig. 6.2. The precise levels of hormones depend on the individual ewe and the method of assay, but the general pattern of cyclic changes is standard. The progesterone level is less than 1.0 ng ml^{-1} at oestrus, rises rapidly between days 3 and 11 of the oestrous cycle, and reaches a maximum of about 3 ng ml^{-1} (Hunter 1980). On about day 15 the progesterone concentration declines rapidly to its minimal level. The concentrations of oestrogens in the peripheral blood are very low. The oestradiol concentration is about 1.5 pg ml^{-1} from day 1 to day 15, rises on the day before oestrus to reach a peak of about 3 pg ml^{-1}, and falls again soon after the start of oestrus. The concentration of LH in the peripheral plasma has a basal level of less than about 2 ng ml^{-1}, but there are pulses of LH which reach a frequency of one every 1–2 hours at the end of the follicular phase. At the beginning of oestrus the LH level rises rapidly to a peak of about 30 ng ml^{-1} which lasts for only about 6 hours.

For Peulh [Fulani] ewes in Niger, the profiles of both progesterone and prolactin are comparable with those reported for sheep in temperate regions (Yenikoye *et al.* 1981; 1982). The levels of progesterone in the luteal phase of the first postpartum cycle of Tabasco [Pelibuey], Dorset and Suffolk ewes in Mexico were studied by Martínez *et al.* (1980). They found that the concentrations of progesterone were greater for the Tabasco ewes than the Dorset and Suffolk ewes, but all the progesterone values recorded were low compared with those of sheep in temperate areas. Madani and Williams (1983) found that North Country Cheviot and Scottish Blackface ewes had significantly lower mid-cycle

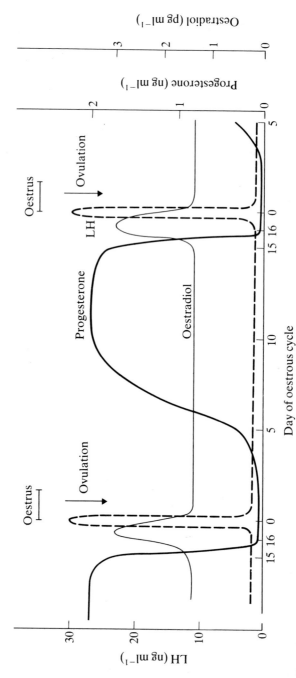

Fig. 6.2 Concentrations of progesterone, oestradiol and LH in peripheral blood during the oestrous cycle of the ewe

progesterone concentrations in an equatorial light environment than in a temperate environment. No studies of other reproductive hormones in tropical ewes have yet been reported in the literature, but with the increasing use of hormone levels in pregnancy diagnosis and to study reproductive failure, this area of study is likely to develop rapidly.

Seasonal variation in reproductive performance of ewes

Oestrus detection by teaser rams and ovulation detection by examination of ovaries in slaughter-houses has shown that some indigenous ewes in the tropics exhibit oestrus in each month of the year. For example, in Rajasthan, India, Mittal and Ghosh (1980) estimated that at least 80% of Marwari ewes exhibited oestrus in each month. Similarly, indigenous ewes in Zimbabwe (Ward 1959) and Javanese Thin-tailed ewes in Indonesia (Obst et al. 1980) do not have a clearly defined breeding season. However, the proportion of indigenous ewes which exhibit oestrus in unimproved conditions in the dry and semi-arid tropics does generally show seasonal variation primarily because of seasonal changes in food supply (Anderson 1972; Molokwu and Umunna 1980). During the summer in Rajasthan, India, the weight of Jaisalmeri ewes was reduced to 60% of their peak weight and the incidence of oestrus was lowered (Sahani and Sahni 1976). Extreme heat stress itself can result in anoestrus (Hopkins and Pratt 1976), but this effect is usually of only secondary importance. In Niger, Oudah [Uda] sheep in an experimental flock show oestrus throughout the year (Fig. 6.3), whereas those in nomadic flocks do not show oestrus in the hot dry season when nutrition is poor. In humid areas such as Indonesia where there is little annual variation in vegetation or climate, sheep do not show seasonality of births (Hardjosubroto and Astuti 1980).

Ewes imported from temperate latitudes to the tropics, and their progeny, show marked seasonal variation in the occurrence of oestrus (Anderson 1972; Williams 1975) and may even have a period of anoestrus. As British hill breeds become adapted to tropical conditions their period of anoestrus becomes shorter (Williams 1975), and as the progeny get older they too have shorter anoestrous periods. Merinos adapted to tropical Queensland, Australia show some seasonal variation, but by examining the ovaries of ewes Murray (1978) concluded that the proportion of ewes ovulating in any month did not fall below 0.78.

In hot countries outside the tropics (for instance in Namibia, Morocco and Iran) ewes show seasonal variation in oestrus even when nutrition is optimal. This seasonality is associated with photoperiod, and the best conception rates are observed when daylength is decreasing, i.e. in the autumn (Matter 1975; Marie and Lahlou-Kassi 1977; Sefidbakht et al. 1978). Daylength is sensed by the pineal gland (Arendt et al. 1981), but the effect of photoperiod is modified by other

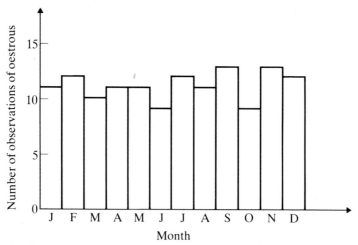

Fig. 6.3 Seasonal variation in the number of Uda ewes showing oestrus in Niger (After Gaillard 1979)

environmental factors. The time of onset of oestrus in the spring is delayed if nutrition is poor (Smith 1965).

Superimposed on the reproductive potential of ewes is the management system which, by controlling mating, may concentrate lambing in certain seasons of the year. Controlled mating is practised in many traditionally managed flocks in the dry tropics. It follows that whereas an even distribution of lamb births throughout the year indicates that ovarian activity is aseasonal, a seasonal distribution of births does not necessarily indicate that ovarian activity is seasonal.

Hormone levels during anoestrus have been measured for ewes in temperate latitudes. Prolactin levels are up to 200 ng ml^{-1} in anoestrus, considerably higher than in cycling ewes (Thimonier 1981). Negative feedback of oestradiol prevents high levels of LH which initiate oestrus in the cycling ewe, and progesterone levels remain less than 0.5 ng ml^{-1} (Pant 1979; Ammar-Khodja and Brudieux 1982).

The introduction of teaser rams into a flock of ewes can induce cycling in ewes which are anoestrous as a result of photoperiod. This may be a useful method of extending the breeding season of ewes in the sub-tropics, but it is doubtful if teaser rams could counter the effects of nutritional anoestrus. For Karakul ewes in South Africa, teaser rams can be used to stimulate oestrus as much as 2 months before the normal breeding season (Boshoff et al. 1977). Merino ewes in Australia ovulate within a few days of the introduction of rams, but in most cases this is a silent heat so that mating does not begin until about 20 days later (Lindsay 1979).

Pregnancy

Fertilisation takes place in the oviducts of the ewe. The fertilised eggs develop to the 8- or 16-cell stage before they descend into the uterus almost 3 days after ovulation. The embryos begin to attach themselves to the uterine wall on about the fifteenth day, although morphological attachments may not be visible until later.

For about 60 days after fertilisation a corpus luteum is necessary to produce progesterone and maintain pregnancy. The presence of one or more embryos in the uterus prevents the production of prostaglandins which in the non-pregnant ewe are secreted and cause the corpora lutea to regress. Later in pregnancy the foetus and placenta secrete significant amounts of progesterone so that the presence of the corpora lutea is no longer necessary (although of course they remain). Progesterone levels in the pregnant ewe therefore remain high; for the first half of pregnancy the peripheral plasma concentration is similar to the maximum observed in the oestrous cycle (i.e. about 3 ng ml^{-1}), and in the second half of pregnancy, considerably higher concentrations of about 15 ng ml^{-1} are found in temperate ewes. Progesterone levels for Peulh [Fulani] ewes in Niger follow a similar pattern but are somewhat lower than these reported for temperate ewes (Yenikoye *et al.* 1981). There is a positive relationship between the number of foetuses carried by a ewe and her progesterone concentration in late pregnancy, particularly if she is poorly nourished (Shevah *et al.* 1975). Therefore, it seems likely that progesterone levels of prolific tropical ewes in late pregnancy may be even higher than those reported for temperate ewes.

Oestrus has been observed in pregnant ewes in India, Egypt and South Africa (Matter 1974; Younis and Afifi 1979; Kandasamy and Pant 1980; Sinha *et al.* 1980). The incidence of this is low and it is seen mostly in the first months of pregnancy. Oestrus is not usually accompanied by ovulation (Williams *et al.* 1956). There is a fear that if mating takes place abortion may result (Sinha *et al.* 1980), but no evidence that it does. A more serious drawback is that ewes which continue to exhibit oestrus after mating with fertile rams may be culled because they are thought to be barren.

Mean gestation lengths reported from India and Africa range from 147 to 153 days. The gestation period is shorter for exotic than indigenous ewes (Sahani and Pant 1978), for ewes with exotic crossbred lambs than those with indigenous lambs (Rao *et al.* 1978), for female than male lambs (Narayanaswamy *et al.* 1975), for lambs with lighter birth weights (Kishore *et al.* 1980), and there are seasonal and year effects (Narayanaswamy 1978).

The hormonal control of parturition is summarised in Fig. 6.4. About 2 days before parturition the concentration of progesterone in the circulating blood falls, while the concentration of oestrogens rises abruptly. These changes appear to occur because of the effects of

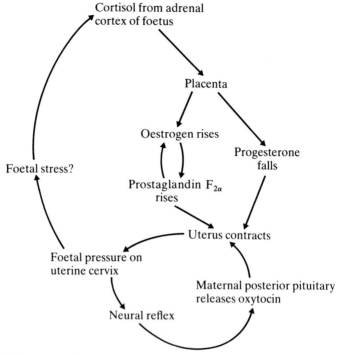

Fig. 6.4 The control of parturition (After Hogarth 1978)

cortisol and prostaglandins on the placenta. The withdrawal of progesterone and the increased concentrations of oestrogens and prostaglandins cause the uterus to contract. Foetal pressure on the uterine cervix results in a reflex release of oxytocin from the ewe's posterior pituitary. Oxytocin synergises with the prostaglandins to increase the intensity of the uterine contractions and thus aids the final step in delivery.

Reports show that although ewes lamb throughout the day and night, there are significant diurnal variations in the frequency of parturitions. This distribution, however, appears to vary with the breed of ewe and the management system (Kaushish *et al.* 1973; Younis and El-Gaboory 1978; Tomar 1979; Bhaik and Kohli 1980).

Puberty and longevity of ewes

Puberty in the ewe is defined as the time when oestrous cycles start. Awassi lambs on a good diet first display oestrus at an average age of 274 days (Younis *et al.* 1978), but Rambouillet crossbred lambs in Rajasthan are about 615 days old before they display oestrus (Kishore *et al.* 1982). This difference is largely due to the different growth-rates resulting from different nutrition. In Egypt the average age of puberty of Ossimi and Barki ewe lambs was 347 days when reared on a high plane of nutrition, but 366 days on a low plane (El-Homosi and El-Hafiz 1982). Females must attain a certain liveweight (which depends on adult liveweight) before they will cycle. Additional variation in age at puberty occurs in the sub-tropics where ewes are seasonally anoestrous so that if a lamb fails to cycle in her first season she will not cycle until the next, several months later.

Longevity is defined as a ewe's age when she dies or is culled. Unlike puberty, longevity receives negligible scientific attention yet it is of critical importance in determining the reproductive rate of a flock of sheep; if there is a high replacement rate for females it is necessary to keep a high proportion of young females in the flock so that the proportion of reproductive females and thus flock offtake are reduced. Age structures of flocks are, however, difficult to determine because there is no way of ageing ewes once they are full-mouthed, unless they are individually identifiable.

Mortality rates in Africa are about 10% per year (Wilson 1982). In addition to natural deaths, ewes may be culled because they become unproductive. Reproductive performance of ewes increases to a maximum at about 6 years of age. Thereafter a ewe's ability to produce lambs is likely to fall more as a result of poor body condition than specific reproductive disorders. Loss of teeth and lameness result in a ewe receiving inadequate nutrition unless she is stall-fed. Mastitis and other udder dysfunctions mean she is no longer able to suckle lambs and she may therefore be culled. Specific diseases such as pneumonia, enterotoxaemia and abscesses were found to be the most frequent causes of culling ewes in Oregon, USA (Norman and Hohenboken 1979), but other causes are probably relatively more important in ewes in the tropics. In Java, ewes are sold when they become difficult to mate because of the fat tail (Mason 1978). There are breed differences in longevity (Vera *et al.* 1978) and, contrary to popular opinion, comparison between systems shows that there is a positive relationship between productivity per year and longevity (Hohenboken and Clarke 1981). In a study with Caucasian ewes, Nikolaeva (1974) found that inbreeding shortened reproductive life. Assuming that the total replacement rate of adult ewes is 20% per year, the average reproductive life of a ewe is approximately 5 years, a value similar to that of 4.3($\pm$0.25) years observed for Rambouillet ewes in India by Tomar and Mahajan (1980). The lambing interval of these Rambouillet ewes was long so that a typical ewe had only three lambings in her lifetime!

Increasing the longevity of ewes increases flock productivity because of the higher proportion of reproductive ewes in the flock and because the individual productivity of older ewes is greater than that of young ewes. However, if genetic selection is practised the annual genetic gain will be low if the generation interval is long. Decreasing the age of puberty, on the other hand, results in both an enhanced productivity and an enhanced rate of genetic gain.

Male reproduction

The reproductive organs of the ram are physiologically similar to those of other male mammals. Sperm are produced in the testes which lie in the scrotum outside the abdomen, and are stored in the coiled epididymis attached to each testis. Before mating, sperm move into the ductus deferens where they are mixed with diluents secreted by the accessory glands. The semen then moves into the penile methna from where it is deposited in the ewe's reproductive tract.

The hormones FSH, LH and testosterone are all involved in the control of reproduction in the ram (Courot and Ortavant 1981). Testosterone produced by the testes regulates spermatogenesis, the function of the accessory glands and behaviour. Testosterone production is under the control of FSH and LH (also known in the male as interstitial cell stimulating hormone) produced by the anterior pituitary gland.

A ram seeks out ewes in oestrus and they form a harem which follows him (Bourke 1967, Mattner et al. 1967). When there are two or more rams in one flock, the old rams dominate the young rams (Hulet et al. 1962) and may prevent the subordinate rams from breeding. If the most dominant ram is infertile this behaviour could have a serious effect on flock fertility. The ability of a ram to breed may partly be assessed by physical examination including general health, and the penis, prepuce and testes (Ott and Memon 1980). In addition, his libido and semen must be evaluated; infertility may be associated with morphological abnormalities in sperm (Arora and Pant 1979). Rams are more successful at mating if they have had heterosexual contact before or subsequent to puberty (Roux and Barnard 1974). Some ewes with fat tails are difficult to serve (Goot et al. 1980; George 1982) and copulation may not take place unless the shepherd assists by holding the ewe's tail to one side.

The number of ewes that can be mated by one ram is large. If management is suitable, a ewe to ram ratio of 200:1 is satisfactory (Allison 1975). Under extensive conditions ratios of between 30 and 50:1 are more usual (Doney et al. 1982), and Devendra and McLeroy (1982) state that a ratio of only 10 or 20:1 is common for tropical sheep. In smallholder systems of production the ratio may be even lower; Devendra (1979) reported that small flocks of sheep and goats in Southeast Asia contain only two to eight animals, of which one or two are males.

Reports of puberty in rams, defined as age at first ejaculate, vary from 132 days for Tabasco [Pelibuey] × Dorset lambs on the high plateau in

Mexico (Valencia et al. 1977) to as old as 738 days for 3/4 Rambouillet 1/4 Malpura lambs in Rajasthan, India (Tiwari and Sahni 1981). The quantity and quality of ejaculates collected from pubescent rams are poor with a high incidence of dead and deformed spermatozoa (El-Wishy 1974).

In the tropics some workers have found that the quantity and quality of semen produced by rams depend on the season (Nivsarker et al. 1971; Saxena et al. 1979), while many others have found no significant effect of season. It appears that exotic breeds such as the Rambouillet and Polwarth give semen of lower concentration and worse sperm motility than indigenous tropical rams in some seasons of the year (Johari 1973), thus resulting in lower overall fertility. It is possible to improve the quantity and quality of semen produced by exotic rams by keeping them in a cooled room during the day and giving supplementary feed (Kaushish and Sahni 1976), but in practice it is difficult to provide adequate cooling by housing (Tiwari and Sahni 1974; Dwivedi 1977). In India, feeding of a herbal preparation is used to improve libido and semen quality (Singh et al. 1978).

Reproductive diseases

In addition to diseases which lower the overall performance of sheep in the tropics, some diseases specifically cause foetal mortality, abortion and ram infertility. Vibriosis (caused by *Campylobacter fetus intestinalis*), salmonellas (particularly *Salmonella abortus-ovis* and *S. dublin*), *Listeria monocytogenes*, chlamydiae, sheep ticks (*Ixodes ricinus*) carrying tick-borne fever, Border disease, mycotic abortion, 'Q' fever (*Coxiella burnetti*) and *Toxoplasma gondii* may all cause foetal mortality and abortion in ewes. *Brucella abortus,* however, is not an important cause of abortion in sheep. Infertility in rams can result from infection with infectious epididymitis (caused by *B. ovis* or *Actinobacillus seminis*). The epidemiology, symptoms, diagnoses and control of these diseases are discussed by Watson (1979). Structural defects of the genital organs are uncommon in sheep (Arthur et al. 1982).

The magnitude of the losses causes by these reproductive diseases in the tropics is not known, but is certain to be very much lower than the losses caused by more general factors such as poor ewe nutrition and lamb mortality.

Reproductive rate

Reproductive rate (RR) can be defined as the total number of lambs weaned per ewe of reproductive age per year. This is related to litter size (S), lamb mortality (M) and lambing interval in years (I) such that

$$RR = S(1-M)/I \qquad [6.1]$$

This is equivalent to the type of formula used for once-a-year lambing, $RR = S(1-M)(1-B)$, where B is the proportion of barren ewes, i.e. those failing to lamb. Alternatively, the reproductive rate for the whole flock, i.e. lambs weaned per sheep in the flock per year (RR') is given by

$$RR' = S(1-M)P/I \qquad [6.2]$$

where P is the number of ewes of reproductive age expressed as a proportion of the total flock.

Other definitions of reproductive performance are found in the literature. There are several variations as to the exact definitions, but the most useful are:

Fertility = number of ewes lambing/number of ewes available for mating.
Prolificacy = litter size = number of lambs born alive/number of ewes lambing.
Fecundity = fertility × prolificacy = number of lambs born alive/ewe available for mating.
Lambing rate = number of lambs born/ewe available for mating, often expressed as a percentage.
Weaning rate = number of lambs weaned/ewe available for mating, often expressed as a percentage.

Litter size (S)

Table 6.1 shows litter sizes for ewes in the tropics. Most values of S are between 1.00 and 1.50, so that the percentage of ewes having twins is generally between 0 and 50%. (Values of litter size less than 1.00 arise where the number of pre- or perinatal deaths is greater than the number of twins born in a flock.) Some breeds of tropical sheep such as the West African Dwarf, Priangan and Barbados Blackbelly are highly prolific and regularly give litters of two or more lambs. These breeds are found mainly in humid and sub-humid climates and are kept in small flocks either completely or partly housed. They receive individual attention from their owners, particularly with respect to feeding of ewes and care of lambs. In addition to having a large litter size, these breeds have early sexual maturity, short lambing intervals and the males have high libido. It seems that, although the heritability of reproductive traits is low, long-term selection in a favourable environment has resulted in high prolificacy (Mason 1980a).

Litter size depends primarily on the number of eggs shed by the ewe, i.e. her ovulation rate. Secondary factors are the proportion of eggs fertilised, losses of embryos and foetuses causing reduction of multiple foetuses, and perinatal deaths. For D'man ewes in Morocco, Lahlou-Kassi and Marie (1981) found that on average the ovulation rate was 2.5,

Table 6.1 *Litter size (S) of sheep in the tropics*

Breed	Location	S	Source
Abyssinian	Ethiopia	1.05	Wilson 1982
Australian Merino	Australia	1.18–1.32	Kennedy et al. 1976
Awassi	Iraq	1.02	Asker and El-Khalisi 1965
Awassi	Egypt	1.12	El-Wishy et al. 1971
Awassi	Israel	1.09	Finci 1957
Awassi	Iraq	0.96	Tešanović 1979
Baggara	Sudan	1.14	Wilson 1982
Barbados Blackbelly	Venezuela	1.45	Bodisco et al. 1973
Barki	Egypt	1.05	Aboul-Naga 1976
Blackhead Persian	Venezuela	1.04	Bodisco et al. 1973
Bornu	Mali	1.05	Wilson 1982
Corriedale	Bolivia	1.14	Choque and Cardozo 1974
Corriedale	Cuba	1.28	Santa-María et al. 1979
Criollo	Venezuela	1.13	Bodisco et al. 1973
Desert Sudanese	Sudan	1.30	Suliman et al. 1978
Dorper	Botswana	1.10	Botswana 1980
East Java Fat-tailed	Indonesia	1.56	Mason 1978
Fulani	Chad	1.07	Dumas 1980
German Mutton Merino	Israel	1.67	Morag et al. 1973
Javanese Thin-tailed	Indonesia	1.36	Mason 1978
Lohi	Pakistan	1.23	Ishaq and Mumtaz-Ali 1959
Lohi	Bangladesh	1.52	Rahman and Huq 1976
Masai	Kenya	1.02	Wilson 1982
Maure	Chad	1.01	Dumas 1980
Maure	Chad	1.07	Wilson 1982
Mayo-Kebbi	Chad	1.64	Dumas 1980
Najdi	Saudi Arabia	1.23	Ramadan et al. 1977
Nilgiri	India	1.08	Raman et al. 1981
Nungua Blackhead	Ghana	1.41	Asiedu and Appiah 1983
Nungua Blackhead	Ghana	1.13	Ngere and Aboagye 1981
Ossimi	Egypt	1.14	Aboul-Naga 1976
Ossimi	Egypt	1.10	Aboul-Naga 1978
Ossimi	Egypt	1.14	El-Wishy et al. 1971
Ossimi	Egypt	1.11	Labban and Ghali 1969
Pelibuey	Mexico	1.20	Castillo Rojas et al. 1972
Pelibuey	Mexico	1.17	Gonzales Reyna and Alba 1978
Priangan	Indonesia	1.70	Mason 1978
Priangan	Papua New Guinea	1.48	Holmes and Leche 1977
Rahmami	Egypt	1.23	Aboul-Naga 1976
Rambouillet	Bolivia	1.25	Choque and Cardozo 1974
Rambouillet	India	1.00–1.38	Kaushish and Sahni 1977
Romney Marsh	Colombia	1.02	Naranjo and Sabogal 1978
South African Merino	South Africa	1.10–1.37	Heydenrych 1977
Soviet Merino	India	1.00–1.38	Kaushish and Sahni 1977
Soviet Merino	India	1.00	Raman et al. 1981
Suffolk	Egypt	1.08	Aboul-Naga 1978
Suffolk × Ossimi	Egypt	1.15	Aboul-Naga 1978
Targhee	Bolivia	1.43	Choque and Cardozo 1974
Tswana	Botswana	1.01	Botswana 1980
Uda	Niger	1.07	Haumesser and Gerbaldi 1980
West African	Venezuela	1.43	Bodisco et al. 1973

Table 6.1 (Cont)

Breed	Location	S	Source
West African	Venezuela	1.12	González-Stagnaro et al. 1980
West African Dwarf	Ghana	1.29	Asiedu and Appiah 1983
West African Dwarf	Ivory Coast	1.00–1.31	Berger and Ginisty 1980
West African Dwarf	Chad	1.24–1.53	Dumas 1980
West African Dwarf	Senegal	1.12	Fall et al. 1982
West African Dwarf	Ghana	1.64	Ngere and Aboagye 1981
West African Dwarf	Togo	1.49	Amegee 1983b
Yankasa	Nigeria	1.23	Molokwu and Umunna 1980

but only 42% of eggs shed developed to full-term foetuses. For Merino ewes in Australia the major source of wastage is non-fertilisation (Entwistle 1972). Ovulation rate varies between breeds of sheep, increases with age of ewe up to about 6 years, and for seasonally breeding ewes is greatest in the first half of the breeding season (Hafez 1974). For individual ewes, ovulation rate has a reasonably high repeatability (Bindon and Piper 1979). There is no evidence that monozygotic twins (those originating from a single ovum) are produced by sheep.

Perinatal deaths occurring at full-term parturition can cause significant losses, which can be greatly reduced by good management. In commercial flocks in Brazil, Selaive et al. (1980) reported that 20% of lambs were stillborn, and for Merinos in South Africa, Marais (1974) gave a value of 6%. In areas such as semi-arid West Africa where multiple births are not desirable the weaker twin may be deliberately killed. Small lambs are usually weaker than larger lambs, yet larger lambs are more likely to suffer dystocia. Lambs born in multiple births are more likely to have complicated lambing presentations than singles, and together with their smaller size, this results in a much higher perinatal death-rate for multiple than single births. As ewes become more experienced, their ability to give birth and to care for their newborn lambs improves so that perinatal mortality decreases with parity. The magnitude of perinatal mortality depends to a large extent on the system of management, and in areas with prolific breeds of sheep, management practices have evolved to reduce perinatal losses to a low level.

Lamb mortality (M)

Lamb mortality between birth and weaning at, say, 150 days is typically between 10 and 30% in traditionally managed flocks in the tropics. Lamb mortality is significantly affected by year. For instance, the mortality of Djallonké [West African Dwarf] lambs in Senegal was 54% in 1978 compared with an overall mean of 33% (Fall et al. 1982), and Valencia et al. (1978) reported a mortality of 70% in Mexico.

Factors affecting lamb mortality are the age of lamb, litter size, birth weight, season, nutrition of the ewe and parity of the ewe. Lamb mortality declines with age; a lamb is most likely to die soon after birth, and the older it gets the better its chance of survival (Ward 1959). Lambs born in large litters have higher mortality rates than single lambs; in India the mortality of Bikaneri lambs was 17% for triplets, 6% for twins and only 1% for singles (Seth *et al.* 1972). In Uganda the mortality rates for twins and singles were 28 and 16% respectively (Trail and Sacker 1966). Part of this effect must be due to birth weight because bigger lambs have a better chance of survival, but it must also result from the better attention and the more milk that the ewe can give to a single lamb.

Lambs born in unfavourable seasons of the year have worse mortality rates than those born in favourable seasons (Malik *et al.* 1980). The rainy reason is generally the worst in the semi-arid and sub-humid tropics because then lambs suffer from both cold stress and from a high disease challenge. Fall *et al.* (1982) found a positive correlation between monthly rainfall and pre-weaning mortality such that there was an increase of 0.014% in mortality for each additional 1 mm of monthly rainfall. In other areas such as tropical Australia, heat stress is thought to be a serious cause of lamb mortality unless shade is provided (Squires 1981). Ewes which are undernourished give birth to small lambs and have a low milk yield so that their lambs have high mortality. The parity of ewe affects lamb mortality as young ewes are inexperienced at looking after lambs and they have a low milk yield. Suliman *et al.* (1978) found that the mortality of Desert Sudanese lambs was 20% for ewes with one pair of permanent incisor teeth, 18 and 5% for ewes with two and three pairs of permanent incisors respectively, and only 3.5% lamb mortality for older ewes. Parity and age are obviously closely related, but there must be an age limit for each ewe beyond which she loses condition and her ability to rear lambs declines.

Lambing interval (I)

The time between successive lambings is known as the lambing interval. It is the sum of the duration of pregnancy (about 150 days) plus the service period which depends on the duration of postpartum anoestrus. The frequency distribution of lambing intervals for a flock is skewed (Wilson *et al.* 1981) with the majority of intervals between 7 and 12 months, but some over 20 months. Values in the literature of lambing interval give flock means of between 218 days for Morada Nova sheep (Teixeira *et al.* 1980) and 408 days for Mandya sheep in India (Purushotam 1978). However, because of the asymmetric distribution of intervals, the use of the mean to describe the flock average is inappropriate because it is unduly affected by the few very long intervals in the flocks. The median or mode intervals are more appropriate.

Long lambing intervals arise because ewes have either a long postpartum anoestrous period, they are mated but do not conceive, all the embryos die, or the ewe aborts. Under good management, ewes which do not breed successfully are removed from the flock so that the proportion of ewes with long intervals is low. Lambing interval decreases with parity up to 4, suggesting that young ewes which are still growing take longer to regain condition after lambing. For Masai sheep in East Africa (Wilson *et al.* 1981) and Djallonké [West African Dwarf] sheep in Senegal (Fall *et al.* 1982) the lambing interval is affected by season, presumably as a result of food availability.

Proportion of reproductive ewes in the flock (P)

The value of P is affected by the age of females at puberty, the length of reproductive life, lamb mortality, the ratio of rams to ewes and many other factors. Traditional flock owners maintain their flock structures to maximise output while keeping risk to a minimum. Rams and other categories of sheep kept in the flock may contribute to flock economy through the production of hair or wool. In the semi-arid parts of Africa P ranges from 48 to 64% (see Table 2.3) with a mean of $57(\pm 1.8)$%. In a survey of sheep production in the Syrian steppe, Thomson and Bahhady (1983) reported that on average 73% of the flock were breeding ewes, 22% yearling females and 5% rams. Taking into account the proportion of lambs in the flocks, these figures give a value of P similar to those in Africa. In traditional systems of management in Sri Lanka, Ravindran *et al.* (1983) reported that P was 56% during the lambing season, and rose to 71% in the breeding season.

Magnitude of reproductive rate

Values of reproductive rate can be calculated using eqns [6.1] and [6.2]. Putting in typical values of $S = 1.0$ for breeds in the dry tropics and 1.6 for prolific breeds in the humid tropics, $M = 0.15$, $I = 0.75$ year and $P = 0.55$ gives RR as 1.1 lambs per ewe per year for the dry tropics and 1.8 lambs per ewe per year for prolific breeds in the humid tropics. Here, RR' is 1.6 and 1.0 lamb per sheep per year for the semi-arid and prolific breeds respectively. There are few estimates of the reproductive rate of traditionally managed flocks in the tropics, but in Niger, Haumesser and Gerbaldi (1980) calculated RR as 1.02 lambs per ewe per year, and calculations based on the data of Wilson (1980) for sheep in the Sudan, Ethiopia, Mali and Kenya give reproductive rates between 0.8 and 1.1 lambs per ewe per year. These values are comparable with those calculated from the components.

The components of reproductive rate (S, M, I and P) are all interrelated and this must be considered when attempts are made to increase the reproductive rate. For instance, a 10% increase in litter size will not directly result in a 10% increase in RR because the higher litter size will be accompanied by higher lamb mortality and longer lambing intervals unless management is also improved. Similarly, the breeding of ewe lambs at a young age reduces P, but the full impact of the reduction in P is not seen in RR' because young ewes tend to have small litters, high lamb mortality and long lambing intervals.

Effect of nutrition on reproductive rate

Nutrition has a large effect on the reproductive rate of sheep. In many cases the reproductive performance of ewes is lowered because the overall quality of available food is low, but in other circumstances specific deficiencies (for instance of protein, minerals or vitamins) or excesses (e.g. oestrogens) may cause problems.

Severe undernutrition of ewes increases the duration of post-lambing anoestrus, thus increasing the service and lambing intervals. Ewes which are heavier at mating have larger litters than lightweight ewes (Reddy et al. 1979). In temperate countries flushing is practised; this means giving the ewes extra food in the 3 or 4 weeks before the rams are introduced, and for several weeks afterwards (Coop 1966). Flushing increases the ovulation rate of ewes, particularly at the beginning and end of the breeding season, so that twinning rate increases by up to 30%. Both the absolute level of nutrition and its rate of increase with time are thought to be important. The response to flushing may be small for prolific breeds; flushing Forest [West African Dwarf] ewes in Ghana did not improve their lambing rate (Asamoah-Amoah 1977). Also, in many extensive systems twins are usually undesirable, so that attention centres on encouraging early ovulation after lambing by adequate feeding of the postpartum ewe rather than increasing the ovulation rate by flushing.

Nutrition during early pregnancy is thought to be relatively unimportant (Honmode and Patil 1975), although both extremely high and very low levels of feeding can result in embryonic loss (Doney 1979). Partial loss of embryos results in reduced litter size, but loss of all embryos means that the lambing interval is increased. As pregnancy proceeds the ewe's nutritive needs rise exponentially and for a ewe carrying twins the ME requirement at the end of pregnancy is approximately twice her maintenance requirement (ARC 1980). For a ewe carrying twins, the ME requirement rises to almost two and a half times that of the maintenance requirement. Unless the ewe is fed a good-quality diet her food intake will be inadequate and she will not satisfy her nutritive needs in late pregnancy, with the result that her body condition will deteriorate. This situation will be aggravated because the voluntary food intake of a ewe may be depressed for a few days around lambing. Supplementary feeding during late pregnancy may increase the birth weights of lambs, and thus reduce peri- and postnatal mortality (Edey 1969). McArthur (1980) demonstrated that, in Afghanistan, selective feeding of weak ewes produced economic benefits.

Little is known about the effect of protein *per se* on reproduction (Doney 1979). In certain parts of the world specific mineral deficiencies reduce the productivity of sheep. Kategile et al. (1978) found that an iodine supplement increased the lambing percentage of Blackhead Persian ewes in Tanzania by up to 17%. The effect of iodine supplementation on semen quality has also been investigated (Kaushish and Sahni 1976). Zinc and selenium are also thought to be related to reproductive performance. In Egypt, metabolic profile tests showed that infertile ewes had lower serum calcium, inorganic phosphorus, magnesium, copper and iron levels than fertile ewes (El-Sherif and Salem 1978). Excess salt can cause a depression in reproductive performance. In India, Singh and Taneja (1979) found that if drinking-water contained 1% saline or more, then lambing percentage and lamb birth weights were depressed. The effects of mineral nutrition are discussed in more detail in Chapter 4. Research on the effects of vitamins on the reproduction of tropical sheep has generally been inconclusive, probably because other nutritional effects are more serious limiting factors.

Oestrogenic activity in forage plants and its association with reproductive problems in farm animals has been reported from countries throughout the world (Bickoff 1968). However, it is difficult to assess the magnitude of this problem. Temporary infertility results when ewes graze on oestrogenic pasture at the time of mating, and permanent infertility occurs after prolonged grazing on clover (Adams 1979). Many of the substances in plants which have oestrogenic activity are called pro-oestrogens because they themselves are inactive but are converted by the sheep into active substances (Moule et al. 1963).

Nutrition affects the growth of ewe lambs and thus the age at which they reach puberty (El-Homosi and El-Hafiz 1982). Early growth, however, has little effect on the reproductive performance of mature ewes (Allden 1970).

Modern technologies for reproduction

Artificial insemination

Artificial insemination (AI) began to be used regularly for cows in the USA and UK in the 1940s. Now it is a highly successful technique for exploiting semen from superior bulls, but is less successful in sheep.

Semen is generally collected from rams using an artificial vagina, although electro-ejaculation is also practised. The artificial vagina has a water jacket maintained at about 41 °C. Semen is collected as the ram mounts a teaser animal or dummy and ejaculates into the vagina (Fig. 6.5). Several false mounts may be used to increase sperm output. A comparison of methods to collect semen in Mexico (Hernández et al. 1976) showed that the volume of semen and semen quality were not affected by the method of collection, but use of an artificial vagina gave higher sperm concentrations than electro-ejaculation.

Semen is evaluated in terms of its volume and the concentration,

Fig. 6.5 Collecting semen from a Pelibuey ram, Yucatan, Mexico (Photo: S. M. Broom)

motility and morphology of spermatozoa. Ram semen has a lower volume but higher concentration of sperm than bull semen. Typical values for rams are semen volume 0.7–2.0 cm^3, concentration 2.0–6.5 × 10^9 cm^{-3}, motility 70–90% and abnormal sperm 5–15% (Memon and Ott 1981). Sperm motility and morphology are usually estimated under a microscope, and sperm concentration may be determined using a haemocytometer, calibrated spectrophotometer or electronic counter.

An enormous amount of work, particularly in India, has been conducted on methods of preserving ram semen for AI. Methods of preservation have been reviewed by Memon and Ott (1981). Semen is diluted using an extender to allow a greater number of ewes to be inseminated. The extender must provide a suitable environment for the sperm, giving them a source of energy and protecting them against temperature, pH and osmotic shocks. Most extenders are based either on milk or egg-yolk. Cow's milk is superior to ewe's milk; it is heated to remove certain proteins and an antibiotic is added. For freezing, egg-yolk is more commonly used and a variety of other components, including sugars and glycerol, may be added. Semen can be diluted by a factor of at least 10 without a drop in conception rate, provided that a suitable extender is used.

If semen is stored above its freezing-point, sperm motility is considerably better if it is kept cool than above 30 °C (Sahni and Roy 1972a). Most diluted semen is therefore kept at 5 °C. However, it is difficult to maintain semen at 5 °C when transporting it in a hot country; Tiwari and Sahni (1976a) reported the effectiveness of different types of transporting container.

Storage of semen for periods longer than 1 or 2 days is possible only if it is frozen. The techniques for freezing ram serum are still being refined, but some general aspects are well known. Immediately after collection the semen is cooled slowly, is extended at 5 °C and glycerol is added to protect the sperm against the otherwise lethal effects of freezing. The extended semen is put into either ampoules, straws or pellets and is rapidly cooled to -196 °C above liquid nitrogen. Rapid thawing of the semen immediately before use gives maximum viability and, once thawed, sperm cells do not remain viable as long as those which have never been frozen.

In AI, semen is usually deposited in the cervix of the ewe. Practical details of the insemination technique are given by Sorensen (1979). For temperate sheep, the optimum time of insemination is mid-oestrus, i.e. 15 hours after the onset of oestrus. For Awassi sheep in Israel too, the highest conception rates were obtained when the ewes were inseminated 8–24 hours before the end of oestrus. Oestrus synchronisation of ewes allows insemination of all the ewes in a flock at a predetermined time (Colas and Courot 1979), thus avoiding problems of oestrus detection.

The success of techniques of semen collection and preservation is most often reported as percentage sperm motility rather than as conception rate or lambing percentage. However, in Brazil, Dutra *et al.* (1980) reported conception rates of up to 71%, and in India lambing rates as high as 86% from AI have been reported (Roy *et al.* 1962). The low number of reported successes with AI in sheep suggests that most attempts give unsatisfactory results. The success rate is better for fresh than frozen semen, for ewes inseminated twice not once, and the time of insemination is crucial (Sahni and Tiwari 1973).

The use of AI for sheep is not widespread in the USA, Australia or the UK. The main reason is that sheep are maintained on extensive systems and receive little individual attention. Oestrus detection or synchronisation is difficult and the labour costs would be prohibitive. In the USSR, on the other hand, where flocks are run on state farms and labour costs are relatively unimportant, AI is widely practised. In Ireland and France AI for sheep is also widespread. In less developed countries AI faces enormous problems. It requires skilled technicians, relatively sophisticated apparatus and a degree of infrastructure (transport, organisation, etc.) which is rarely found. The potential advantages of AI are the extensive dissemination of superior genotypes, reduced investment in large numbers of rams and its use in selection programmes. However, the present benefit from AI is minimal, and its future potential even when technical problems have been solved is likely to be limited. In certain circumstances, for instance in importing semen for development of breeds on research stations, AI may be useful, but it seems unlikely that its widespread use in most tropical countries will develop. Therefore the diversion of considerable national resources into improving techniques associated with AI must be criticised; it appears that research is being conducted for the prestige rather than for practical purposes.

Oestrus synchronisation

Synchronisation of oestrus and ovulation allows AI to be used more efficiently. The most commonly used method of oestrus synchronisation in sheep is a vaginal sponge impregnated with progestagen which is administered for 14 days, and following withdrawal, a single dose of pregnant mare serum gonadotrophin is given (Britt 1979; Combellas et al. 1980). An alternative method, using subcutaneous implants, gives similar results although the interval from removal of progestagen to oestrus and the duration of oestrus are both shorter with implants than sponges (Faure et al. 1980). Synchronisation is also possible by daily injection of progesterone (Dhanda and Arora 1977) and by feeding melengestrol acetate (Tripathi 1977). The techniques of oestrus synchronisation thus seem well developed and have been demonstrated to work satisfactorily on research farms in Africa, Asia and South America (e.g. Steele 1983c), but in the field conception rates after oestrus synchronisation are usually poor unless nutrition is good (Westhuysen et al. 1981).

Ovum transfer

Ovum transfer, also known as embryo transfer, in farm animals has been reviewed by Trounson and Rowson (1976) and Willadsen and Polge (1980). It is a highly sophisticated technique which, together with superovulation allows the rapid multiplication of valuable genotypes. In sheep the technique does not yet have a high rate of success (Armstrong and Evans 1983), although it has been used in India to multiply wool-producing genotypes (Zanwar 1981).

Superovulation

In an attempt to increase the lambing percentage of ewes, the ovaries may be stimulated with exogenous gonadotrophins to increase ovulation rate (Hunt et al. 1971). Attempts to superovulate ewes in India have met with variable success (Pandey et al. 1972b; Sengupta et al. 1978), and in addition, lambs born in multiple births have less chance of survival than those born in single births (Seth et al. 1972). Therefore superovulation is likely to be used only in conjunction with embryo transfer, not as a technique to increase litter size.

Short lambing interval

Within the tropics the aseasonality of breeding means that an 8-month lambing interval is possible provided that nutrition and management are satisfactory (Naude and Grant 1979). In the sub-tropics where seasonal anoestrus results from photoperiod, hormone treatment can be used to stimulate oestrus in the non-breeding season (Fletcher *et al.* 1980), following techniques devised in temperate areas (Carpenter and Spitzer 1981). These hormonal techniques work reasonably well except for the postpartum ewe.

Typically the period between lambing and conception of poorly fed ewes is about 180 days if no hormonal treatment is given (Sahni and Roy 1972b). If fed intensively, Djallonké [West African Dwarf] ewes will conceive at an average of 43 days after lambing (Berger and Ginisty 1980). Similarly, Orji and Steinbach (1980) reported that Nigerian Dwarf [West African Dwarf] ewes grazing improved pastures and given concentrates first showed oestrus at an average of 55 days after lambing. Oestrus can be induced by treating ewes with intra-vaginal sponges (as for oestrus synchronisation) as early as 17 days after lambing (Niekerk 1979). The fertility resulting from this treatment is very low, but increases if the period between lambing and treatment is increased. Honmode *et al.* (1971) successfully induced parturition in indigenous and crossbred Indian ewes by inserting sponges 30 days after lambing. Early postpartum breeding of Nungua Blackhead ewes in Ghana (Brown *et al.* 1972) reduced the lambing interval to 6 months and the annual lambing rate was increased to 150% of the normal. In Israel, Awassi and Assaf ewes which lambed twice in a year produced 30% more milk than those which lambed only once, even though earlier conception lowered the yield per lactation (Eyal *et al.* 1978).

These techniques require a high standard of management, and in particular, good nutrition, if they are to result in higher lambing percentages. Frequent lambing puts more strain on the ewe, so that unless she is fed a superior diet she loses weight (Sahni and Tiwari 1974). Aseasonal production of lambs results in high lamb mortality (Labban and Ghali 1969) and poor growth-rates (Ganesakale 1975) for those lambs born in unfavourable seasons of the year, so that the overall effect of frequent lambing must be carefully evaluated before it is advocated.

Reducing the age of puberty

The age at which puberty is reached depends on the nutrition of the lamb. Some flock owners do not allow their gimmer lambs to mate immediately they reach puberty, but there is no evidence from traditional systems that early first lambing results in diminished reproductive capacity (Wilson and Durkin 1983). The gonads of pre-pubertal sheep respond to exogenous hormones, but animals stimulated in this way do not commence spontaneous oestrous cycles (Hunter 1980). It is unlikely that hormonal stimulation of lambs will be of practical use in the tropics.

Pregnancy diagnosis

Pregnancy diagnosis allows management to take account of the varying needs of the ewes, particularly the diversion of scarce resources (e.g. food) to pregnant ewes. Some methods of pregnancy diagnosis are relatively simple. For example, the failure of a ewe to return to oestrus gives a good indication that she is pregnant. A harness with a marking raddle fitted to the rams allows the shepherd to see which ewes have been mated in his absence. Raddle marks are not permanent so that they must be supplemented with some recording system. The use of raddles with teaser (sterile or aproned) rams allows identification of ewes on heat without mating taking place.

Physical examination of ewes in late pregnancy is simple and quick, but has low accuracy. By gently palpating the ewe's abdomen, the lamb can be felt, and the udder is firm and enlarged if the ewe is pregnant. A rectal probe can be used to palpate the uterus of ewes from about 70 days of pregnancy and is reliable when used by an experienced operator (Plant 1980).

More sophisticated methods of pregnancy diagnosis were reviewed by Richardson (1972). She found that only three methods (vaginal biopsy, ultrasonic foetal pulse detection, and radiography) gave accuracies of over 80%. Vaginal biopsy and radiography require sophisticated apparatus not often available in the tropics. Foetal pulse detection is more practicable and is reviewed by Thwaites (1981) and Wani (1981). Sounds coming from the foetal heart and umbilical vessels are detected by a probe placed on an area of bare skin near the udder. This method can accurately predict whether or not a ewe is pregnant if used after 60 days of pregnancy (Aswad et al. 1976; Wani and Sani 1981), and the rate of lamb heartbeat can be used to predict the date of lambing. Attempts to use this method to detect litter size have so far been unsuccessful.

In the last 10 years ultrasonic scanning instruments developed to measure carcass fatness have been used for pregnancy diagnosis. Trapp and Slyter (1983) found that two commercial scanners compared favourably with other methods – when used between days 69 and 112 of pregnancy they gave an accuracy of pregnancy diagnosis of about 90%.

Laparotomy can be used to detect pregnancy as early as 30 days (Pacho 1973). It requires a skilled operator for accurate diagnosis and to avoid causing embryo deaths, and it is difficult to perform in fat ewes.

The concentrations of reproductive hormones can be used to determine pregnancy. Provided that the date of oestrus and mating is known, an analysis of plasma progesterone 17 days later will determine whether or not a ewe is pregnant (Robertson and Sarda 1971). When the dates of mating of individual ewes are not known, progesterone analysis of blood samples taken at set intervals can be used to determine pregnancy with an accuracy of 96% (Tyrrell et al. 1980). For lactating ewes, progesterone levels in the fore-milk can be used to diagnose pregnancy (Ayalon and Shemesh 1979). Analysis of hormones demands sophisticated laboratory facilities, but if these are available pregnancy diagnosis is quick and efficient.

Parturition induction

In an attempt to sychronise lambing, parturition may be induced by injection of hormones or their synthetic analogues such as dexamethasone or flumethasone (Bosc 1972). This practice has been used to combat prolonged pregnancy of Karakul ewes in South Africa (Roux and Wyk 1977). However, the interval between injection and parturition depends on the stage of pregnancy (Aswad *et al.* 1974), and if ewes are injected too early, the lambs die (Webster and Haresign 1981). There is therefore little scope for parturition induction in tropical ewes at present.

Chapter 7

Growth and meat production

Growth is usually measured by the increase in liveweight of an animal, and it is accompanied by changes in body form and composition known as development. If the diet is optimal the growth curve is sigmoid in shape. Growth-rate increases until the point of inflection is reached in lambs at the age of between 1 and 5 months (Owen 1976). Thereafter the animal continues to increase in weight, but its growth-rate declines as mature weight is approached. In practice the diet is not optimal and the early part of the growth curve may contract. The increase in weight of Masai lambs in Tanganyika is shown in Fig. 7.1. The rate of increase in weight becomes progressively slower as the lambs get older, indicating that the point of inflection occurs at, or soon after, birth.

Pre-weaning growth

Some examples of growth rates of lambs before weaning are shown in Table 7.1. They range from 20 g d^{-1} for Karakul lambs in Iran to over 200 g d^{-1} for Awassi, Chios and Cyprus fat-tailed lambs in Cyprus. Much of this variation in growth-rate is the result of the environment in which the lambs are reared.

The principal factor affecting the growth-rate of a lamb is its nutrition. This is a composite measure of food intake mediated through the milk yield of the ewe, availability of creep feed and pasture quality. Because the lamb derives the majority of its nutrient requirements from milk there is a strong relationship within flocks between milk yield and early lamb growth. For example, in Rajasthan, India, Singh *et al.* (1973c) found that the correlation coefficient between total milk yield to 75 days and lamb growth-rate was 0.47. The efficiency of conversion (unit weight of lamb per unit weight of milk) varies from about 17% (El-Sherbiny *et al.* 1972; Lawlor *et al.* 1974) to over 30% (Mahajan and Singh 1978). The lower values probably indicate the true efficiency of conversion of milk, while higher apparent values are observed if the food conversion efficiency is measured over a longer period when the lambs ingest solid food in addition to milk.

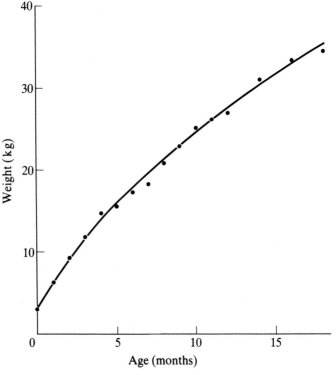

Fig. 7.1 Increase in liveweight of Masai lambs in Tanganyika (Data from French 1942)

Although lactation has a high priority for use of nutrients compared with other body processes, the nutrition of the ewe does affect milk yield and thus lamb growth (Butterworth et al. 1968). The number of lambs reared (rather than litter size) has a large effect on the growth-rate of lambs. The growth-rate of single lambs is faster than those of twins or triplets (Seth et al. 1972; Ugalde Orta 1978). Fall et al. (1982) reported that twin lambs born to West African Dwarf ewes in Senegal were 27% lighter at 61 days of age than singles. This difference arises because of the competition between the lambs for their dam's milk and the small size of twin lambs at birth. Provided that both lambs survive, the sum of the liveweight gains of twins is, however, usually considerably greater than the liveweight gain of single lambs (Butterworth et al. 1968), presumably because the ewe produces more milk for two lambs than for one, as a consequence of the greater appetite of two lambs.

Lambs which are heavier at birth grow more rapidly than lightweight lambs. Heavy lambs are usually singles or are produced by ewes which are in good body condition and which have high milk yields. Also a large

Table 7.1 Growth-rates of lambs before weaning

Breed	Location	LWG (g d^{-1})	Age at weaning (d)	Source
Awassi	Cyprus	240	35	Lawlor et al. 1974
Awassi	Iraq	165	120	Eliya and Juma 1970
Baggara	Sudan	140	91	Wilson 1976
Barbados Blackbelly	Trinidad	102	21	Rastogi et al. 1979
Barbados Blackbelly	Trinidad	129	84	Rastogi et al. 1979
Blackhead Persian	Brazil	117	112	Figueiredo et al. 1982
Blackhead Persian	Trinidad	103	21	Butterworth et al. 1968
Blackhead Persian	Trinidad	188	84	Butterworth et al. 1968
Blackhead Persian	Trinidad	125	21	Rastogi et al. 1979
Blackhead Persian	Trinidad	136	84	Rastogi et al. 1979
Chios	Cyprus	230	35	Lawlor et al. 1974
Chios	Cyprus	183	42	Mavrogenis 1982
Chokla × Rambouillet	India	156	21	Sahni and Tiwari 1975
Chokla × Rambouillet	India	118	84	Sahni and Tiwari 1975
Corriedale × Ile-de-France	Brazil	240	30	Surreaux et al. 1980
Cyprus Fat-tailed	Cyprus	230	35	Lawlor et al. 1974
Deccani	India	116	28	Gupta et al. 1974
Deccani	India	98	84	Gupta et al. 1974
Deccani × Merino	India	105	28	Gupta et al. 1974
Deccani × Merino	India	102	84	Gupta et al. 1974
Deccani × Rambouillet	India	118	28	Gupta et al. 1974
Deccani × Rambouillet	India	99	84	Gupta et al. 1974
Desert Sudanese	Sudan	157	42	Pollott and Ahmed 1979b
Desert Sudanese	Sudan	96	112	Pollott and Ahmed 1979b
Jaffna	Sri Lanka	45	91	Ravindran et al. 1983
Jaisalmeri × Rambouillet	India	160	21	Sahni and Tiwari 1975
Jaisalmeri × Rambouillet	India	108	85	Sahni and Tiwari 1975
Javanese Thin-tailed	Indonesia	151	91	Hetzel et al. 1982
Javanese Thin-tailed × Suffolk	Indonesia	173	91	Hetzel et al. 1982
Karakul	Iran	21	—	Farid et al. 1976
Madras Red	India	100	119	Ganesakale 1975
Malpura	India	89	28	Bohra et al. 1979
Malpura	India	77	84	Bohra et al. 1979
Malpura × Rambouillet	India	153	21	Sahni and Tiwari 1975
Malpura × Rambouillet	India	117	84	Sahni and Tiwari 1975
Mandya	India	81	119	Ganesakale 1975
Masai	Kenya	73	153	Wilson et al. 1981
Morada Nova	Brazil	119	112	Figueiredo et al. 1982
Muzaffarnagari	India	122	80	Ali et al. 1980
Nali	India	90	84	Gaur et al. 1977
Polwarth	India	89	61	Johari 1972
Rambouillet	India	162	21	Sahni and Tiwari 1975
Romney Marsh	Mexico	193	30	Ugalde Orta 1978
Romney Marsh	Mexico	156	60	Ugalde Orta 1978
Santa Inês	Brazil	152	112	Figueiredo et al. 1982
Uda	Nigeria	118	91	Buvanendran et al. 1981
West African	Trinidad	116	21	Rastogi et al. 1979
West African	Trinidad	141	84	Rastogi et al. 1979
West African	Venezuela	122	—	Combellas 1978

Table 7.1 (Cont)

Breed	Location	LWG (g d^{-1})	Age at weaning (d)	Source
West African × Dorset Horn	Venezuela	171	—	Combellas 1978
West African Dwarf	Nigeria	48	30	Adeleye and Oguntona 1975
West African Dwarf	Senegal	73	61	Fall et al. 1982
Yankasa	Nigeria	109	91	Buvanendran et al. 1981

birth weight may indicate that a lamb will achieve a heavy adult weight and has a large potential for growth.

Age appears to have no consistent effect on the daily liveweight gain of suckled lambs in the first 3 months of life so that the relationship between weight and age is approximately linear (Bhadula and Bhat 1980). Although the lamb's potential for growth increases with age up to about 70 days (Pálsson 1955) the achievement of growth is limited by the availability of milk and solid food. As the lamb gets older its capacity for ingesting solid food increases, but the supply of milk from its dam declines from 2 or 3 weeks after parturition. Under conditions of poor nutrition the daily liveweight gain of the lamb may therefore decrease as it gets older.

There are large differences between years and seasons within years in growth-rate. In Senegal, the causes of variation between years are the annual rainfall affecting pasture quality, the disease situation and changes in management (Fall et al. 1982). Surprisingly, these authors found that there were strong negative correlations between lamb weight and rainfall. In Senegal the month of birth also had a significant effect on lamb weight; lambs born in July at the beginning of the rainy season had a growth-rate to 61 days of 58 g d^{-1} compared with 83 g d^{-1} for those born in September at the end of the rains. This effect is probably because of the effect of nutrition of the ewe resulting in differences in milk yield, especially noticeable in semi-arid areas. Malik et al. (1978) reported that in Rajasthan, India, lambs born in the rainy season (between July and September) were 18% heavier at 30 days than lambs born between February and April.

Sex affects the birth weight of lambs; ram lambs are heavier than ewe lambs (Marai 1972) but the effect is small, and the growth rate of the sexes is similar up to about 5 months of age (Eliya and Juma 1970; Wilson 1976). Thereafter ram lambs grow faster than ewe lambs.

The genotype of the ewe affects the early growth of her lambs because, by her mothering ability and milk yield, she determines the environment in which the lambs are reared as well as contributing to their genotype. Thus the breed of ewe has more effect on the growth of the young lamb than does the breed of ram. Because some breeds are better able to withstand harsh conditions than others, there may be a genotype–season interaction in the weight of lambs at weaning (Singh et al. 1981b).

The mortality of lambs before weaning is closely related to the weaning weight of the survivors (Fall et al. 1982). For Djallonké [West African Dwarf] lambs the weight of survivors at 4 months was 140 g less for each additional 1% mortality.

Weaning

Weaning is defined as the time at which lambs cease to receive milk. In many systems lambs are never separated from their mothers, and in others lambs are weaned at an age of between 2 and 6 months. A lamb can surive solely on solid food from a very young age. There have even been experiments in which lambs have been weaned at 10 days of age (Pond et al. 1982), although their growth-rates were extremely poor.

The growth of lambs is temporarily slowed down after weaning (Fig. 7.2). The younger the lamb at weaning, the more severe the check in growth. The time of weaning may have little effect on the overall growth

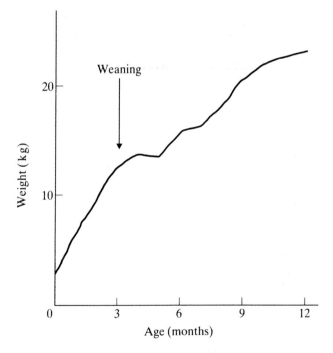

Fig. 7.2 Growth curve of Nali lambs in India, weaned at 3 months (After Malik and Acharya 1972)

of lambs. Makarechian et al. (1973) weaned lambs in Iran at 45, 60 or 70 days of age, and found that by the time they were 195 days old there was no significant difference in weight between the groups. Similarly, no beneficial effects on lamb growth as a result of delaying weaning have been observed for German Mutton Merino lambs in Egypt (El-Shaffei et al. 1975), Malpura and Chokla lambs in India (Tiwari and Sahni 1976b), Dorper lambs in South Africa (Rudert 1976) or Desert Sudanese lambs (Pollott and Ahmed 1978). On the other hand, Awassi lambs weaned on to feedlots in Iran weighed more at 195 days if weaned at 60 than 90 or 120 days (Bhat et al. 1978), presumably because the nutrition in the feedlot was superior to their pre-weaning nutrition.

The overall efficiency of converting pasture (and presumably also concentrate) into lamb is considerably greater if lambs are weaned at 4 weeks than at 12 weeks (Geenty and Sykes 1981). This is partly because the direct utilisation of solid food by the lamb is more efficient than the two-stage process of consumption by the ewe and subsequent consumption and conversion of milk by the lamb.

Early weaning may be desirable where the lambing interval is less than 1 year. For Egyptian ewes lambing every 8 months, Aboul-Naga et al. (1980) found that ewe weight (and condition) at mating was significantly affected by the duration of suckling period. These authors found some indication that reproductive performance of ewes was better when their lambs were weaned at 42 days than at 70 days.

Post-weaning growth

The potential for growth of lambs after weaning is determined primarily by their genotype and sex: under optimum conditions large breeds and male lambs grow faster than smaller breeds and females. However, the fulfilment of this potential depends on the nutrition and health of the lamb. Other factors, such as litter size and age of dam, which are important influences before weaning are of little consequence after weaning (El-Kouni et al. 1974). Year and season influence growth by their effect on food supply.

Table 7.2 shows the growth-rates of lambs after weaning. The values can be classified into two distinct categories: lambs given no supplementary food grew at rates of up to 100 g d^{-1}, whereas those receiving supplementation grew at between 100 and 200 g d^{-1}. In a direct comparison with Malpura sheep, those grazing pasture without supplementary food grew at an average of 56 g d^{-1}, while those receiving a protein and energy supplement grew at 112 g d^{-1} (Bhatia et al. 1981). The relationship between liveweight gain and quality of food is discussed in Chapter 4. Provided that the sheep has not reached its mature liveweight, the better the diet, the faster it grows. The data of

Table 7.2 *Growth rates of lambs after weaning, with (Y) or without (N) supplementary food*

Breed	Location	LWG (g d^{-1})	Suppl. food	Period (days)	Source
Awassi	Egypt	144	Y	470–533	Galal et al. 1975
Awassi	Iraq	150	Y	122–210	Alwash et al. 1983
Awassi	Saudi Arabia	93	Y	126–203	Pritchard and Ruxton 1977
Barki	Egypt	128	Y	470–533	Galal et al. 1975
Barki	Egypt	175	Y	600–670	Younis et al. 1975
Jaffna	Sri Lanka	37	N	182–365	Ravindran et al. 1983
Javanese Thin-tailed	Indonesia	137	Y	91–154	Hetzel et al. 1982
Javanese Thin-tailed × Suffolk	Indonesia	200	Y	91–154	Hetzel et al. 1982
Lohi	India	102	Y	91–182	Malik and Acharya 1972
Madras Red	India	35	N	119–182	Ganesakale 1975
Malpura	India	12	N	92–182	India, CSWRI 1982
Malpura	India	56	N	90–247	Bhatia et al. 1981
Malpura	India	112	Y	90–247	Bhatia et al. 1981
Malpura × Karakul	India	139	Y	90–180	Prasad and Singh 1982
Mandya	India	35	N	119–182	Ganesakale 1975
Merino	Egypt	148	Y	470–533	Galal et al. 1975
Najdi	Saudi Arabia	143	Y	84–126	Ramadan et al. 1977
Nali	India	116	Y	91–182	Malik and Acharya 1972
Sonadi	India	37	N	92–182	India, CSWRI 1982
Sonadi	India	102	Y	90–174	Sehgal et al. 1983
Sonadi × Dorset	India	137	Y	90–174	Sehgal et al. 1983
Sonadi × Suffolk	India	135	Y	90–174	Sehgal et al. 1983
Sonadi × Karakul	India	149	Y	90–180	Prasad and Singh 1982
Uda	Nigeria	51	N	91–182	Buvanendran et al. 1981
West African	Venezuela	77	N	212–282	Stagnaro 1983
West African	Venezuela	172	Y	212–282	Stagnaro 1983
Yankasa	Nigeria	53	N	91–182	Buvanendran et al. 1981

Galal *et al.* (1975) and Younis *et al.* (1975), obtained in Egypt show that sheep can grow rapidly even when considerably more than 1 year old.

In practice the food supply is not constant in terms of either quantity or quality during the year, so that growth-rate in all but the humid tropics shows a seasonal variation. In the dry season the vegetation may be so poor that lambs lose weight. The resulting growth curve for individual lambs is typically jagged. Where weights of lambs born in different seasons are combined to form an average growth curve, this seasonal effect is not seen. Figure 7.3 shows the increase in weight of gimmers on a seasonal breeding system in South Africa over a period of 3 years. Each year during the winter (May–August) the animals lost weight, but subsequently gained weight as pasture quality improved.

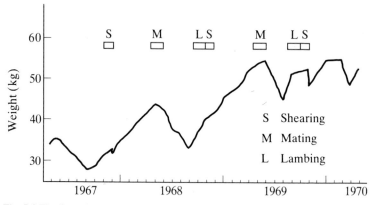

Fig. 7.3 The liveweight of Döhne Merino ewes (initially two-tooth) on natural pasture in South Africa, over a period of 3 years (After Reyneke and Fair 1972)

Compensatory growth

During a period of undernutrition an animal ceases to grow and may lose weight as it utilises its body tissues to maintain essential bodily functions. Subsequently when the food supply is better, the animal regains weight more rapidly than would be predicted from its pre-stress growth-rate (e.g. Elliott and O'Donovan 1969). This rapid growth is known as compensatory growth and is seen in all but very young animals. It results from the higher food intake, the lower metabolic rate and the lower energy cost of depositing tissue in an undernourished animal compared with a well-fed animal.

In an experiment in Northwest Egypt, weaned Barki lambs aged about 7 months were allocated to different nutritional treatments ranging from 100% to only 30% of recommended feeding levels (Younis *et al.* 1975). These treatments were continued until the sheep were about

20 months old, then they were finished for 70 days on a diet containing concentrates. The liveweights of the groups, their growth-rates during the finishing period and dressing percentages are shown in Fig. 7.4. The lower the quality of the diet given to the sheep between the ages of 7 and 20 months, the greater their capacity for subsequent growth on the high-quality diet. Nevertheless, although the sheep exhibiting compensatory growth grew twice as fast as the well-fed animals, this was not enough to allow them to catch up. The two worst-fed groups began the finishing period weighing less than 25 kg, and after the 70-day period reached only 35–40 kg. In contrast the two best-fed groups started at over 40 kg and reached 50 kg.

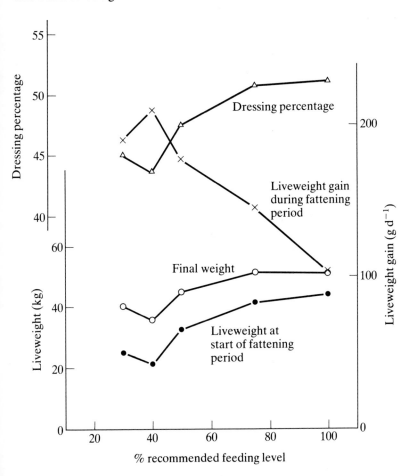

Fig. 7.4 The effect of feeding level from 7 to 20 months of age on performance during a subsequent 70-day finishing period (Data for Barki lambs in Egypt, from Younis *et al.* 1975)

In a study with Targhee and Corriedale rams in the USA, Osman and Bradford (1967) found no significant differences between individual rams in the degree of compensatory growth exhibited by their offspring. There may be differences between dissimilar breeds in this respect – those breeds which evolved in areas with seasonal undernutrition being able to recover better after a period of stress.

Because of compensatory growth the effect of a period of undernutrition is less severe than would otherwise be expected, and sheep are able to utilise food when it is available. The cost of providing good-quality food during the dry season is high compared with the wet season. Therefore compensatory growth allows growing sheep to make good use of cheap food during a period of abundancy, and minimises the effect of a period of inadequate feeding.

Breed

Breeds differ in their mature size and thus in their potential growth-rate, in the priority with which nutrients are diverted to growth, and in body conformation.

The relative merit of fast- and slow-growing breeds depends on the system of production. On a high level of feeding, large breeds with a good potential for growth can be exploited and they grow rapidly without becoming too fat. This rapid growth is advantageous for commercial finishing schemes. On the other hand, where the available nutrition is poor, a rapid growth potential is no advantage, and lambs of small breeds grow as well or better than lambs from large breeds. Early maturity is an advantage in these conditions because the lambs reach an acceptable body composition for slaughter at a reasonably young age. These principles of matching genotype to production system evolved for breeds and systems in Europe. The same principles also apply to tropical sheep.

If the progeny of ewes of a small breed are to be used in an intensive finishing system, their growth-rate can be increased by crossing them with a ram from a rapidly growing breed. Rams used in this way are known as terminal sires. In the UK the Down breeds, such as the Suffolk, are used as terminal sires. Similarly, provided that the nutrition is reasonable, the post-weaning growth of indigenous tropical lambs can be enhanced by mating local ewes with rams of imported breeds (Trail and Sacker 1969).

Within breeds, there are differences between individuals in their performance. Bessa et al. (1980) found significant differences between Morada Nova rams in the growth of their offspring. In the UK the differences between individuals within breeds are much greater than the differences between breeds, and this is likely to be the case also for tropical breeds in finishing units.

Breed differences in food conversion efficiency are small. Breeds developed for meat production tend to be more efficient than undeveloped breeds or those bred for wool production. Thus in a comparison in South Africa in which Bapedi, South African Merino and South African Mutton Merino (SAMM) lambs were fattened on a variety of levels of feeding, the SAMM had the best food conversion efficiency from 63 days to slaughter at approximately 40 kg. However, because the SAMM had the greatest potential for growth, they were slaughtered at an earlier age than the other two breeds, and thus their total requirement of energy for maintenance was lower (Hofmeyer et al. 1976).

Sex and castration

Sex has a consistent effect on growth-rate. Both on a restricted diet and when fed *ad libitum*, male lambs grow faster than female lambs (Badreldin 1951; Dass and Acharya 1970; Pollott and Ahmed 1979a; Wilson et al. 1981; Stagnaro 1983). Female lambs lay down more fat on a given diet so have a lower food conversion efficiency and a higher dressing percentage than male lambs (France, IEMVT 1978). Male lambs may be castrated to prevent them from breeding or becoming aggressive. Several workers have found that castration has no significant effect on liveweight gain (Younis et al. 1972; Ahmed et al. 1975; Al-Mallah et al. 1979), but Silva et al. (1980) found that for Polwarth and Corriedale lambs reared by their dams on natural pasture in Brazil, castration at 4 weeks of age decreased the growth-rate from 150 to 99 g d^{-1}. The dressing percentage of the castrated lambs was higher than that of entire lambs, but this difference was not enough to compensate for the difference in growth-rate. In some circumstances where the production of small fat carcasses is required, castration may be advantageous, allowing lambs to be finished at a younger age. However, in general, rapid growth is desirable and methods other than castration are used to prevent ram lambs from breeding indiscriminately.

Hormones

Several hormones influence tissue activity and growth (Frandson 1981). Together they influence the metabolism of food and food intake. Growth hormone or somatotrophic hormone produced by the anterior pituitary has a general growth-stimulating effect. Its primary effect seems to be the stimulation of protein synthesis from amino acids. Insulin, secreted by the pancreas, controls blood-sugar levels by facilitating the entry of glucose into cells where it is used to produce energy. The thyroid is concerned with the regulation of metabolic rate, and the thyroid hormones are complementary to growth hormone in the regulation of growth. Thyroxine accelerates the maturation of the body, but has little effect on growth. The adrenal corticoids are concerned with the regulation of the electrolyte balance and carbohydrate metabolism.

Growth promoters

Many attempts have been made to improve the growth-rate and food conversion efficiency of growing sheep by two types of growth promoter. The first type are anabolic agents which have hormonal properties and act on the metabolic processes (Heitzman 1980). These include resorcylic acid lactone (zeranol or Ralgro), and synthetic oestrogens, e.g. hexoestrol and diethylstilboestrol. The second type consists of rumen-active anaboles which modify rumen fermentation and protein synthesis in the small intestine; monensin is in this category.

The effects of hormones on growth and body composition have recently been reviewed by Galbraith and Topps (1981). Zeranol and other hormonal chemicals are generally applied as implants 30–100 days before slaughter. Experiments conducted in Rhodesia (Grant and Naude 1977) and Brazil (Figueiró et al. 1980) did not demonstrate any significant improvement in lamb performance as a result of zeranol implantation. On the other hand, in Colombia, Bautista et al. (1976) found that zeranol implanted into Romney Marsh and Corriedale males 224 days before slaughter increased their liveweight gains by about 12.0% for the castrated and 4.5% for the entire males, and on average zeranol increased the hot carcass yield from 37.9 to 38.4% (Bautista and Gonzalez 1976). In Brazil, Hall et al. (1977) found that zeranol increased the rate of liveweight gain of entire males by 20%. In temperate countries, too, some experiments have shown small but statistically significant gains in liveweight gain and food conversion efficiency, while others have not demonstrated significant advantages of zeranol implantation (Vipond and Galbraith 1978; Wiggins et al. 1980; McKenzie 1981; Thompson et al. 1982).

Implants of diethylstilboestrol increased the growth-rate of fattening lambs in Egypt over a period of 29 weeks, but the resulting increase in carcass weight was not statistically significant (Shehata et al. 1978). Implantation of Awassi lambs soon after birth with stilboestrol increased their growth-rates over 34 days (Fox and Husnaoui 1967). A combination of trienbolone acetate and oestradiol has given improvements in performance (Coelho et al. 1978), and increased the proportion of protein and water in the carcass and decreased the proportion of fat. The greatest potential hazard of anabolic agents is that meat will contain residues which affect humans. It is difficult to detect these residues and to study the health hazards they may cause. Implants of anabolic agents are absorbed over several months and it is therefore unlikely that there are sufficient residues in edible tissues at slaughter to cause harm (Heitzman 1980).

A possible new method of hormonal manipulation which avoids the problem of residues in meat was investigated by Spencer et al. (1983). They immunised lambs against somatostatin and found that this treatment enhanced growth-rate, suggesting that stomatostatin normally has an inhibitory effect on growth.

Monensin is given as a food additive. An experiment to study its effect on lambs in the Ivory Coast showed that it increased growth-rate and food conversion efficiency, and decreased mortality during finishing (Mignon and Diague 1981). However, reports from other parts of the world do not consistently show a beneficial effect; in the USA, monensin significantly decreased the intake of food by lambs and reduced their growth-rate, an observation attributed to the low palatability of diets containing monensin (Sharrow et al. 1981). Excess monensin in the diet of ruminants results in severe scouring and death (Heitzman 1980).

A variety of other antibiotics have been used to increase the productivity of livestock kept in intensive systems. It is widely accepted in the USA that antibacterial drugs fed as supplements increase the growth rate and food conversion efficiency of pigs and poultry, as well as reducing morbidity and mortality. However, there is concern that indiscriminate feeding of antibiotics to animals results in the development of resistant strains of harmful micro-organisms which may be difficult to control in both animals and man (Hays and Muir 1979). Little work has been done on the effects of feeding antibiotics to sheep. Feeding chlortetracycline to lambs in India had no consistent effect on the digestibility of hay (Bidarkar and Mudaliar 1981). A daily supplement of oxytetracycline can increase the growth-rate of lambs (Chorey et al. 1965), but unless the animals are in an intensive system an economic benefit is unlikely to be obtained. If the antibiotic is withdrawn the mortality of lambs can be very high.

Composition of growth

As a lamb gets older and heavier, its body composition changes. The viscera, skin, head and feet grow relatively slower than the carcass tissues, so that the ratio of carcass weight to total liveweight increases as the lamb matures (Prescott 1979). Within the carcass, the ratio of fat to muscle and the ratio of muscle to bone both increase with age. These changes in carcass composition are described by Berg and Walters (1983).

The level of nutrition affects the composition of the body. The essential tissues (brain, heart, liver, intestines, bones, etc.) have the highest priority for nutrients, so that even if the level of nutrition is very low these tissues still maintain their weight. Lean tissue or muscle has a high priority, and although each unit of lean deposited carries a specific amount of 'essential fat', large quantities of fat are produced only if there are surplus nutrients.

Because the potential growth decreases as a lamb proceeds from weaning to maturity, a plane of nutrition which satisfies only the needs for essential tissue and lean growth at weaning, will result also in fat growth in an older lamb. The higher the level of nutrition or the lower the potential body weight, the more fat there is in the lamb's carcass at any given age or at any given body weight.

When weight loss occurs as a result of undernutrition, all tissues are depleted, but the relative effect on fat is greater than on muscle, and bone resists depletion to an even greater extent than muscle and fat. Thus the ratio of meat to bone is low in old animals in poor condition as well as in young animals (Chawdhary 1978).

Compared with temperate mutton breeds, tropical sheep tend to lay down more intramuscular fat and internal fat, and less subcutaneous fat (Gaili 1979). Fat-tail sheep deposit fat in their tails rather than in the carcass (in a similar way to the deposition of fat in the hump by *Bos indicus* cattle). The fat tails of finished Barki sheep weigh over 1 kg (Galal *et al.* 1975). The tail of Najdi lambs slaughtered at 18 weeks accounted for 6% of carcass weight (Ramadan *et al.* 1977), and in Iran tail fat accounts for up to 20% of carcass weight (Parvaneh 1972).

Fat tails are considered a delicacy in some parts of the world, but are not desired in modern carcass classification systems. Fat-tail sheep may be docked (i.e. the tail removed from the young lamb) because this is thought to improve growth-rate and food conversion efficiency. Docking may reduce weight gains to weaning (Juma *et al.* 1973), but improves post-weaning performance (Sefidbakht and Ghorban 1972; El-Karim 1980).

Apart from differences in size and distribution of fat, there is no evidence that tropical and temperate breeds differ in either carcass composition or conformation (e.g. Amegee 1981).

Optimum age for finishing and slaughter

After weaning the growth-rate of lambs is increased if they are given supplementary food to bring them more rapidly to a marketable weight with acceptable carcass quality. This process is known as finishing or fattening. In some areas of the tropics the duration of finishing is dictated by the seasonal availability of food, but where lambs are fed on stored food such as cereals and cereal by-products decisions must be made concerning the optimum age to begin finishing, the duration and the feeding level. In Iraq, Al-Mahmood *et al.* (1976) conducted an experiment to study the effect of age on finishing. They began to finish Awassi lambs at 5, 7, 9 and 11 months of age, and killed them at a uniform weight of 50 kg. There was no consistent relationship between the age at which finishing began, and the rate of liveweight gain or food conversion efficiency. Thus, a recommendation can be made that finishing should take place when lambs and food are available.

For lambs fattened on a uniform diet, as age and weight increase so growth-rate decreases (Coetzee 1973) and the fat content of the body increases. Even though fatter lambs have higher dressing percentages, prolonging the finishing period results in the inefficient use of food to produce fat carcasses, while too short a period gives small lean carcasses and does not make good use of the growth potential of the lamb. In the UK it is recommended that lambs are slaughtered when they reach about 50% of mature weight (MLC 1975). This recommendation is appropriate also for the tropics if the nutrition is moderately good. However, where nutrition is poor lambs can be grown to heavier weights without getting too fat.

In an experiment in Nigeria, Adeleye (1982) fed West African Dwarf rams, initially aged 10–12 months for a period of 84 days. The rations they received contained either 0, 33, 50 or 67% concentrate, and the remainder was grass hay. His results are shown in Table 7.3. The rate of liveweight gain and food conversion efficiency were both better in the higher-quality diets and, despite the higher cost of these diets, they resulted in cheaper production costs per unit liveweight gain. High levels of concentrate were associated with fatter carcasses and slightly higher dressing percentages, but also with significantly increased lean carcass weights.

This and other experiments show that even unimproved breeds of sheep respond well to intensive feeding. The practicality of doing this depends on the availability and cost of suitable foods and store lambs. The high cost of cereals prevents their use as sheep food in most less developed countries, but alternative foods such as crop and industrial by-products have potential. Because of the problems with fluctuating supplies of inputs, systems which have low capital costs are more likely to be successful than those which are capital intensive. Thus peasant systems are generally more successful than commercial finishing systems (Sandford 1983).

The body condition of an animal is a more consistent guide to the optimum time of slaughter than its weight or age. Experienced evaluators are able to estimate carcass measurements reasonably

Table 7.3 *Feedlot performance of West African Dwarf ram lambs on grass-concentrate diets in Nigeria*

	% concentrate in diet			
	0	33	50	67
Initial weight (kg)	17.00	16.96	16.96	16.92
Rate of LWG (g d^{-1})	51	76	111	150
Food conversion efficiency	0.086	0.14	0.22	0.32
Food cost (Naira/100 kg food)	8.00	13.00	16.00	19.00
Production costs (Naira/100 kg LWG)	93.00	94.00	74.00	59.00
Carcass weight (kg)	11.6	12.3	14.5	16.7
Dressing percentage	55	54	57	58
Lean in carcass (kg)	7.8	8.1	9.8	11.5
Fat in carcass (kg)	0.18	0.56	1.08	1.42

Source: After Adeleye 1982.

accurately by visual appraisal of live animals (Lewis et al. 1969). Condition scoring is done by feeling the amount of fat on the animal: on the back of thin-tailed sheep, and the tail as well for fat-tailed sheep. Although condition scoring gives an estimate of the carcass quality of lambs, the method is very subjective and there are considerable differences between the assessments of different evaluators (Everitt 1962). A more accurate method is to measure the back-fat depth using an ultrasonic probe. This gives a very good estimate of the fat depth, and can give a reasonable estimate of the total fat in the carcass (Gooden et al. 1980). However, if insufficient measurements are made, the ultrasonic probe may not be as accurate as condition scoring by an experienced operator (Alliston and Hinks 1981).

Slaughter

Pre-slaughter handling affects the quality of the carcass. Sheep may lose condition if they are transported long distances to slaughter and receive little food. In India for instance, sheep intended for consumption in the cities have to walk 160–320 km before they reach the market (Taneja 1978). Transport on lorries or trains can cause bruising and oedema which lower the quality of carcasses and possibly result in condemnation. Losses and shrinkage are slightly higher when sheep are transported by lorry or train than on foot (Sandford 1983). Tranquilising drugs are available to minimise the stress suffered by transportation, but are rarely used for sheep.

Mann (1978) discusses the slaughter facilities used in the tropics, and categorises them into simple rural slabs, public slaughter-houses and factory abattoirs. In remote rural areas butchering is carried out by individuals, rarely on authorised premises. A clean area with a hoist and a water supply are all that is needed at such a slaughter-slab. Details of the design of slaughter-houses and slabs are given by Eriksen (1978).

Many countries now have slaughter-houses in most towns. These are designed primarily for the slaughter of large ruminants, but may also have a unit for sheep and goats. A factory abattoir may be located near a city to supply the requirements of the modern local market and to export meat. An example of a factory abattoir is that run by the Kenya Meat Commission at Athi River near Nairobi in Kenya. Such an abattoir requires a high degree of co-ordination of livestock supply with abattoir management and marketing of meat.

In most religions, there are rules governing the slaughter of animals. For instance, Islamic law states that only healthy animals can be slaughtered for food, a sharp knife must be used to cut the major blood vessels in the neck at one stroke, the animal must be blessed in the name of Allah and the carcass must be properly bled. Bachhil (1980) describes the usual Islamic procedure of 'halal' in India, in which the head is jerked back to rupture the spinal chord.

In westernised slaughter-houses, sheep are stunned before killing. Stunning is done by either a captive bolt, a controlled blow on the head, an electric shock or anaesthesis by carbon dioxide (Lawrie 1979; Mitchell 1980). After stunning the carcass is bled by cutting the carotid artery and jugular vein. The heart and lungs continue to function and help to pump blood out of the body. Failure to remove the blood gives the meat an undesirable appearance and encourages the growth of micro-organisms.

The head, feet, skin, excess fat, viscera and offal are removed from the carcass. This is known as dressing. The edible meat and bones are then cut up for sale. In Europe, traditional butchering gives cuts each of which contain parts of several muscles and usually some bone, and which are sold at different prices determined by demand. In modern abattoirs, carcasses may be deboned while still warm, and sold in vacuum packs partly for hygienic reasons, but also to reduce costs and give a more uniform product. However, in most parts of the tropics there is traditionally little discrimination between different parts of the carcass.

The dressing percentage of a carcass is the ratio of dressed carcass weight to liveweight, expressed as a percentage. An alternative term is 'carcass yield'. The value of dressing percentage recorded depends on the access of the animals to food and water before slaughter, whether the hot or cold carcass weight is used, and on the dressing procedure, i.e. what parts of the body are removed (Berg and Butterfield 1976) as well as the type of animal. Great care must be therefore be taken when comparing values obtained by different workers. Some examples of dressing percentages reported in the literature are shown in Table 7.4.

In the tropics more of the carcass is eaten than in developed temperate countries. Almost all the offal is cleaned and consumed by humans. In most parts of rural Africa and Asia meat is sold at a price which is either uniform or agreed by bargaining (e.g. Osman and El-Shafie 1967). Figure 7.5 shows mutton being sold in Nepal. There is little discrimination between different cuts of meat, and even between species. Sheep and goat meat are both called 'mutton' in India. It is difficult to tell apart the carcasses of sheep and goats after the head has been removed (Cuq et al. 1978).

Meat quality

The consumer assesses the quality of meat initially by its appearance and subsequently by its palatability or eating qualities – particularly tenderness and flavour. Consumers buying fresh meat are encouraged by a good colour and in westernised shops, little excess fat. In particular, yellow fat is said to be undesirable (Cuthbertson and Kempster 1979).

Table 7.4 *Dressing percentages (D) of lambs at slaughter*

Breed	Location	D	Age	Source
Awassi	Egypt	49	17 months	Galal et al. 1975
Awassi	Iraq	42	7 months	Al-Tawash and Alwash 1983
Bangladesh	Thailand	50	—	Falvey and Hengmichai 1979b
Barki	Egypt	46	17 months	Galal et al. 1975
Coimbatore	India	38	12 months	Singh et al. 1973b
Criollo	Brazil	43	Lambs	Figueiredo et al. 1983
Criollo	Brazil	39	Old ewes	Figueiredo et al. 1983
Desert Sudanese	Sudan	52	14 months	Osman et al. 1970
Indian [Muzaffarnagari?]	India	41	Lambs	Bachhil 1980
Indian [Muzaffarnagari?]	India	43	Adults	Bachhil 1980
Javanese Thin-tailed	Indonesia	49	2 permanent incisors	Obst et al. 1980
Kelantan	Thailand	47	—	Falvey and Hengmichai 1979b
Kelantan × German Mutton Merino	Thailand	47	—	Falvey and Hengmichai 1979b
Kelantan × Polwarth	Thailand	42	—	Falvey and Hengmichai 1979b
Malpura	India	42	3 months	Basuthakur et al. 1980

Table 7.4 *(Cont)*

Breed	Location	D	Age	Source
Malpura	India	45	6 months	Nivsarkar and Acharya 1982
Malpura	India	53	6 months	Prasad *et al.* 1983
Malpura × Suffolk	India	50	6 months	Prasad *et al.* 1983
Mandya	India	49	7 months	Bidarkar 1982
Mandya × Somali	India	60	7 months	Bidarkar 1982
Merino	Egypt	43	17 months	Galal *et al.* 1975
Najdi	Saudi Arabia	45	4 months	Ramadan *et al.* 1977
Rahmani	Egypt	42	7 months	Darwish *et al.* 1973
Rahmani	Egypt	50	18 months	El-Serafy *et al.* 1976
Rahmani	Egypt	59	36 months	El-Serafy *et al.* 1976
Santa Inês	Brazil	46	Old ewes	Bellaver *et al.* 1980b
Sonadi	Brazil	51	6 months	Prasad *et al.* 1983
Sonadi × Suffolk	Brazil	50	6 months	Prasad *et al.* 1983
South African Merino	South Africa	38	20 kg	Cloete *et al.* 1975
South African Merino × South African Mutton Merino	South Africa	42	20 kg	Cloete *et al.* 1975
Tswana	Botswana	46	2 permanent incisors	Owen and Norman 1977
West African	Venezuela	44	9 months	Stagnaro 1983

Fig. 7.5 Sheep meat on sale in Nepal. Note poor hygiene and absence of meat grading.

However, in some parts of the world fat is highly desirable. Lyne-Watt in 1942 reported that the Kikuyu in Kenya fattened lambs to produce about 10 litres of fat for use as a food and for toilet, and Epstein (1982) stated that in heavy male lambs of the Awassi breed the tail fat weighs up to 8 kg. In developed countries such as the USA, Europe, Australia and New Zealand, carcasses are graded into categories (e.g. Kirton and Colmer-Rocher 1978; Kempster et al. 1982) depending on their weight, age of sheep, sex, fat cover and conformation (shape). Other attributes may also be included in the grading: kidney and channel fat, marbling, colour, texture of meat and fat, and eye muscle area (Calder 1982).

The tenderness of meat is an important quality where meat is cooked quickly. Older sheep are believed to be less tender than lambs, but there is no evidence that this is so in the USA (Kirton 1982). Tenderness can be improved by hanging the meat in a cool place for a few days between slaughter and cooking. However, rapid chilling or freezing soon after slaughter makes meat tougher, an effect called cold shortening. Chilling facilities are rare in the tropics and meat is cooked and consumed soon after slaughter. Cooking methods, such as boiling for several hours, improve the tenderness of meat. Only in a few instances, such as in the highlands of Ethiopia is meat eaten raw or quickly cooked, and the meat used in this way comes from young animals.

The flavour of meat is rarely classified, yet can have a profound effect on its acceptability. The flavour of meat may be unacceptable if it is unfamiliar. The flavour depends on the species from which the meat originates, and is carried largely in the fat. Meat from entire males slaughtered at an advanced age has a stronger flavour than that from other classes of animal. Grazing lambs on pure swards of the legumes *Glycine wightii* and *Dolichos axillaris* resulted in meat which was considered to have a characteristic and objectionable odour and flavour (Park and Minson 1972). Contamination of a carcass during and after slaughter, particularly with male urine, can give the meat an unpleasant flavour. Freezing can alter the flavour of meat and make it less desirable. For instance, fresh sheep meat imported from Africa is much preferred in Saudi Arabia over frozen meat from Australia (Anteneh 1982).

In the tropics the quality of meat deteriorates rapidly unless it is preserved. In rural markets most meat is sold on the day that the animal is slaughtered, but there are both traditional and modern methods of preserving meat which are used to match the meat supply to the demand. In other words, if it is necessary to transport meat from one area to another or if the rate of meat production is greater than consumer demand, then meat is preserved. A traditional way of preserving meat is by smoking. This is done by hanging strips of meat in the fireplace so that each time the fire is used the meat is smoked. Commercial smoking followed by treatment with vegetable oils gives a product which can be kept for several months. Alternatively, meat can be preserved by dry salting which prevents the growth of micro-organisms. The salt is removed by soaking before cooking. Modern abattoirs rely on the chilling of meat for short-term storage, and freezing if the meat is to be kept for more than a few days.

Chapter 8
Wool production

Breeds of sheep in the tropics are either 'wool sheep' which grow fleeces, or 'hair sheep' which do not. Both types have evolved from wild sheep which have an outer coat of coarse, nearly straight fibres and an undercoat of fine wool. Some more primitive breeds still have coats very like those of their wild ancestors, but in many wool breeds the undercoat has been greatly increased, while the outer coat has been partly or entirely modified into finer fibres.

In some breeds there is no obvious visible distinction between the two coats, but they can still be distinguished histologically by examining the follicles from which the fibres grow. The outer coat grows from those follicles which develop earliest in the foetus (primary follicles) and which have sweat glands and erector muscles attached to them, as well as large sebaceous glands. The secondary follicles, from which the undercoat grows, have smaller sebaceous glands and no sweat glands or erector muscles. Thus the two kinds of follicle can be distinguished at all ages by their accessory structures. The number of secondary follicles per primary follicle is called the S/P ratio, and is an important parameter for wool biology.

In wild sheep, moulting of the coat occurs as in other mammals. The undercoat is mainly cast in spring, and the outer coat in autumn, by which time the new undercoat has grown in. Thus the sheep is never bare, but has a thin, open coat in summer and a warm, thick covering in winter. In domestic fleece-bearing sheep, the tendency to moult has been suppressed, but some primitive breeds still moult, as does one well-known British mutton breed, the Wiltshire Horn. In other breeds the tendency to moult is expressed as the partial peeling off of the fleece following a period of stress, and as the shedding of individual fibres, particularly kemp fibres.

In some tropical breeds, evolution under domestication has taken an opposite course, with extreme reduction of the undercoat, while the outer coat of coarse hair remains, giving a smooth glossy coat. This is presumably an adaptation to tropical heat, possibly assisted by deliberate selection. Some of these hair sheep grow quite a long undercoat as lambs, which is cast at about a year old and replaced by such short fine fibres that they are difficult to see on the animal. Burns (unpublished) found that West African Dwarf sheep in Ghana shed and replaced their outer coat continuously, so that tips of regrowing hairs were present in every month of the year. It appears that the evolution of hair sheep may have occurred independently in the long-legged (Sahel) sheep such as the Uda and Yankasa breeds in which the rams have argali-type horns, and in the West African Dwarf sheep in which the rams have mouflon-type horns (Burns 1968). The two also differ in fibre type array (see below). Symington (1959) found that Blackhead Persian sheep in South Africa grew fleeces when exposed experimentally to a light regime similar to that of temperate countries.

The evolution of fleece-bearing sheep is the result of deliberate human selection. Wool has been used in textile manufacture at least since the Bronze Age. Three main types of fibre are found in fleeces:
1. Kemps are fibres which cease growing at intervals, and are shed. They contain a hollow core, the medulla, throughout their length except at the extreme ends (base and tip). Kemp fibres grow in primary follicles and closely resemble the outer coat fibres of wild sheep.
2. Hairs are continuously growing fibres which have medulla either in part or, more rarely, throughout their length. They are usually grown in primary follicles, and do not occur in primitive fleeces (Ryder 1969).
3. Fine wool, which typically has no medulla but may have slight traces; the fibres grow indefinitely and are not shed regularly. They grow in secondary follicles and also in primary follicles of fine-fleeced breeds.

Hair and wool grow from the follicle bulbs below the skin, and as the component cells are pushed towards the skin surface, they die and become keratinised. Thus the fibre is composed of long-chain protein molecules (keratin) linked together by bisulphide and salt linkages. A non-medullated fibre is mainly composed of spindle-shaped cortical cells (the cortex) covered by a thin layer of flattened cells, the cuticle. The medulla consists of a network of keratinised material with hollow spaces. The keratin of the medulla is chemically different from that of the cortex and cuticle.

Wool textiles

In this context the term 'wool' refers to whole fleeces which may include kemp and hair as well as fine wool.

One of the great advantages of wool is its versatility, with different types being suited to different purposes. There is therefore no one 'best wool'. In order to understand this, it is necessary to appreciate the basic methods of wool processing. As some readers may be entirely new to

168 *Wool production*

wool biology, and even to fleece-bearing sheep, these processes will now be briefly described.

With very few and minor exceptions (such as certain felts made by pounding and beating wet fleeces), all wool has to be spun into yarn before it can be made up into textiles, either by weaving or knitting. Starting with the fleeces as shorn from the sheep, the first process is sorting into batches of similar quality, i.e. fineness, staple length, strength, medullation and colour. The purpose of manufacture (fine worsteds, coarser cloths, carpets, etc.) determines how exact sorting must be: the uniformity of the yarn is much affected by the uniformity of the fibres composing it. Hand-spinners (Fig. 8.1) sometimes wash fleeces first, and pick out any vegetable matter or other obvious contamination, before sorting rather roughly, mainly for colour, fineness and staple length. In the wool textile industry scouring (washing) takes place at some stage between blending (mixing a large quantity of the wool from different fleeces together) and spinning.

The fact that the process of blending is normally an integral part of

Fig. 8.1 Hand-spinning of wool, Nepal

wool manufacturing is important for sheep breeders, because often similar yarns can be made by blending fleeces from various breeds of sheep, or different grades within a breed. Thus in a market economy a manufacturer producing a particular type of yarn may at different times make it from various blends of raw wool according to their cost. This means that in many countries there is little incentive to improve the quality of wool grown, because if it thereby becomes more expensive the manufacturer may cease to buy it. Much research on wool growth and improvement has involved Merino sheep, particularly in Australia, not only because wool is the main export, but because almost all is used for fine worsted clothing materials which require fine and regular fleeces. However, in many countries there is now a trend towards trying to standardise fleece type and usage according to breed; especially in the socialist countries and some tropical countries such as India. Thus certain breeds are selected to produce high-quality carpet wool, others for apparel wool, others for special felting wools, etc.

In the wool trade, fleeces are classified into four main groups: fine (or Merino), crossbred, lustre longwool and carpet. The term 'crossbred' originated to describe wool produced by crosses between Merino and Longwool or Down breeds in Australia, but includes virtually all wool which does not fall into any of the other three categories. The quality of wool is often described by its 'quality count', originally defined as the number of hanks of yarn, 560 yd long, which can be spun from 1 lb of wool. A more accurate measure of quality is given by fibre diameter. Approximate counts and corresponding fibre measurements are shown in Table 8.1.

In preparing the wool for spinning there are two alternative processes, either worsted or woollen carding. In the worsted process, only long (4 cm or more) fibres can be used, and any shorter ones are combed out mechanically and described as 'noil', which can be transferred for use in the woollen process. Only raw wool is used, and it is the method of choice for making superior woven cloths from fine wool, for clothing materials. Soundness, uniformity of fibre diameter and adequate staple length are important for worsted spinning. The woollen process can use both long and short fibres, and also recycled wool and noils; it produces yarn suitable for blankets, knitwear, hosiery and many other purposes.

The main difference between worsted and woollen yarns is that the former are stronger and can, if required, be given more twist during spinning, giving added strength. The short fibres in the woollen process cannot be twisted so hard, but the yarns are softer to handle and give a more fluffy surface in the finishing process called 'milling'. Milling raises the individual fibres at the surface of the woven cloth, giving a smoother appearance and obscuring the weave. In felt-making, milling is carried further until the fibres are tangled together in a more solid mass, which can be perfected only with certain wools known as felting wools. These felts are important for press cloths used in paper-making and other

Table 8.1 *Approximate quality counts and diameter ranges of wool fibres*

Wool type	Quality count (s)	Fibre diameter (μm)
Merino	80–60	18–24
Crossbred	60–48	25–30
Lustre longwool	48–36	30–45
Carpet	40–30	40–50

Source: After Onions 1962.

special industrial purposes. Hand-knitting wools and carpet yarn may be made either by the worsted or the woollen process, but more commonly the latter.

The processes of sorting, blending and spinning are those of most direct interest to the wool grower, as they are most directly dependent on the quality of raw wool. The subsequent processes include dyeing, weaving and finishing. Wool quality affects dyeing mainly in two ways. The presence of medullated fibres leads to uneven colour because the medulla, being hollow, does not take the dye. This is more noticeable in pale shades than when the dye is a dark colour. Coarsely medullated fibres such as kemps also tend to be brittle, with low breaking strength and elasticity. Darkly pigmented fibres, which may be black or brown (and also 'yellowed' wool, see p. 186), obviously cannot be dyed clear or pastel shades, but they may take very dark colours satisfactorily.

The dark pigment in fibres is called eumelanin; yellow pigment, which is probably phaeomelanin, is present in white wool fibres and gives them a creamy or even definite tan colour, according to its concentration. The creamy tendency does not interfere with dyeing, but the wool may have to be bleached if pure white yarn is required. Coloured wool is normally of much less value than white wool, although sometimes natural coloured wool is used instead of dyed wool to produce patterned textiles. This is the case, for instance, with the traditional black or brown patterns on white blankets or rugs produced in many parts of North Africa and elsewhere (Fig. 8.2). At the present time there is a revival of interest in coloured wool for hand-spinning, weaving and knitting, although sometimes natural colour fades more than dyes do if exposed to sunlight. The deliberate establishment of coloured flocks for the production of pigmented wool may be worth considering, especially as crossbreeding of local sheep with Merinos and some other improved wool breeds often produces a proportion of coloured offspring.

Assessment of adult fleeces

Detailed assessment of fleeces requires both considerable skills of a highly specialised kind, and sophisticated laboratory facilities. All countries in which wool production is of major importance have

Fig. 8.2 Blanket made in the traditional way from naturally coloured wool, Nepal

therefore established wool biology laboratories and testing stations, from which the non-specialist can seek advice and assistance. However, anyone involved in research and improvement work which uses fleece-bearing sheep, even if only crossbreeding for meat production, needs to have at least a general knowledge of fleece assessment methods. These will therefore be briefly described, indicating which assessments can be made fairly easily on the farm, and which require special facilities.

In assessing parameters such as fibre length, diameter or medullation, only a very small sample of wool can be used, and as these assessments are laborious it is desirable to use as few samples from each fleece as possible. One method is to form a composite sample by taking small staples from different parts of the fleece, blending them thoroughly and using sub-samples from the composite sample for the various assessments.

Carter (1943) showed that in Merinos the mid-side position, over the last rib, is closely representative of both follicle and wool characteristics and this position has been much used in sampling. In a study of certain Indian breeds, however, Acharya et al. (1972) found that the back sample was the best indicator of fleece quality. In fibre diameter, length and medullation, the correlations between the composite samples and the back samples were 0.75, 0.66 and 0.83 respectively. Where there are obvious differences between different regions of the fleece, it may be necessary to sample them separately; it could, for example, be misleading to blend a sample from a hairy breech into an otherwise fine fleece.

In assessing a flock or breed, a number of animals should be examined to determine the general fleece type, which, in unimproved breeds, will probably be either medium or carpet type. In typical carpet-wool fleeces there is a definite long outer coat of medullated fibres, and a distinct undercoat of short, fine wool which may be scarce or plentiful. Kemp is nearly always present, though not desirable. All fleeces without a distinct double coat would initially be classed as intermediate or 'medium' wool unless they were exceptionally fine. Fleeces with much hairiness (heterotype hairs) are often classed as carpet wools even if they have no double coat, because this is their main manufacturing outlet.

Assessment of individual fleeces includes those given under the following headings.

Fleece weight

Greasy fleece weight is the weight as shorn, without any treatment, and is the easiest trait to record on the farm. It is highly correlated with clean fleece weight, which is the weight after scouring out grease, suint, vegetable matter and dirt. The clean fleece weight as a percentage of greasy fleece weight is called the yield. In some countries, a higher price is paid per unit weight for greasy fleeces with a high yield than for those with a low yield. Washing the sheep before shearing by immersing them in clean water increases the yield, provided they are shorn within a day or so of washing and drying, but the increased price seldom pays for the extra labour.

Fibre types

In the adult sheep only three fibre types are distinguished: kemp, hair and fine wool.

Kemps are coarse fibres which, unless pigmented, appear chalky white due to the broad medulla within them. They are often fairly short and strongly crimped, but sometimes long kemps occur, which may be distinguishable from coarse hairs only when they are shed. The distribution of kemps over the fleece area is such that the fewer the kemps the more they are restricted to the dorsal and posterior areas. If kemp is absent from the britch it is unlikely to be present elsewhere, except on the extremities. Kemp grade is best determined on the animal, but may be assessed from the complete shorn fleece. It is usually graded on a scale from K0 (no kemp) to K5 (kemp plentiful all over), but the same scale may be applied to each region (e.g. shoulder, back, britch) separately to give a more detailed description.

Hairs are long medullated or partly medullated fibres which do not shed, but they often become thin and non-medullated at the time when the kemps on the same animal are shedding. The term 'heterotype hair' refers to the fact that such fibres are like wool fibres in their non-medullated parts.

Wool in this context means non-medullated fibres which are usually fine, but in some breeds such as Longwools may have quite large diameters.

If it is deemed necessary to determine the proportion of fibres of each type, this may be done by sorting a small representative sample by hand, against a black background such as black velvet, all fibres which show any chalkiness to the naked eye being counted as hairs, unless they are kemps. Some fibres counted as wool may actually contain fine medulla, but this is not of great practical importance. The results are expressed as percentages of each type. The whole of the selected sample must be counted, and must contain at least 200 fibres for a fine fleece, and more if there is much variability. In the laboratory, medullation may be more accurately assessed by use of high magnifications and by immersion of the wool sample in clearing oils such as xylol. However, some of these methods do not indicate the number of medullated fibres, but only the percentage of cut pieces which are medullated. This is not very meaningful for sheep breeding, as the important question is how many follicles are producing partly or fully medullated fibres, and which follicles are involved. The same objection applies to assessing the proportion of different fibre types by weight instead of by number.

Fibre parameters

These include staple length, fibre length distribution and fibre diameter distribution.

Staple length is important in determining whether or not the wool is suitable for worsted processing. It is simply measured against a rule, the staple being held taut but not severely stretched; it therefore roughly measures the length of the longest fibres, but if there is much crimp their length is underestimated. Staple length is easily ascertained on the farm, and may be measured on the newly shorn fleece, on a small sample cut at other times, and even on the animal.

Fibre-length distribution is obtained by measuring the length of every fibre in a representative sample, the fibre being stretched just sufficiently to remove any crimp. This may be done by holding the ends of each fibre and stretching it along a rule, but this is slow and laborious. In the laboratory quick and accurate estimates are obtained by using fibre-measurement machines. So much breakage occurs during manufacture that fibre-length variation is relatively unimportant in most wools.

Fibre-diameter measurements require laboratory facilities, and the methods vary according to whether diameter variation along individual fibres is of interest, or only the average variation within a staple. Medullation and proportion of pigmented fibres are often assessed at the same time. Machines for diameter measurement are available.

Pigmentation

Pigmentation may be either general, with large coloured areas or many spots and patches, or incidental with scattered coloured fibres or a few small spots in an otherwise white fleece. Both can be assessed on the farm, but incidental pigmentation may also be assessed with diameter measurements as mentioned above, or by counting the number of pigmented fibres in a small representative sample. General pigmentation may be noted as coloured animals to be culled, where white fleeces are required, or described in detail, for example for studies on the genetics of pigmentation and/or the establishment of pigmented flocks. Photographic records or detailed diagrams showing the markings of individual animals are useful in such work. 'Canary colouring' is dealt with in a later section.

Soundness

'Soundness' refers to the strength of wool. Weak, easily broken wool is called tender, but only if it is abnormally weak for its diameter. In sound wool, tensile strength is correlated positively with diameter, but negatively with amount of medulla. Tenderness occurs when the growth-rate of many of the wool fibres slows down drastically or ceases, and can also be caused by bacterial or fungal action in the fleece. Soundness can be roughly assessed by taking a small staple, holding one end in each hand and either pulling it until it breaks, or 'twanging' it with one finger while holding it taut. Measurement of soundness involves testing the tensile strength of individual fibres by stretching them to breaking point under special conditions. Under worsted processing, tender wool produces excessive amounts of noil.

Assessment of lambs' birthcoats

Halo hair grades

In many lambs at birth and up to about 8 weeks old, long coarse fibres are seen to protrude from an otherwise short coat, forming a fuzzy 'halo' round the lamb. These are halo hairs (HH), the birthcoat kemps. Their distribution and quantity give a rough indication of the amount of kemp and coarse hair likely to follow in the adult fleece. It can, however, occasionally be misleading, particularly, for example, in lambs with coarse plain arrays (see below) which are not followed by kemp or hair. The HH grade is determined by visual assessment of the density of HH (Dry 1955), from grade I (no HH) to VII (dense mat of HH), and their distribution over the body, from grade 1 (britch only) to 7 or F (full coverage).

Fibre-type arrays

The fibre-type array technique can be used to predict many important features of the adult fleece from the coat of a young lamb. It thus makes possible selection (e.g. against kemp) at a very early age. The technique is based on a study of the effects of a phenomenon known as pre-natal check, which occurs in almost all sheep and Angora goats, but has not so far been found in any other species (Duerden 1932; Duerden and Spencer 1930; Dry 1975). It is important to understand that it is a check to follicle and fibre size, not to the general growth of the lamb.

It has long been recognised by zoologists that in almost all mammals the hair follicles which develop earliest in the foetus grow the coarsest fibres in the adult coat, while the latest-developed follicles grow the finest fibres. Between the two extremes there may be a continuous reduction in length and coarseness, but more often there is a discontinuity, leading to a coarser outer coat produced by the early follicles and a fine undercoat grown by the later follicles, as in wild sheep. In some domestic breeds of sheep this primitive order of fibre size is retained or only slightly modified, the early follicles producing kemps or hairs, and the later follicles growing fine wool. However, in many wool breeds this orderly progression is interrupted, rendering some early fibres as fine as, or even finer than, some which develop later.

In foetal lambs the first follicles begin to be formed about 60 days after fertilisation. They appear first on the head, and then further back and more ventrally, but within any one location the order of development of the follicles is the same. The first formed follicles are called the primary centrals (PC), because typically two more primary follicles are formed, one on each side of each PC. These are the lateral primaries (PL).

In the main fleece areas, fibres are beginning to grow in all the primary follicles by 105 days foetal, at which time their sweat glands and erector muscles develop, on the ental side (i.e. that on which the angle of slope of the follicle forms an acute angle with the skin surface) of each primary. There is then an interval of 6 days before the earliest secondary (S) fibres begin to appear (Side 1964). The secondary follicles develop in a cluster between or on the ectal side of the primary follicles, opposite the sweat glands, and with the three primary follicles form a trio group. Figure 8.3 shows a trio group in the skin of a Romney Marsh lamb. Duo groups and solitary primaries also occur at low frequencies.

At about the time when secondary fibres begin to grow (110 days foetal), the growth-rate of some or all the primary fibres, and sometimes the early secondary fibres, is suddenly reduced, affecting both length and diameter. This check to fibre growth persists until birth, after which the fibre may or may not resume its former coarseness and growth-rate.

Thus the primitive order of fibre size corresponds to the order in which foetal follicles develop, fibre size decreasing directly with the age of the follicle in which it grows. The pre-natal check reduces the size of some early starting fibres relative to that of some which grow later, and this effect may be permanent for the follicle concerned. According to Dry (1975) the pre-natal check varies between sheep in intensity and timing, and its effects also depend on the strength and innate tendency for wool growth in the individual animal. A sheep with a strong tendency to grow long coarse fibres is said to have a strong base, while one with a weaker tendency is said to have a weak base. It is probable that the pre-natal check has little or no effect on total keratin production, but greatly affects the distribution of that production between follicle types.

In fibre-type array work, medullation is not significant unless it is coarse enough to give the fibre a chalky-white appearance to the naked

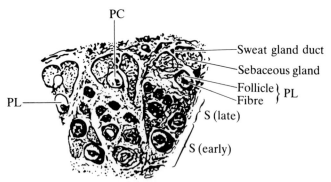

Fig. 8.3 Trio group of follicles from a Romney Marsh lamb. The section was taken at a level a little below the sebaceous glands of the majority of the secondary follicles. The follicles are: PC, primary central; PL, primary lateral; and S secondary. In this group there are 3P and 17S, thus the S/P ratio is 5.7 (After Burns and Clarkson 1949)

eye. For verbal accuracy the terms 'chalky' and 'fine' are preferred to 'medullated' and 'non-medullated'. Thus, by definition, medullation in fibre-type array field work is judged by the naked eye (Dry 1975), although in some research work it may be necessary to use a microscope in order to determine the exact extent of medullation.

Fibre-type array technique

If a small tuft is cut from the birthcoat of a lamb about 7 weeks old, and the fibres are laid out in a row with the oldest (earliest developed) fibres to the left and progressively younger fibres to the right, the whole array of fibres from the tuft can be seen. They will of course have come from numerous trio groups, but the fibres on the left will be from PC follicles, the next from PLs, then the products of early and later secondary follicles to the right. The age of each fibre is judged, not by its size, but by the nature of its pre-natally grown tip.

In a coarse birthcoat, for example from a lamb of a carpet-type fleeced breed, the earliest fibres will be very coarse, heavily medullated throughout and probably show a slight thinning about half-way or less from the tip to the base. These are the halo hairs. Next to them may be somewhat similar fibres in which the thinning is pronounced, forming a distinct neck and making the tip of the fibre appear sickle-shaped; the medulla continues throughout part or all of the neck region as well as through the rest of the fibre in these the supersickles (SS).

Sickle fibres (Sk) may be similar to the SS except that the medulla is absent from the neck region but present below it. Alternatively, they may be fine sickles, with no return of the medulla below the neck, although it may be present or absent in the sickle tip. The smaller the sickle tip the later the fibre, until the sickle tips are lost altogether and the following fibres have wavy or curly tips like sickle necks without the sickles. These are called curly tips (CT), and the more curls or waves there are in the tip the earlier the fibre started to grow. Hairy-tip curly tips (HTCT) have a medulla in part of the pre-natal tip, thus resembling SS fibres which have lost the sickle-shaped tip. In all the above fibre types the fibre becomes at least slightly coarser after birth, and the significant fine neck and tip regions are formed pre-natally.

The last fibre type, called histerotrichs, start to grow at or soon after birth and have no pre-natal tip; in conformity with the primitive order they are small fine fibres which seldom contain any medulla.

Dry and his colleagues named the various arrays according to the position and extent of manifestation of the pre-natal check, from Plateau (the least checked) through Saddle, Ravine and Valley, which all show some sickle fibres more checked than some early CT, to Plain, in which there is no 'rise' (increase in coarseness or medullation) after the latest sickles, which are fine and non-medullated post-natally. Plain arrays with many HH are called 'Coarse plain'. Figures 8.4 and 8.5 show a Plateau and a Coarse plain array respectively.

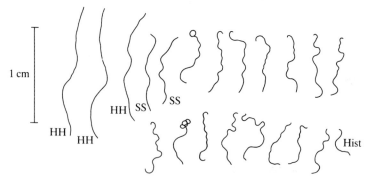

Fig. 8.4 Plateau array, showing 3 halo hairs (HH), 2 supersickles (SS), 15 curly tips (unlabelled) and 1 histerotrich (Hist) (From Burns 1966)

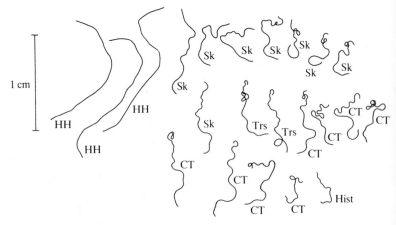

Fig. 8.5 Coarse plain array, showing 3 halo hairs (HH), 8 sickles (Sk), 2 transitionals (Trs), 8 curly tips (CT) and 1 histerotrich (Hist) (From Burns 1966)

Full fibre-type array studies are time-consuming and involve sorting tufts of no fewer than 200 fibres, determining the array and possibly counting the various types. With some types of fleece, certain short cuts may be taken in judging the future fleece characteristics of the lamb as follows:
1. In medium- and fine-fleeced sheep, the presence of numerous fine sickle fibres (i.e. sickles with no post-natal medullation) indicates that kemp is very unlikely to occur in the adult fleece, even if the birthcoat is hairy (high HH grade). If there is no post-natal medullation in the late sickles and early curly tips, both hair and kemp are likely to be absent from the fleece.

2. In coarse-fleeced breeds, whether carpet type or intermediate, kemp later than the first generation (birthcoat kemp) will be less plentiful if sickles are present (Saddle array) than if they are not (Plateau array).

Review of literature

In this section some of the more important and interesting research reports from tropical and sub-tropical countries will be reviewed, concentrating on fairly recent publications. Estimates of fleece parameters are summarised in Table 8.2. It will be noted that the majority of relevant papers come from either India or Egypt.

India is exceptional among tropical countries in having a national co-ordinated scheme for sheep improvement. In this scheme both fundamental and applied science have been encouraged, and have complemented each other. Precisely defined objectives, in terms of fleece weight and quality, have been attained in experimental flocks, and the problem now is to obtain large-scale extension to the peasants' flocks. The difficulties seem to be concerned with husbandry rather than genetics. The use of non-tropical Merino breeds was probably unwise, but these were no doubt the only available types.

The methods used in India have been largely based on Australian research results, derived almost entirely from work with Merinos, which is discussed in a later section. The success of the wool work in India appears to indicate that the determinants of wool production are similar over a very wide range of sheep breeds, and their crosses with Merino breeds not very closely related to the Australian types.

In Egypt, the approach to wool improvement has been a different one, partly due to the small number of sheep breeds, but also because of the influence of Dr R. A. Guirgis, who as a colleague of the late Dr F. W. Dry, became interested in the possible application of pre-natal check theory and fibre-type array techniques to wool improvement in Egypt. This work has confirmed the predictive value of fibre-type arrays in regard to fleece fineness, uniformity, medullation and kempiness, and as a means of selecting Barki lambs which have reduced kemp in their adult fleeces, while retaining the protective hairy birthcoat in lambs.

Egyptian work has also demonstrated maternal influence on manifestation of pre-natal check, as was also reported in British breeds by Burns (1972) and Burns and Ryder (1974). Unfortunately, in most countries the tendency is to import fine-wooled rams to cross with coarse-wooled females so maternal influence may tend to enhance coarseness in the crossbreds. However, the information may be of use in later generations in encouraging strict culling of females with coarse (Plateau or Saddle) arrays as lambs.

Table 8.2 Mean values of wool parameters of some tropical and sub-tropical breeds

Breed	Greasy fleece weight of ewes (kg) [% yield]	Fibre (F) or staple (S) length (cm)	Diameter (μm)	Medullation (%)	Source
Arabi	1.8 [55]	F 17.4	26.2	10.9	Ashmawy and Al-Azawi 1982a, b
Awassi	2.0	—	—	—	Ghoneim et al. 1974
Awassi	2.0 [78]	F 9.8*	39.7*	53.7*	Pritchard et al. 1975
Awassi	—	F 12.8	35.7	14.6 MI†	Guirgis et al. 1978
Awassi	2.0 [53]	F 19.9	28.8	9.6	Ashmawy and Al-Azawi 1982a, b
Barki	—	S 11.4	33.9	14.0 MI	Guirgis 1973
Barki	1.9	S 9.8	—	—	Marai 1975
Barki	0.9	—	—	—	Aboul-Naga 1977
Barki	3.5 [52]	F 14.6 S 13.4	30.9	—	El-Oksh et al. 1979
Chokla‡	—	S 10.2	28.9	10.3	Gupta et al. 1972
Chokla‡	—	—	32.6	37.6	Chatterjee and Arora 1974
Chokla‡	0.9	S 6.5	29.8	23.0	Arora et al. 1975b
Chokla‡	1.2	S 4.8	26.9	18.8	Mahajan et al. 1980a, b
Chokla‡	—	S 4.2	23.9	12.5	Pant et al. 1980
Deccani‡	—	S 5.9	35.1	23.0	Daflapurkar et al. 1979
Hamdani	2.0 [47]	F 18.6	30.0	9.4	Ashmawy and Al-Azawi 1982a, b
Israeli Improved Awassi	2.7	—	—	—	Wallach and Eyal 1974
Jaisalmeri‡	—	—	36.5	—	Chatterjee and Arora 1974
Karadi	—	F 23.5 S 19.3	48.1	19.4	Ghoneim et al. 1973
Karadi	—	F 14.8	40.7	16.7 MI	Guirgis et al. 1978
Karadi	1.6 [52]	F 19.9	28.8	9.6	Ashmawy and Al-Azawi 1982a, b
Libyan Barbary	2.7	F 12.7 S 13.1	32.5	3.0	Labban 1973
Magra‡	—	—	41.7	—	Chatterjee and Arora 1974
Magra‡	1.4	—	—	—	Mittal and Pandey 1975
Magra‡	—	—	30.8	37.0	Sattar et al. 1978
Malpura‡	—	—	47.9	—	Chatterjee and Arora 1974
Malpura‡	0.5	S 6.7	39.0	79.7	Mahajan et al. 1980a,b
Malpura‡	0.6 [69]	S 5.0	40.6	89.7	Kishore et al. 1978
Najdi	2.2	F 17.5*	66.5*	87.8	Pritchard et al. 1975
Nali‡	—	—	42.1	—	Chatterjee and Arora 1974
Nali‡	1.0	F 11.2 S 6.9	—	—	Ram et al. 1978
Nali‡	0.9	—	—	—	Tomar 1978
Nali‡	—	S 4.5	37.0	80	Pant et al. 1980
Nali‡	—	F 9.2 S 6.2	30.9	37.8	Singh et al. 1981a

Table 8.2 (Cont)

Breed	Greasy fleece weight of ewes (kg) [% yield]	Fibre (F) or staple (S) length (cm)	Diameter (μm)	Medullation (%)	Source
Nilgiri‡	0.8	—	—	—	Bhuvanakumar and Subramanian 1981
Ossimi	2.1	S 11.6	—	—	Marai 1975
Patanwadi‡	0.45	S 5.7	29.6	19.1	Daflapurkar *et al.* 1979
Sonadi‡	—	—	61.1	—	Chatterjee and Arora 1974

* Calculated from data given in the source.
† MI is medullation index.
‡ Indian data refer to wool obtained at 6-monthly shearings; in other countries wool is shorn annually.

India

Breeding experiments

In India in 1974, two previous all-India research schemes were merged to form the All India Co-ordinated Research Project on Sheep Breeding. This aims at developing, for both fine and carpet wool, and mutton production, improved strains and breeds of sheep, each adapted to the particular environment and husbandry system of the region in which it is developed. According to Turner (1978), 'superior' rams were produced either by selection within indigenous breeds, or by crossing with exotic breeds. Some indigenous breeds were maintained by within-breed selection without crossbreeding. These include Sonadi and Malpura for coarse carpet wool in very hot dry areas, and Chokla, Magra, Nali and Jaisalmeri for good-quality carpet wool. Originally, all pigmented rams were culled, but in 1975 the rule was relaxed, when the advantages of having some naturally coloured wool for local use was recognised.

The breeding objectives were precisely defined (India, CSWRI 1981) as:

(a) to evolve new fine wool breeds of sheep suitable for different agro-climatic conditions, capable of producing annually 2.5 kg of greasy fleece of 58–64s quality, and weighing 35 kg at 2 years of age;

(b) to evolve new mutton sheep breeds suitable for different agro-climatic regions and weighing 30 kg at 6 months under an intensive individual feedlot system;

(c) to create a superior carpet-wool breed producing 2 kg greasy fleece annually with an average fibre diameter of 30 μm and 20% medullation.

The results of crossbreeding local ewes with several exotic breeds of ram have been recorded in a standard manner so that corresponding data are accumulating at all centres. These data for all fleece-bearing breeds and crosses include greasy fleece weight, clean yield, staple length, fibre diameter, medullation percentage (apparently by number) and fibre density (number per unit area). Greasy fleece weight per kg body weight is also calculated. At most centres, two exotic breeds (Merino and Rambouillet for wool, or Suffolk and Dorset Horn for wool and mutton) are mated to one or more local breeds to produce 1/2, 3/4 and 5/8 (exotic) crossbreds; 3/4 and 5/8 lambs involving two exotic breeds are also produced. At each centre local breeds are bred pure for contemporary comparison. At the Central Sheep and Wool Research Institute in Rajasthan, two new breeds have been developed by crossing coarse-wooled breeds with Rambouillets: the Avivastra for fine wool and the Avikalin for carpet wool.

In many cases it has been concluded that there is no advantage in proceeding beyond 50% exotic genes, and a policy of interbreeding first crosses and their descendants, with the use of selection indices, is being adopted. In several centres it has also been found that there is no significant difference between the descendants of the two fine wooled breeds and the two crossbred lines are being merged. The same applies to the two mutton breeds. In some of the 'mutton' flocks, crosses to Romanov are being tested, with a view to increasing multiple births. The carpet-wool qualities of the Romanov do not seem to have been discussed, but could combine well with the Down fleeces to produce a superior carpet wool.

There are few reported results of selection within indigenous breeds or crosses between them, although there are indications (e.g. Taneja 1974; Chatterjee and Arora 1974) that assessment of local breeds is not being neglected. In general, their wool production is so much below that of selected exotic crossbreds that the prospects of rapid genetic improvement without crossbreeding are not encouraging. However, there are three breeds, all derived from exotic crosses in the past, which merit such attention. These are the Kashmir Valley with average fibre diameter quoted as 18–20 μm, no medullation and fleece weights around 5.5 kg (rams) or 3.8 kg (ewes); the Hissardale with fleece weights of 2–3 kg, fibre diameter around 22 μm and negligible (0.6%) medullation; and the interesting Nilgiri which has exceptionally high medullation (18%) for its fine (average 21 μm) fibre diameter. In any case, in the interests of maintaining genetic resources, it is hoped that no Indian breed will be crossed out of existence which, if AI of sheep becomes widely used, could become a real danger, although not at present imminent.

It is not possible to review in detail the very numerous papers which have been produced over the years in the Co-ordinated Project. Some data on fleece weight and fibre-quality measurements are given in Table 8.2. The following brief review of literature from India concentrates on conclusions rather than on extensive data.

Narayan (1960) in a histological study showed that Chokla, Malpura,

Marwari and Jaisalmeri sheep retain the primitive order of follicle size, with the primary follicles conspicuously larger than the secondaries. Medullation was prominent in the primary fibres of all four breeds, and in the secondary fibres of all except the Choklas. The S/P ratios were very low (1.8) except in the Chokla (2.3) which in consequence had a higher follicle density. The Chokla fleeces were markedly finer and less medullated than those of the other breeds. A report by Bhatnagar *et al.* (1974) stated that the density of primary follicles was increased from 110 cm^{-2} at 6 months of age to 154 cm^{-2} at 60 months. The S/P ratios at the two ages were 3.7 and 1.9 respectively. As all primary follicles are present before birth it is evident that in this study follicles were not correctly identified at the later age, many secondary follicles being mistaken for primaries.

Kapoor *et al.* (1972) studied the copper and sulphur contents of diet and wool fibre in relation to fibre characteristics of Chokla rams. They found that wool copper was strongly correlated with wool sulphur, and both were inversely correlated with medullation percentage and fibre diameter. Wool sulphur, but not copper, was inversely correlated with staple length. Canary stain grade was inversely correlated with diameter, crimp frequency and copper content.

Arora *et al.* (1978) reported that crossbreeding Chokla and Nali sheep with exotic breeds such as the Merino and Rambouillet, caused deterioration in fibre tenacity and elongation. Krishniah *et al.* (1980) examined skins from Malpura, Sonadi and their first crosses with Dorset and Suffolk breeds. The physical and chemical characters of the wool and skin did not differ significantly between groups, but on visual appraisal the best skin and wool were obtained from Dorset × Malpura sheep. All the crossbreds produced better skin and wool than the native breeds. Krishnappa (1980) found that Corriedale × Deccani crosses were significantly superior to Deccanis in all wool traits except staple length at first shearing.

Singh *et al.* (1980c) reported that the relationship between medullation and fibre diameter was linear in both Chokla and Nali wool, and fibre length was not related to diameter. Malik (1976) found a highly significant genetic correlation (0.86) between yearling body weight and fibre diameter in the Chokla breed, while there was a significant negative correlation (-0.20) between body weight and crimp frequency. Malik and Chaudhary (1975) reported that fleece weight in Choklas was significantly correlated genetically with staple length (0.57), fibre diameter (0.67) and 'purity' percentage, presumably the percentage of non-medullated fibres (-0.48). Fibre diameter and staple length had a positive genetic correlation of 0.48. Estimated heritabilities (h^2) included: greasy fleece weight, 0.8(±0.16); staple length, 0.05(±0.13); crimps cm^{-1}, 1.60(±0.98); average fibre diameter, 0.86(±0.30); and percentage purity, 0.58(±0.24).

Nivsarkar and Acharya (1980) reported the h^2 of greasy fleece weight at first shearing in Malpura and Sonadi sheep to be 0.16(±0.05). Greasy fleece weight was greater in the Malpuras than the Sonadis, and was significantly correlated with body weight at 6 months of age, both genetically and phenotypically. The selection index which would give the best improvement for market weight and greasy fleece weight at first shearing was $I = 0.902B + 4.415F$, where B is body weight at 6 months and F is greasy fleece weight.

Mahajan et al. (1980a) found that in Rambouillet crosses with Malpura and Chokla ewes, heterosis for greasy fleece weight ranged from 7 to 27%, and from 4 to 18% for greasy fleece weight per kg body weight. Mohan and Acharya (1980) calculated individual and maternal heterosis for greasy fleece weight and fleece quality data in various grades of crosses between Rambouillet and Chokla, Malpura and Jaisalmeri. The data are too complex to quote in full, but individual heterosis for greasy fleece weight was large for all F_1s, while maternal heterosis was negative for most traits.

Age and season

Gupta et al. (1972) studied the effect of age (from 6 to 42 months) and season on fibre parameters in Chokla sheep. Staple length decreased and fibre diameter increased with increasing age of sheep. Fibre density also increased with age up to 36 months. Staple length and fibre diameter were significantly greater in the autumn than in the spring, while fibre density tended to be greater in the spring. The percentage of medullated fibres was not affected by age but was affected by the season of shearing, and there was an age–season interaction. Mittal and Pandey (1975) found a significant increase in the fleece weight of Magra sheep from 1.2 kg at 1–2 years to around 1.5 kg at 3–4 years of age, followed by a slight decline. The season of shearing had no significant effect on fleece weight.

Singh et al. (1980a) reported that fleece weight in autumn ranged from 0.39 kg for Coimbatore ewes to 1.63 kg for Corriedales, and in spring from 0.32 to 1.50 kg. The crossbreds had intermediate fleece weights. The season of shearing and the year had significant effects on fleece weight in some groups. Tomar (1978) recorded the greasy fleece weights and body weights of Nali ewes shorn twice yearly from 6 months of age to 5 years. Both traits increased significantly with age, but, whereas body weight had doubled by the time the sheep were 4 years old, fleece weight had increased by only 38%. Fleece weight per kg of body weight decreased until the sheep were 3 years old, and then remained unaffected by age. Greasy fleece weight was significantly correlated with age (0.37) and liveweight (0.57). Of the total variation in fleece weight, 42% was attributed to variation in liveweight and 8% to variation in age.

Ghanekar and Soman (1972) found that the fleece weights of both Deccani and Deccani × Rambouillet ewes were lower at the summer shearing (July/August) than at the winter shearing (January/February).

Weights of Deccani fleeces averaged 0.28(±0.01) kg for summer-shorn ewes and 0.36(±0.02) kg at the winter shearing.

Biochemical polymorphs

Singh *et al.* (1973a) recorded the frequencies of three haemoglobin (Hb) types and two potassium (K) types in 211 Magra, 65 Jaisalmeri, 79 Soviet Merino (SM) and 153 SM × Magra sheep. Potassium type had no significant effect on staple length, crimp, fibre diameter or medullation. Kalla and Ghosh (1974) supported this conclusion in regard to medullation in Marwari, Chokla and Soviet Merino ewes, but found that least-squares analysis of the pooled data indicated a possible association between the LK type and the production of finer wool.

Seth *et al.* (1973), using Magra sheep, found that the wool medullation percentage was significantly lower in HbB ewes (34%) than in HbA ewes (46%), despite the significantly higher body weight of the HbB ewes. In HbAB females it was 35%, close to the value for HbB animals. Singh *et al.* (1976) confirmed the lower medullation percentage in HbB Magra sheep, and reported a significant negative regression of crimp frequency on blood potassium concentration. In a study of 726 sheep, including Nilgiri and two imported Merino breeds (Stavropol and Grozny), Krishnamurthy and Rathnasabapathy (1980) found only two significant associations between Hb type and wool parameters. Sheep of type HbA had significantly longer staples (9.6 cm) than those of type HbAB (8.8 cm) and HbB (8.2 cm), and significantly more crimps per cm (3.9, 3.1 and 3.0 respectively).

Taneja (1974) compared the weight of wool produced per unit area of body surface in the progeny (both LK and HK) from LK × LK matings, and in the progeny (all HK) of HK × HK matings. In both Marwari and Magra breeds, the LK progeny produced more wool than the mid-parent value, but the HK progeny did not. Taneja concluded that wool production could be increased by using only LK breeding stock, and culling any HK segregates in each generation. In contrast, Singh *et al.* (1979) found no consistent relationships between Hb or K types or blood sodium levels with fleece weight, crimp frequency, staple length, fibre diameter or medullation percentage.

Kalla and Ghosh (1975) typed 725 adult ewes for blood potassium, Hb and erythrocyte-reduced glutathione (GSH). Breeds included were Marwari, Chokla, Magra, Pugal, Jaisalmeri and Soviet Merino × Marwari. Mean wool production per kg body weight was higher in animals of the HbA type than the other Hb types. The only other significant correlation (−0.13) with a wool parameter was between wool production per kg liveweight and erythrocyte GSH level in the Marwari breed.

Shukla *et al.* (1978) in a study involving Soviet Merino, Rambouillet, Nali, Rampur Bushair and their crosses, found a significant correlation with wool quality traits in only one crossbred group, in which GSH level was negatively correlated with staple length (−0.70) and medullation percentage (−0.66). Murugaraj *et al.* (1980) reported that fibre diameter was significantly negatively correlated with GSH concentration, irrespective of GSH type (high or low) in Grozny Merino sheep.

Canary colouring of wool

A problem for wool growers in tropical countries is a condition known as canary colouring or yellowing of wool. This is particularly important in India where more than a third of the total clip is affected. The yellow stain cannot be scoured out, and adversely affects dyeing and other properties of the wool so that its market value is greatly reduced. Intensive research on the causes of yellowing has been conducted in India, and the conclusions (which are supported by work in other countries) are summarised by Acharya *et al.* (1980), on which the following account is mainly based.

Canary staining occurs mainly during the wet season over the northwestern Indian plains, affecting much of the autumn clip but little of the spring clip. No staining occurs at temperatures below 43 °C unless the relative humidity exceeds 57%. Staining is associated with fleeces of high fibre density, and does not occur in the coarse open fleeces grown in peninsula India, despite the hot humid climate. Thus among Indian breeds susceptibility to staining appears to depend mainly on the variation in fleece density, and this applies even to individual sheep, which may show patchy yellowing with less dense areas remaining white. Low grease content also predisposes to yellowing.

The cause of staining has been found to be bacterial action associated with a high pH (8.5–10.5) and low grease content. These factors lead to alkali damage to the fibre cuticle with consequent chemical changes in the fibre; these reduce tensile strength and alter 'setting', rendering the wool unsuitable for 'permanent pleating' and similar manufacturing processes.

Sweating of Indian sheep is profuse at high ambient temperatures. The suint is alkaline, containing potassium bicarbonate which is converted to the highly alkaline carbonate at the skin surface. This alkalinity favours chromogenic bacteria; bacterial counts on yellow wools are in the order of 480×10^4 mg^{-1} wool, in contrast to counts of 26×10^4 mg^{-1} on white wool.

Merinos and related breeds which depend mainly on evaporation from the respiratory tract when heat stressed, show no yellowing under Indian conditions. The effect of crossing with Indian breeds varies, for example the F_1 Rambouillet × Chokla shows less colouring than the Chokla, but the F_1 Rambouillet × Malpura shows more than the Malpura. The effect apparently depends on the interaction between fleece density and mode of heat dissipation. The increased metabolic heat production which results from a high plane of nutrition can result in increased sweating and consequent canary staining, and reduced water intake can have similar effects due to increased heat stress. Washing the sheep in boric acid solution reduces alkalinity and therefore yellowing, and even washing in water reduces staining by removing suint.

An interesting observation was that a few Merino sheep bred in India had a lower fleece density and a higher suint production than the original imports. It may be that in hot humid zones it is unwise to breed for fine dense fleeces. Selection for increased fibre length can increase fleece weight without increasing fleece density. Practical management suggestions to reduce canary colouring include shearing susceptible sheep just before the hot/humid season, providing shade and water during the day, with night grazing and weekly washing in water.

Pakistan and Bangladesh

Mian (1976) examined the wool fat content and yield of Lohi sheep shorn in the spring and autumn, with an age range of 3–7 years. The fat content ranged from 0.5 to 1.3% and was lower in the spring than in the autumn. The clean wool yield was also higher in the autumn and varied from 57 to 77%. Rahman *et al.* (1971) compared the wool quality of two grades of Lohi × local ewes in Bangladesh. For shoulder samples the following data were obtained: fibre diameter, 24 and 39 μm; percentage of undercoat, 27 and 43; percentage of heterotypes, 17 and 20; percentage of medullated fibres, 55 and 37. Shah *et al.* (1971) determined the percentage of true wool, heterotype, medullated (hair) and kemp fibres in ten samples of Pakistani carpet wool. The degree of medullation increased with increasing fibre diameter.

Near East

In recent years, workers in Egypt have played a major part in developing wool biology in this area. Work in Egypt has concentrated on the Barki breed and its crosses with imported Merinos, while the Ossimi and Rahmani breeds have also received some attention.

The Barki is a coarsewooled breed well adapted to arid grazings. Guirgis (1973) found that in wool samples from yearling animals, medullated fibres comprised 3.9% of the total. When these fibres were discounted, a reduction of 6% in mean diameter and 18% in standard deviation resulted. The main sources of variation between animals were fibre diameter, percentage medullation and percentage of fine and coarse fibres. Kemp score and percentage varied between sites and between animals. Guirgis *et al.* (1979) studied kemp succession in relation to birthcoat fibre-type array. They found that 65% of the arrays were Plateau and the remainder, Saddle, while HH grade averaged 5.5. All the HH, 42% of the SS and 23% of the HTCT were shed as G_1 (first generation) kemp, while few CT and histerotrichs were shed and they contributed most of the persistent fibres. A higher proportion of pre-curly tips (pre-CT) in the birthcoat was followed by increased kemp in the adult fleece. The proportion of kemp decreased with age, but Plateau arrays were followed by more kemp in G_1 and subsequent generations than were Saddle arrays. The amount of G_1 kemp was significantly correlated with the amount of G_2 and G_3 kemp which followed. The authors concluded that, because high HH grade is important for the survival of the lamb at birth, especially under desert conditions, direct selection against HH grade cannot be recommended. Instead lambs should be selected which have a high HH grade with a Saddle array and a relatively high CT/pre-CT ratio; these are the most likely to have an adequately hairy birthcoat with minimal kemp in later fleeces.

Also in the Barki breed in Egypt, Kadry et al. (1980) estimated h^2 and repeatability for greasy fleece weight, percentage clean wool, staple length, fibre length and fibre diameter. They obtained an unusually high estimate of h^2 for fleece weight, 0.84($\pm$0.20), whereas Guirgis et al. (1982) obtained a much lower value, 0.21($\pm$0.94), from a much larger number of sheep (1150 against 153). The two estimates of h^2 for staple length correspond more closely: 0.26($\pm$0.33) and 0.16($\pm$0.82). Two studies report data from Ossimi, Rahmani and Barki sheep reared at six different locations in Egypt. Aboul-Naga (1977) found that locality had a significant effect on the weight of the first fleece, which averaged 1.1 kg for Rahmanis, and 0.92 kg for both Ossimis and Barkis. More detailed data on wool production in the same flocks (Aboul-Naga and Afifi 1977) showed, however, that differences in clean wool production in the different locations were slight, and for greasy fleece weight the location effect was significant for the Rahmani breed only. The age of ewe had a highly significant effect on greasy fleece weight in all three breeds. Heritability estimates for annual greasy fleece weight differed markedly with age, but were generally low; the highest estimates were for the first fleeces of Barki (0.34) and Rahmani (0.19). Repeatability estimates for greasy fleece weight were: Barki, 0.29; Ossimi, 0.11; and Rahmani, 0.04. Estimates of genetic correlations between yearling greasy fleece weight and weaning and yearling body weight were high and positive in Ossimi and Rahmani ewes.

Marai (1975), in a study of the Ossimi and Barki breeds, found significant positive correlations between staple length and fleece weight, and both staple length and fleece weight were positively correlated with body weight. The month of shearing significantly affected all three parameters.

Fleece studies on the Barki breed led Guirgis and his colleagues to conclude that improvement efforts should concentrate on wool production, as the scanty pasturage of the desert environment would be unfavourable to any large increase in meat production. Two types of Merino were imported into Egypt for crossbreeding: the German Mutton Merino and the Hungarian Combing Wool Merino. It appears from the data reported that both these breeds have lighter and more variable fleeces than typical Australian Merinos. In studying the results of crossing these Merinos with the local Barki and Ossimi breeds, much attention has been given to the birthcoat fibre-type arrays. In general, the results have confirmed the relationships between fibre-type array and adult fleece characteristics reported by Burns (1955) and Dry (1975). Finer arrays are followed by finer and more uniform fleeces, with kemp being mainly confined to fleeces following the coarsest arrays (Plateau and Saddle).

Guirgis and Galal (1972) found that kemp was positively and significantly correlated with birth and weaning weight in Barki and Merino × Barki lambs, whereas staple length was negatively correlated with these weights, although its correlation with fleece weight was positive. This appears to imply that kemp was replaced by long hair which determined the staple length. In one of the most significant papers of the series, Guirgis (1977) demonstrated a strong maternal influence on the birthcoats of Merino × Ossimi lambs, those with Merino dams having a significantly greater proportion of finer arrays, and a lower percentage of HH and SS than the reciprocal crossbreds. There was no maternal effect on lamb birth weight, but the latter was significantly correlated with fibre-type array. Lambs with Plateau or Saddle arrays had significantly heavier weights at birth than those with Ravine arrays.

Guirgis (1980) compared the fleeces of Merino, Barki and five grades of their crosses, measuring greasy fleece weight, clean fleece weight, kemp score, mean staple length, mean fibre diameter and medullation percentage on representative samples from the whole fleeces. Except in clean yield, there was a gradation from Barki to Merino, through 1/4, 3/8, 1/2, 5/8 and 3/4 Merino. The 1/2 Merinos had the highest yield, and the average yield of all the crosses was higher than that of the Merino and similar to that of Barki wool. Although mean diameter was reduced by increasing proportions of Merino genes, within-fleece variability in fibre diameter remained high and medullated fibres occurred in both fleece wool and skirtings in all except 3/4 Merino. The author attributed this to the failure of the Merino cross to reduce the size of the primary central follicles, and suggests the use of a Wensleydale cross to overcome this problem. The staple length of all the crossbreds was sufficient for processing on the Noble comb (worsted) system.

El-Sherbiny et al. (1978) studied the effect of light and temperature on the wool growth of German Mutton Merinos, Ossimis and various crosses between them, by comparing wool growth of sheep kept in continuous dim light with that of others kept in sunlight with shade zones. Length growth of fibres was higher in all breeds and crosses in sunlight, but diameter was slightly higher in Merinos and 3/4 Merinos kept in dim light. No significant seasonal rhythm was found under sunlight, but under dim light, wool grew significantly faster in the winter and autumn than in the spring and summer. Also in Egypt, Selim and

Youssef (1971) studied the relationships between clean fleece weight and body weight in thirty-nine German Mutton Merino rams, and concluded that body weight alone could be used in selecting for body weight and fleece weight combined. Although a more complex estimate combining the h^2s of body weight and fleece weight and the estimated economic value of expected genetic gains was considered slightly more efficient, body weight was highly correlated (0.997) with this estimate.

Ragab et al. (1978) reported on some non-genetic sources of variation in fleece weight of German Mutton Merinos in five commercial flocks in Egypt, over the first three shearings. Greasy fleece weight averaged 1.3, 3.7 and 2.6 kg at the three shearings, and was significantly affected by farm and year of birth, while season of birth affected the first two shearings. These factors, together with sex and type of birth, accounted for 42, 57 and 36% of fleece weight variance at the three shearings. In a further study of the same flocks, Abdel-Aziz et al. (1978) calculated the h^2 of greasy fleece weight (for the first fleece) to be $0.32(\pm 0.06)$; it showed significant genetic and environmental correlation with body weight at three ages (birth, 4 months, 1 year).

The effects of various factors on fleece weights of Awassi sheep in Iraq were analysed by Ghoneim et al. (1974). Sex, age, year of shearing and type of birth significantly affected fleece weight. The fleece weight averages (over all ages) were 2.5 kg for males and 2.0 kg for females, and the highest fleece weights were obtained at the first and second shearings. Heritability estimates for fleece weight were 0.47 by paternal half-sib correlation and 0.16 by regression of offspring on dam. Repeatability was 0.32 by intraclass correlation and 0.36 by correlation between pairs of records. The authors recommended that, for improvement of fleece weight, breeding stock should be selected for fleece weight at the first shearing when they are 18 months old. Wool follicle characteristics of Awassi lambs in Iraq were examined by Fayez et al. (1976), at intervals from birth to 6 months of age. Follicle density was significantly affected by age, sex and birth type, but the density of primary follicles was not affected by sex. The S/P ratio at birth was 2.0 and by 2 months of age it had risen to 4.0, almost its adult value. It was greater in singles than in twins, and in males than in females. Fibre diameter was affected by age and birth type, but the diameter of secondary follicles was affected only by sex.

In a detailed study of Karadi and Awassi sheep and their reciprocal crosses in Iraq, Guirgis et al. (1978) obtained wool samples from sheep at 4 and 12 months of age, and measured fibre length and diameter, medullation index, and percentage of fine, coarse and kemp fibres. Samples showing a bimodal distribution of fibre length and diameter at 12 months were common in both breeds, but more frequent in Karadi than Awassi wool. There was significant heterosis in the crossbred lambs for fibre diameter but not for fibre length.

Also in Iraq, Ashmawy and Al-Azawi (1982a) studied the influence of age, location and husbandry system on greasy fleece weight, 'shrinkage per cent' (apparently the reciprocal of yield) and fibre-type ratio in Awassi, Arabi, Karadi and Hamdani sheep. The highest clean fleece weight and undercoat fibre percentage were found in the Arabi, followed by Awassi, Karadi and Hamdani, while kemp was highest (7%) in the Awassi. The same authors (1982b) reported data on fibre diameter, percentage of medullated fibres and fibre length on the same four breeds. The Karadi breed had the longest and coarsest fibres, and the Arabi had the shortest and finest fleece. The environmental factors studied had little effect except on fleece weight and shrinkage, but the percentage of undercoat fibres was reduced by intensive husbandry except in the Arabi breed. No attempt was made to explain the latter effect, but it would appear probable that either the better nutrition caused some follicles, which otherwise contributed to the undercoat, to grow medullated fibres, or else outer coat fibres were lost, for example by rubbing on fences, etc.

Labban (1973) studied the wool characteristics of Barbary sheep in Libya. Greasy fleece weights averaged 2.7($\pm$0.09) kg for ewes and 4.5($\pm$0.14) kg for rams. Fibre diameter from different sampling positions differed significantly and for mid-side samples was 32.5($\pm$0.67) μm, with 97% of true wool, staple length 12.7($\pm$0.33) cm and fibrelength 13.1($\pm$0.20) cm. Pritchard et al. (1975) gave data from mid-side wool samples of Najdi and Awassi ewes in Saudi Arabia. The Najdi ewes were found to have 88% coarse fibres but no kemp; diameter of coarse fibres was 68($\pm$3.4) μm and of fine fibres 20($\pm$0.6) μm. Length of the coarse fibres was 18.3($\pm$0.68) cm and length of fine fibres was 12.7($\pm$0.94) cm, indicating a double-coated carpet-type fleece. The average fleece weight at 1 year old was 2.2 kg.

Goot (1972a) reported on the effect of lactation on some wool parameters in Awassi sheep in Israel. In both ewes and hoggets lactation significantly ($P<0.001$) affected fleece weight, but there was no effect on amount of cotting (tangling and felting), canary stain or hairiness. There were no significant correlations of fleece weight with milk yield or lactation length. In Syria, Mukhamed (1973) described some wool characteristics of three types (Shagra, Absa and Porsha) of Awassi sheep kept in two different climatic zones, semi-desert and littoral. Kemp percentages varied from 10 to 20, and 'undercoat' from 56 to 64. Presumably the remainder was heterotype hair. Undercoat fibre fineness varied from 21.1($\pm$0.19) to 22.6($\pm$0.18) μm.

Nigeria

A wool-growing project was started at Katsina in northern Nigeria in 1958, by crossing Merino rams from South Africa and Rhodesia with two local breeds of hair sheep, the Uda and Yankasa. The F_1 animals carried short fleeces composed of very fine wool mixed with numerous kemps; subsequent research showed that all the primary follicles grew kemps. The fleeces were shed annually, and were rejected by manufacturers because of the excessive amount of kemp. Burns (1967a, b) reported that the S/P ratio was doubled in the F_1 to approximately 8.0, from around 4.0 in both Uda and Yankasa, with no increase in primary follicle density. The 3/4-bred lambs from the back cross to Merino fell into two categories as adults, kempy and non-kempy, suggesting involvement of a major gene. All those with kemp-free adult coats had Plain fibre-type arrays as lambs, although a few had very kempy (high HH grade) birthcoats. Those which developed kempy fleeces all had high HH-grade birthcoats with Plateau arrays. The non-kempy 3/4 Merinos had higher average S/P ratios than the kempy group.

The use of a ram showing 'central checking' (severe checking of the central primary fibres) was recommended as likely to eliminate kemp, and the use of Wensleydale rams was successful in this respect. In the F_1 Wensleydale × hair sheep, the primary follicles grew heterotype hairs instead of kemp, and fleece shedding did not occur. A feature of all the crossbred fleeces, even the kempy ones, was their remarkable softness of handle. The average diameter of the non-kemp fibres of 3/4 Merino animals was actually less than that of the Merino rams used. The possibility therefore existed for developing a fine, soft, speciality wool suitable for hand-knitting, and comparable to the British Shetland wool. Unfortunately, external circumstances prevented this development.

Kazmi and Mathieson (1976) conducted chemical and physical measurements on fleece samples from Katsina sheep including Merino, Yankasa and a 5/8 Merino 3/8 Yankasa crossbred, and concluded that the crossbred wool was only slightly inferior to Merino in most respects.

South Africa

Wool biology in South Africa has a long history, from the early years of this century when Duerden, working on Merino sheep, discovered sickle fibres, and laid the basis from which the work of Dry and his colleagues was developed. However, as research in South Africa is confined almost entirely to the Merino breed, with which we are not dealing in detail, this review will be confined to a few recent papers of general interest.

Attention has been paid to factors affecting the damage by weathering of the tip of the wool staple in the tropical environment. Venter and Edwards (1977) concluded from experimental work, that the degree of weathering showed a linear relationship with the hours of sunshine and maximum daily temperature, but was mainly dependent on the season of wool growth. In order to limit weathering to the extreme tip of the staple, the sheep should be shorn just before the hottest time of year. Venter et al. (1977) studied the effects on fleece weathering of shelter (shade) and of covering the sheep with blankets. The use of blankets was the most effective protection, and even when used for only the hottest 3 months weathering was almost eliminated. Provision of shade had little effect on the tip 1 cm of wool, but weathering of the next 2 cm was reduced by about 35%. The provision of shade was recommended as being more practicable than covering the sheep with blankets.

Two similar reports by Venter (1980a, b) describe the assessment, over a 9-year period, of midrib wool samples from Merino two-tooth ewes and rams selected for good- or poor-quality wool. Soft handle was a requisite for classification as good wool. The effect of selection was to increase fineness, length and feltability of the 'good-quality' fleeces, but to reduce the crimp frequency and resistance to compression, leading to increased susceptibility to weathering of the tip, faults which were not developed in the sheep selected for 'poor-quality' wool. It was concluded that soft handle should not be overemphasised in selecting for wool quality in the South African environment.

Nel et al. (1972) estimated the repeatabilities for greasy fleece weight and staple length as 0.58 and 0.42 respectively. Ewes produced 0.30 kg less greasy wool for each lamb reared in the same year. The age at which ewes were culled (6–8 years) had no effect on the rate of genetic improvement in greasy wool weight. Heydenrych (1977) and Heydenrych and Meissenheimer (1979) gave data on the production and genetic parameters in a flock of South African Merinos. Breeding ewes produced their maximum amount of greasy wool (6.6 kg) at 3 years old, and lactation retarded wool production more than did pregnancy. Type of birth and of weaning had significant effects on wool production, fibre dimensions and the S/P ratio. Age of dam had a significant effect on clean wool production of the progeny. Estimates of heritability gave higher values with data from non-selected sheep than from sheep selected for wool weight at 18 months old, or higher S/P ratio at 3 months old. Females had higher values than males. Genetic correlations with clean wool yield were 0.81 for greasy wool weight and 0.37 for S/P ratio. Body weights at age 42 and 120 days were significantly correlated (about 0.65) with greasy and clean wool weights. Heritabilities for males and females respectively were as follows: greasy fleece weight 0.42, 0.49; clean fleece weight 0.28, 0.41; yield 0.55, 0.61; staple length 0.21, 0.26. There was no significant sex difference in the h^2 of crimp frequency (0.19), and the h^2 of fibre diameter was slightly higher in males (0.45) than in females (0.41).

Poggenpoel and Merwe (1975) derived selection indices for Merino rams and ewes, following consideration of the relative importance of wool-production characters, their heritabilities and correlations among them. It was recommended that selection indices should be applied to the remaining sheep after not more than 10% had been culled for serious defects and faults in breed type. Heydenrych et al. (1977) obtained positive responses in Merino sheep to selection for either high clean fleece weight or high S/P ratio, and found that realised heritabilities showed fair agreement with estimated values. Staple length, fibre diameter and crimp frequency were included in the study.

Reyneke and Fair (1972) found that wool production in Döhne Merinos was depressed in the winter months due to an inherent physiological rhythm combined with low nutritional level. This effect was much greater than that of pregnancy or lactation. Wool production was significantly correlated with staple length (0.73) and fibre diameter (0.74).

South America

Wooled sheep in South America are either mainly of Spanish and Portuguese origin (Criollos) or Merino and Corriedale types with Australian ancestry. Lincoln Longwools, Romney Marshes and British Down breeds are also kept in some parts. The Criollo wool is of carpet quality, and most of it, as well as some finer wool, is used locally. The rest is nearly all exported as raw wool.

In 1975 Mexican wool production totalled 3800 t of greasy and 1550 t of clean wool (Moreno Chan 1976); 80–85% was short wool of 46–64s quality, and the remainder was long, finer wool (56–70s). For Romney Marsh ewes in Mexico, Castañon Canet (1978) reported that annual wool production averaged 13.4 kg of 52s quality, staple length 81.1 mm and fibre diameter 21.0 μm. Fibre diameter was significantly correlated (0.22) with length. Aguilar Corona (1980) presented data on Hampshire × Criollo and Suffolk × Criollo ewes in Mexico. Greasy fleece weight averaged 2 kg and was highest in 3–4-year-old ewes; the wool was of medium quality with short staple length.

Recent papers from southern Brazil give data on wool parameters of several breeds, but there have been few studies on wool production in tropical Brazil. Prucoli (1973) reported that the season of shearing had no effect on fleece weight or quality in Corriedale ewes. Selaive *et al.* (1980), from records of commercial flocks, found that fleece weights of females mated as two-tooth ewes were lower over the next 3 years, by 10%, 10% and 4% respectively, than those of ewes not mated until a year later.

Data on Corriedale sheep kept at an altitude of 3800 m in Bolivia were presented by Iñiguez (1975). The heritabilities (h^2) of fleece weight and fibre length were 0.23 and 0.15 respectively, and the repeatability of fleece weight was 0.62. Corriedales of three different genetic origins, kept at even higher altitudes (4260–5000 m) in Peru were studied by Diez *et al.* (1974). Based on paternal half-sib correlations, the following h^2s were estimated: fleece weight, 0.20; fibre diameter, 0.01; and staple length, 0.03. Also in Peru, Cerrón and Pumayalla (1974) reported that, in Junin hoggs, greasy fleece weight of ewes averaged 3.0 kg and of rams 4.6 kg. Staple length was 4.1 cm for ewes and 11.0 cm for rams.

Liveweight was significantly correlated with greasy fleece weight (0.35 for ewes, 0.10 for rams) and staple length (0.8 and −0.2); greasy fleece weight and staple length were also correlated (0.44 for ewes, 0.35 for rams). The negative correlation of liveweight and staple length is surprising, but it must be remembered that staple length measures the longest fibres, not the mean fibre length.

Advantages of wool production

In those parts of the tropics where only hair sheep are kept, many people feel that wool production in tropical climates is totally inappropriate, mainly because of heat stress and ectoparasites. Vegetable contamination, especially from adherent seeds, is also often a problem. In hot wet climates hair sheep undoubtedly have special advantages; Mason and Buvanendran (1982) advise that 'fleeced sheep should not be imported into a hot humid environment'. Nevertheless, wooled sheep are kept in some such climates, although they are more suited to dry tropics, sub-humid tropical regions where there is a substantial dry season, and the sub-tropics.

Undoubtedly wool production adds to the labour costs and to a small extent to capital costs (shearing machines, wool storage facilities, etc.), but there are two major economic advantages of wool production:
1. There is no biological antagonism between meat and wool production (e.g. selection for rapid genetic growth and large body size will tend to increase fleece weight) so that where nutrition is adequate, any income from wool is a bonus over the income from meat.
2. In arid areas and elsewhere where nutrition is inadequate for mutton production, wool growing can be economic, as wool, unlike meat and fat, continues to grow even when a sheep is losing weight. In such areas, Merinos and their crosses excel as wool producers, at least when bred in sub-tropical and tropical zones such as southern Africa and Australia; it is uncertain whether other types of Merinos, bred in temperate climates, will prove adaptable to heat stress and malnutrition. However, these breeds (Rambouillet, Soviet Merino, German Mutton Merino) are all better meat producers than the tropical Merinos.

In the humid tropics the high fleece density of Merinos is a disadvantage, and in these areas it seems logical to concentrate on carpet wools. Where nutrition is adequate for mutton production, 'crossbred' apparel wool may be the best option, especially as most of the highly developed temperate mutton breeds grow this type of wool. Some of these breeds could be useful for crossing in the sub-tropics and mountainous areas in tropical countries. The place of lustre longwool breeds in the tropics and sub-tropics has been little explored, but where nutrition is adequate for their large size and fairly high fertility, their open (low-density) fleeces and heavy fleece weights should make them suitable crossing breeds for wool production; those with pigmented skins (Wensleydale and Teeswater) may have advantages when exposed to direct sunlight.

It should not be assumed that wool improvement necessarily involves crossing with imported breeds. Selection within local breeds and crosses will involve less loss of adaptation and, although increase in fleece weight may be more limited, such a policy may be more cost effective, especially for the carpet wools. The potential of sub-tropical breeds such as the Awassi and Arabi for crossbreeding in the tropics have not been explored, but they can be expected to adapt to the dry tropics better than breeds of temperate origin.

Maximising wool production

Environmental factors

In order to maximise wool production it is necessary to attend not only to the genetic potential of the sheep, but to relevant environmental factors, and to correct shearing, handling and storage of the wool. Restricted nutrition reduces wool growth and may produce tenderness or even a definite break, the latter particularly in coarse hairy fleeces. In the tropics and sub-tropics, any apparent effect of season on wool growth is probably indirect, due to effect on pasture growth. Pregnancy, and even more, lactation reduce wool growth, which suggests that in desert environments where wool is the only offtake, it could be worth keeping wether flocks; moreover, at 2 or 3 years old, wethers might be sold to other regions for fattening.

Contamination of fleeces with vegetable matter is a problem which often cannot be avoided on natural pastures, but where tropical pastures are specially sown for sheep, grass species with adhesive seeds should be avoided.

Incorrect handling of the fleeces at shearing can greatly reduce the price received for the wool. In shearing, double cuts must be avoided, and the fleeces must be kept clean and dry, and any soiled locks removed. It is good practice to cut off soiled wool from the sheep the day before shearing, at a place other than the shearing area. Fleeces of widely different quality (e.g. fine wool and carpet wool) should be kept separate, and any pigmented animals should be shorn last to prevent contamination of white fleeces by coloured fibres.

Genetic improvement

Genetic improvement of wool can take two main directions: (a) within type, keeping the type of wool already being produced but aiming to eliminate faults and increase clean fleece weight; and (b) changing the type, for example from coarse to fine. Improving the existing type can usually be achieved by selection without crossbreeding, but if the fleece type is to be drastically changed, crossbreeding is usually necessary. However, in the tropics and sub-tropics there are many breeds, described as coarse-wooled, which in fact have a relatively small proportion of medullated fibres, and the elimination of these can bring their fleeces within the medium ('crossbred') apparel wool class; whereas selection for increased medullation in the primary fibres would be necessary to convert them into good carpet wool. Such fleeces do not fall exactly into any of the usual trade grades, and offer valuable opportunities for changing fleece type by within-breed selection.

In countries where wool is or has been of major commercial importance, farmers and shepherds have established breeds with characteristic fleece types. Even within the Merino breed there are different strains, the fleeces of which can be easily identified by anyone experienced in appraising Merino wool. Genetic control of many wool-production characteristics has been demonstrated by a large amount of scientific experimental work, although most of it has been concerned with Merino sheep and closely related breeds, and still requires to be more fully tested on other types of sheep.

As with appraisal of shorn wool, fleece quality on the live animal was, and usually still is, assessed by hand and eye. It is thus clearly possible to improve wool production without sophisticated techniques, and this is particularly true where gross faults, such as kemp or pigment are concerned. Even for Merino types (assuming skilled classers are available as in Australia), Turner (1956) concluded that 'extensive measurements on ewes, other than greasy fleece weight are not warranted on the grounds of expense, but on rams a study of the relationship of the fleece components is justifiable, at least on sires used extensively'. However, pragmatic assessments do have limits and can actually be misleading. In Merinos, appraisers long believed that high crimp frequency was directly related to fineness, but scientific studies have demonstrated that this correlation can be very low. Merino breeders therefore wasted a great deal of selection pressure, over many years, selecting for a character which is of no practical or commercial significance. Similarly, lack of understanding of genetic principles can lead to very misguided breeding policies.

The fleece components most commonly measured are fleece weight (greasy and clean), fibre length, diameter and medullation, and staple length. In the Australian Merino and its close relatives, medullation is negligible, and Turner (1956) defined the components of clean wool production per animal (W) by the formula

$$W = S \cdot R \cdot A \cdot F \cdot N \cdot \rho \qquad [8.1]$$

where

S = area of smooth body surface, $\propto$ (body weight)$^{0.67}$;
R = wrinkling factor;
A = mean cross-sectional area of fibres;
F = mean fibre length;
N = mean fibre density (number of fibres per unit area of smooth skin); and
ρ = specific gravity of wool.

This formula may need modification when applied to non-Merino fleeces (e.g. to allow for medullation and perhaps for extreme variation in fibre length) but it does very neatly summarise the components available for measurement.

Fleece weight

The simplest form of wool improvement is to select for increased greasy fleece weight, which involves merely weighing the fleeces at shearing, and favouring those sheep with the heaviest fleeces. Clean fleece weight is strongly correlated with greasy fleece weight, although this correlation obviously varies with the yield, so that direct selection for clean fleece weight is more accurate, although seldom worth while on ewes. Experimental work has established that sheep which produce more clean wool per head are also producing more wool per unit of food consumed (Turner 1977).

Increased clean fleece weight can be produced only by increasing some or all of fibre length, diameter and density. In general, length and diameter are positively correlated with clean fleece weight, but both are also correlated with increased medullation; fibre density is negatively correlated with diameter, and this tends to improve quality in fine and medium wools, but thus may partly counter increased fleece weight. As none of the correlations are absolute, complex interactions occur, and the results obtained in one breed may not be applicable to others. The heritability of fleece weight is usually fairly high, typically 0.4–0.6 for clean fleece weight (Turner 1956). Body weight has been reported to account for up to 20% of the variation in clean fleece weight, but selection for clean fleece weight (in medium Merinos) produced no marked change in body weight; selection for body weight produced increases in clean wool weight of the order of 0.02 kg kg^{-1} body weight (Turner 1956).

A large number of studies on Merinos and other non-medullated fleece types have demonstrated that diameter is by far the most important determinant of quality, with average fibre length next. This probably holds true for medium wools and even carpet wools, although the required diameters and lengths would be different, and would be bimodal in carpet wools. Diameter is almost the only determinant of quality count.

The wrinkling factor, R, need not be considered in non-Merino breeds, which do not have artificially 'developed' (wrinkled) skin. It seems to be clearly established that, for improvement of fine and medium wool, the correct policy is to select for fleece weight while setting upper limits on diameter, but conducting detailed measurements only on élite rams. A minimum staple length may have to be set for worsted spinning wools, and if kemp is a problem, it can most quickly be reduced by culling lambs with Plateau or Saddle arrays in their birthcoats; if hairy birthcoats are not required, the elimination of all such lambs will eliminate most kemp without the need to study fibre-type arrays.

Heritabilities and repeatabilities have been determined in fine and medium breeds for all the measurable parameters of the fleece, and although they vary greatly, most are reasonably high. Correlations between these parameters show great variability, so that it is possible to select successfully for various combinations, such as long and fine or long and coarse. Medullation percentage is positively correlated with coarseness, so that the coarsest fibres in a fleece are always the most medullated, but the threshold for medulla formation varies greatly. In Longwool breeds, fibres of 60–65 μm diameter may be free of medulla (though they contain a metacortex) (Marcet and Onions 1963), whereas in some hair sheep, and also in the Wiltshire Horn, some fine fibres growing in secondary follicles contain a medulla.

Carpet wool

As already explained, carpets (Fig. 8.6) are often made from blends of coarse and medium wool, and the exact proportions vary according to the type of carpet and the manufacturer's preference. It is generally agreed that both long hairy fibres and bulky fine fibres are essential. Although the presence of medullated fibres is regarded as important for carpet wools, their actual function in the finished product is not understood. The keratin which forms the network between the medullary spaces contains much less sulphur than does the keratin of the cortex, and is resistant to attack by chemicals that dissolve the cortex. This may be relevant to resistance to abrasion in carpet wools. In Britain, some wools with very little medullation are favoured for carpets because of their resilience. Resilience and the ability to withstand hard wear are the two special requirements for carpet wool, which are less important for other uses. Measurement of these qualities is not easy, but resilience may be roughly assessed by squeezing a handful of wool into a tight ball and watching it spring back on release.

The quality of carpet wool has received relatively little attention from wool biologists, most of whom have concentrated their efforts on apparel wools. Shepherd (1959) stated that the coarse outer fibres are required to give strength, but a fine bulky undercoat is also necessary to

200 *Wool production*

Fig. 8.6 Carpet weaving, Ethiopia

give fullness and resilience to the yarn. Burns and Chaudhary (1965) defined standard requirements for carpet wool as follows: it should contain 85% true wool by count, or 65% by weight, and 15% non-kempy medullated fibre (hair and heterotype) by count, or 35% by weight. Kemp is not required, and is a defect. The average diameter of all the fibres should be between 37 and 41 μm, but the diameter distribution is more important than the mean. The range of diameters should be from 11 to 90 μm, with about 85% of the fibres in the range 21–60 μm. Thus they confirm the desirability of bimodal diameter distribution. This standard has been adopted in the Indian work already described, with the additional requirements of not more than 2% kemp, and staple length of about 10 cm (Taneja 1978).

Many of the famous Indian carpet wools have very little undercoat, but when exported to Britain are blended, for example, with Cheviot wool, to provide resilience and bulk.

The development of the Drysdale breed from the N-type Romney and

Cheviot in New Zealand focused the attention of Dry and his colleagues on carpet qualities (e.g. Nash 1964), and Drysdales are now recognised as probably the best carpet-wool breed in the world.

It seems strange that in India, a carpet-wool breed is being developed from Merino crosses. Merino wool lacks resilience and the high S/P ratio of the breed is neither necessary nor desirable in carpet wool. Probably the main attraction of Merino crosses is the rapid genetic gain in fleece weight obtained, but this results largely from increased fleece density, which is not desirable in hot humid climates, and is conducive to canary yellowing. Selection for increased fibre length would be more logical, especially where twice-yearly shearing is practised. Breeds which appear to have potential for development in this direction include, for example, the Hamdani, Karadi and related types in the Near East, and many of the Indian breeds, among which the Nilgiri with its fineness plus medullation could be important for crossbreeding with other indigenous breeds. The coarser-fleeced type of Awassi could be useful in bringing more undercoat to some of the Indian (and other) breeds which have a very low S/P ratio. If any temperate breed can improve carpet-wool production in the tropics or sub-tropics it must surely be the Drysdale or one of the later versions, such as the Turkidale.

Chapter 9

Milk and other products

Milk production

In all systems of sheep production, milk is essential for lamb survival and growth. In some areas of the world ewe's milk is also used for family consumption or sold. Table 9.1 shows the milk produced for human consumption from sheep in different parts of the world. Data for individual countries are included if production exceeds 50 000 t y^{-1}. In both Africa and Asia, sheep produce about 5% of total milk production. In Asia, dairy sheep are mainly in the Near East (Syria, Iran, Afghanistan, Iraq, Saudi Arabia and Turkey) or China and Mongolia. In Africa, production of milk from sheep is limited to the northern parts: little milk is produced from sheep south of the Sahara. European production is about 3.6 million t y^{-1}. In much of Europe, sheep were originally triple-purpose animals, producing wool, milk and mutton (Ryder 1983a). Production of milk is now limited to the eastern and southern parts of Europe and accounts for less than 2% of total milk production. In North, Central and South America, Australia and the Pacific, human consumption of sheep's milk is negligible.

It is interesting to note that the total world consumption of sheep's milk (8.2×10^6 t y^{-1}) is slightly greater than that of goat's milk (7.7×10^6 t y^{-1}). On average, less than 2% of all the milk consumed by humans is from sheep (Table 9.1). However, in developing countries which, as defined by FAO (1983a) include most of the tropics, sheep produce about 4% of all milk consumed by humans (Table 9.2). In Syria 46% of milk is produced by sheep, and in four other countries in the Near East the proportion is over 20%.

Management of dairy ewes

Systems of production

The most common systems of management of ewe flocks producing milk are similar to those for dairy herd of cows, namely:

Table 9.1 *Worldwide distribution of milk production from sheep*

Area	Milk produced by sheep (1000 t y^{-1})	Sheep milk as a proportion of total milk (%)
World	8163	1.7
Asia	3699	5
Turkey	1229	21
Iran	731	27
China	504	6
Syria	450	40
Afghanistan	241	28
Iraq	175	12
Saudi Arabia	100	24
Mongolia	54	21
Europe	3634	2
France	1131	3
Italy	623	5
Greece	591	35
Romania	340	8
Bulgaria	325	13
Spain	230	3
Yugoslavia	147	3
Portugal	90	11
Poland	77	0
Africa	694	5
Algeria	166	20
Sudan	130	9
Somalia	99	18
Ethiopia	58	8
Mauritania	54	25
USSR	100	0
S. America	36	0
N. and C. America	0	0
Oceania	0	0

Source: Data from FAO 1983a.

1. A milking and suckling system in which lambs are given access to their dams for limited periods, and in which the ewes are milked once or twice each day.
2. A milking system in which the lambs are removed from their dams soon after birth, and are subsequently not allowed to suckle. This occurs only in intensive dairy flocks where there are adequate facilities to rear lambs independently of their dams, either by fostering, feeding milk back to lambs or giving an artificial diet. The artificial rearing of lambs has been reviewed by Treacher (1973).

In some parts of the world ewes are milked only after their lambs have been weaned. In this case, the flock is kept primarily for a product other

Table 9.2 *Origin of milk in less developed countries (excluding milk from other species, e.g. camels)*

Species	Production (1000 t y^{-1})	% of total
Cow	75 258	66
Buffalo	28 385	25
Goat	5 619	5
Sheep	4 408	4
Total	113 670	

Source: From FAO 1983a.

than milk (either meat or wool), and milk is obtained as a by-product from the high-yielding ewes in the flock after weaning or if their lambs die.

Both in subsistence and commercial ewe dairying, milking and suckling (system 1) is the most common. Allowing the lambs to suckle increases the total milk production of dairy ewes (Lawlor *et al.* 1974), although the extra milk is consumed by the lambs. Many different suckling regimes are practised. Lambs may be present with their dams during the night or for limited periods after each milking. For the lambs to have a beneficial effect on milk yield they must have access to their dams at least twice each day. The longer they are present with the ewes the more milk they obtain and the less is available for human consumption (Folman *et al.* 1966). Management of milking and suckling systems therefore depends on a careful balance between the nutrient requirements of the lambs and human demands. The decision will depend on the availability and cost of alternative food supplies for the lambs. Provided that good-quality creep feed is available, the growth-rate of lambs is independent of the time allowed for suckling (Louca 1972).

Practicalities of milking

Sheep must be trained to stand quietly while being milked. Training is much easier if ewe lambs are used to being handled or are hand-reared. Initially, ewes must be restrained, often by a collar, but once they have accepted the milking routine they should co-operate if they are fed during milking. In traditional Bedouin systems in the Near East where Awassi ewes are milked in the open, they are tied in rows, head to head for milking.

Milking is facilitated by raising the ewes above ground level. For hand-milking it is usual to lubricate the teats with oil (lanolin, lard or proprietary udder cream) or human saliva in place of the lamb's saliva. In most areas ewes are milked from behind (Fig. 9.1), but depending on the shape of the udder, and particularly the direction of the teats, ewes may be milked first from one side then from the other (Mills 1982).

Fig. 9.1 Milking an Awassi ewe in Turkey (Photo: M. L. Ryder)

Sheep can be milked by machine, although this practice is not widespread in the tropics or sub-tropics. Machine-milking is limited to large dairy flocks where there is a reliable supply of power and water. Unless the flock is large, machine-milking is no faster than hand-milking, although it is less tiring for the milker. The basic milking machine for ewes is similar to that for cows, but the teat liners must fit the ewe's two teats (which vary greatly in size between individuals, and are often small), the cups should be lighter than for cows, and the operating vacuum should be about 45 KPa with about 100 pulsations min^{-1} (Mills 1982). Richards (1957) described a rotary parlour for simultaneously milking 128 sheep.

It is common practice in Israel to milk each ewe two or three times at each milking (Eyal 1972). The secondary milkings, known as strippings, take place after an interval of up to 20 minutes. The proportion of milk obtained in the primary milking is known as the fractionation. For Chios ewes, strippings account for about 16% of the total milk obtained, but have a higher fat content than the milk obtained in the primary milking so that their economic value is higher than their relative volume (Cyprus, Agricultural Research Institute 1981). However, for low-yielding ewes the quantity of the strippings is so low that the time and effort of a second milking is not justified. Sagi and Morag (1974) showed that the fractionation depends on udder conformation, suggesting that fractionation may be improved by selecting ewes with favoured udder conformation as heritability of udder conformation is medium to high

(Gootwine et al. 1980). Alternatively the udder may be supported on a mechanical arm during milking to improve fractionation (Sagi 1978).

Some authors claim that milk yield is enhanced if ewes are milked three times each day, but in practice milking is done twice daily. Experiments have shown that intervals between milkings of 8 and 16 hours, and 10 and 14 hours produce as much milk as intervals of 12 and 12 hours (Jatsch 1977).

Weaning causes a steep drop in milk production (Eyal 1972). The magnitude of this drop is greater the earlier in lactation that weaning takes place. Some Awassi ewes may completely dry off after weaning, but the majority continue to lactate. In Israel the normal practice is to wean lambs at 2 or 3 months of age and to continue to milk the ewes for a further 3-5 months (Morag et al. 1970).

Health of lactating ewes

Lactating sheep are no less prone than other sheep to diseases, and these can lower milk production. For instance, experimental dosing with *Ostertagia circumcinta* larvae can reduce milk production by 0.27 kg day^{-1} (Leyva et al. 1982). In addition some diseases are associated with lactation. Mastitis is less common in the ewe than in the cow, but is usually much more severe when it does occur. It is associated with several species of organism, especially *Staphylococcus* spp, and *Pasturella* spp.

Lactating sheep can suffer from metabolic disorders. Grass staggers or hypomagnesaemia is associated with a fall in the concentration of magnesium in the blood, and acute cases result in rapid death. Grass staggers rarely occurs on rough pasture, but mainly on improved pastures or where the animals are stall-fed. It can be prevented by ensuring that the ewes receive a continuous supply of dietary magnesium. Hypocalcaemia or lambing sickness is seen soon after lambing. It is associated with a low concentration of calcium in the blood, and is precipitated by sudden changes in diet. Other metabolic disorders may occur in lactating ewes, and the causes of these are often unclear. In general, metabolic disorders are found with sheep on improved production systems and are rare under traditional management.

Milk yield

Some milk yields of sheep in the tropics are listed in Table 9.3. They range from 0.06 kg d^{-1} for Shropshire ewes in the Philippines to over 2 kg d^{-1} for Assaf ewes in Israel. The higher yields are comparable with the yields of dairy ewes in Europe (Flamant and Morand-Fehr 1982) and dairy goats in the tropics (Devendra and Burns 1983). There are many factors which account for the variation in reported values of milk yield of ewes. The most important of these will now be discussed.

Table 9.3 Milk yields of sheep in the tropics and sub-tropics

Breed	Location	Yield (kg d^{-1})	Period measured (d)	Source
Assaf	Israel	1.34	156	Eyal et al. 1978
Assaf	Israel	2.23	100	Gootwine et al. 1980
Avikalin	India	0.47	92	Kishore et al. 1983
Avivastra	India	0.45	92	Kishore et al. 1983
Awassi	Iran	1.16	207	Kaschanian 1973
Awassi	Lebanon	0.68	150	Bhattacharya and Harb 1973
Barki	Egypt	0.71	84	Aboul-Nasa et al. 1981a
Beni Ahsen	Morocco	1.31	28	Kabbali 1977
Blackhead Persian	Trinidad	0.63	84	Butterworth et al. 1968
Chios	Cyprus	1.27	209	Louca 1972
Chios	Cyprus	1.28	90	Mavrogenis 1982
Chokla	India	0.46	90	Sahni and Tiwari 1975
Chokla × Rambouillet	India	0.51	90	Sahni and Tiwari 1975
Corriedale	Kenya	1.36	70	Bush 1965
Cyprus Fat-tailed	Cyprus	1.00	158	Louca 1972
Jaisalmeri	India	0.40	90	Sahni and Tiwari 1975
Malpura	India	0.50	90	Sahni and Tiwari 1975
Malpura × Rambouillet	India	0.60	90	Sahni and Tiwari 1975
Merino	India	0.70	112	Mallikeswaran et al. 1980
Merino	Egypt	0.85	126	El-Sherbiny et al. 1972
Middle Atlas	Morocco	1.90	28	Kabbali 1977
Morada Nova	Brazil	0.21	92	Bellaver et al. 1980a
Najdi	Saudi Arabia	1.35	84	Ramadan et al. 1977
Nilgiri	India	0.65	112	Mallikeswaran et al. 1980
Nilgiri × Merino	India	0.67	112	Mallikeswaran et al. 1980
Ossimi	Egypt	0.75	126	El-Sherbiny et al. 1972
Ossimi	Egypt	0.78	84	Aboul-Naga et al. 1981a
Pelibuey	Mexico	0.53	84	Castellanos Ruelas and Zarazúa 1982
Priangan	Indonesia	0.35	84	Atmadilaga 1958
Rahmani	Egypt	0.77	84	Aboul-Naga et al. 1981a
Santa Inês	Brazil	0.18	92	Bellaver et al. 1980a
Sardinian	Italy	0.86	44	Dattilo and Congiu 1979
Shropshire	Philippines	0.06	73	Villegas and Cruz 1960
Tunisian Barbary	Tunisia	1.06	35	Sarson 1972
West African	Venezuela	0.60	56	Combellas 1980
West African × Dorset Horn	Venezuela	0.69	56	Combellas 1980
West African Dwarf	Nigeria	0.42	70	Adu and Ngere 1979

Method of measuring milk production

Aboul-Naga et al. (1981b) and Peart (1982) discussed methods of measuring milk yield of ewes. It is difficult to estimate milk yield by hand-milking a ewe unless she is accustomed to being milked, because the milker may not stimulate milk let-down. This problem can be overcome by injecting oxytocin before milking. Alternatively, if suckling is restricted to certain periods of the day, milk production can be estimated by weighing the lambs before and after suckling. Moore (1967) found that for non-dairy ewes, hand-milking following oxytocin injection generally gave higher estimates of milk yield than weighing lambs.

Stage of lactation

In dairy ewes, milk production rises during the first few weeks to a peak which may occur anywhere between the second and seventh week of lactation (Eyal 1972). Thus the shape of the lactation curve for a dairy ewe is similar to that for a cow, but is of smaller magnitude. However, where nutrition is less than optimal, daily milk yield declines throughout lactation (Doney et al. 1983). This appears to be the case for most non-dairy breeds in the tropics and sub-tropics. For example, the lactation curve of Jaisalmeri ewes in India (Fig. 9.2) shows no peak, and milk yield declines almost linearly throughout the 12-week lactation.

Breed

Traditional dairy breeds include the Awassi (Near East), Chios (Cyprus) and Sardinian (Italy). These breeds have higher milk yields, longer lactations and better milk let-down than non-dairy breeds. The average annual milk yield of Awassi sheep in Palestine has been increased from about 60 kg in 1930 to about 350 kg in milk-recorded flocks in 1962, as a result of selective breeding and improved feeding (Epstein 1977). It would appear that increases in milk yield by selection in other breeds could also be achieved should this be desired. The heritability of milk yield is about 0.30 (Flamant and Casu 1978), but the heritability of milk let-down seems to be very low. An alternative approach to increase the milk yield of dairy sheep in the tropics is to cross with a European dairy breed such as the East Friesian which in a typical lactation of 260 days gives 500 kg milk (Gall 1975), a daily average of 1.9 kg. In Israel, the Assaf breed has been developed in this way by crossing the Israel Improved Awassi with the East Friesian.

Milk let-down

Non-dairy breeds of sheep are difficult to milk by hand because the oxytocin release mechanism allowing let-down works only when stimulated by the lamb. This problem is also encountered in non-dairy breeds of cattle. In some dairy breeds of sheep the lamb's presence is needed at milking to stimulate let-down, but some highly developed dairy breeds such as the Awassi can be milked in the absence of their lambs. Even in dairy ewes accustomed to machine-milking the milk generally flows in two waves and so that the teat cups may be removed

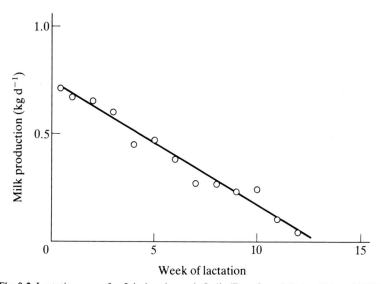

Fig. 9.2 Lactation curve for Jaisalmeri ewes in India (Data from Sahni and Tiwari 1975)

after the first wave, thus allowing time for the alveolar milk to be ejected, before putting them on for a second time (Gall 1975).

Nutrition

The nutrition of the ewe during pregnancy and lactation is one of the most important factors affecting milk yield of ewes in temperate conditions (Peart 1968). There are, however, many contradictory reports about the effect of nutritional level during pregnancy on the subsequent milk yield of ewes: changes in nutritional level cause changes in body condition and lamb size which themselves affect lactation. For example, Mallikeswaran et al. (1980) reported that there was a positive correlation between a ewe's weight at lambing and her subsequent milk yield. For Blackhead Persian ewes, the level of feeding in the last 6 weeks of pregnancy affected milk yield even if ewes received identical feeding during lactation (Butterworth and Blore 1969). Similarly, for native ewes in Egypt, an increased level of feeding during late pregnancy and lactation gave an increase in milk yield and persistency (Aboul-Naga et al. 1981a), but was not economical. Reviewing the literature, Eyal and Folman (1978) concluded that very high levels of nutrition during pregnancy have a beneficial effect on milk production only when the ewes are in poor condition. For ewes which are already fat, too heavy feeding in late pregnancy may be detrimental.

For dairy ewes in Israel, Eyal and Folman (1978) found that peak food consumption during lactation reached 4 kg DM d^{-1}, and individual animals producing 4–5 kg milk per day consumed 5–5.5 kg DM daily. Ewes on a high level of feeding maintained their body weight, while those on a standard level lost weight and subsequently had a worse reproductive performance. Milk production beyond the third month was not significantly affected by the level of nutrition. The effects of undernutrition in early lactation are a reduction in milk yield (which is seen as poor lamb growth if the lambs are suckling) and a fall in body condition. Sacker and Trail (1966) reported that the pre-weaning growth of East African Blackheaded lambs was low in the dry season and they attributed this to the low milk production of their dams. In many sedentary systems of sheep production, in order to protect very young lambs, their dams are tethered near the house for a few days after lambing and thus receive little food. Studies by Coop *et al.* (1972) in New Zealand showed that food restriction for 3 days after lambing had no effect on lamb growth, suggesting that post-weaning tethering does not have a deleterious effect. However, if the ewes are in poor body condition at lambing, the effects of nutritional stress after lambing may be more serious.

The protein requirements of lactating sheep in the tropics are not well defined. Robinson (1978) concluded that for temperate sheep there is little response in milk yield to dietary crude protein levels above 11 or 12%. This is similar to the level recommended by Eyal and Folman (1978). In many tropical systems the forage contains less than the recommended level of protein, and in tropical Australia, addition of urea to drinking-water has been used to increase the liveweight gain of Merino lambs; the authors attributed this to increased milk production by the ewes (Stephenson and Hopkins 1978).

Number of lambs suckled

For non-dairy ewes the number of lambs suckled has a greater effect on milk yield than level of nutrition during pregnancy or lactation (Treacher 1978). Whether this is true for tropical ewes which may be subjected to very severe nutritional stress is not known. Most studies show a 30–50% increase in milk production of twin-suckling ewes over the production of single-suckling ewes, and a further small increase if the ewe is suckling triplets. Figure 9.3 shows the effect of number of lambs suckled on the lactation curve of Scottish Blackface ewes. Early in lactation the greater appetite of twins stimulates a higher milk production than a single lamb, but as lactation proceeds this difference in milk production becomes smaller. Where ewes are milked and suckled, the number of lambs has much less effect; Goot (1972b) found that dairy ewes in Israel produced only 6% more milk if they were suckling twins than single lambs.

Parity and age of ewe

It is generally accepted that milk yield increases with parity (number of lactations) up to a maximum found in the third to sixth lactation, and thereafter declines (Treacher 1978). The peak lactation yield is at least

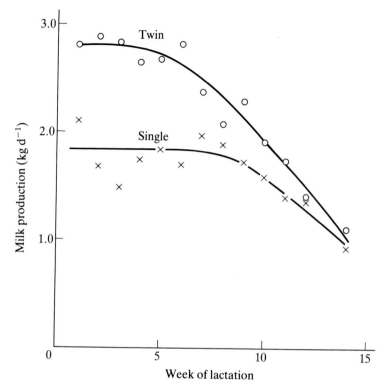

Fig. 9.3 Lactation curves for Scottish Blackface ewes suckling either single or twin lambs (Data from Peart et al. 1979)

25% greater than the first lactation yield. In most studies the effect of age is confounded with parity. Old ewes may have very poor lactation yields depending on their feeding and management.

Milk composition

Milk is an emulsion comprising water, fat, protein, lactose and minor constituents including minerals and vitamins. Some values of the composition of milk from ewes in the tropics and sub-tropics are shown in Table 9.4. There is a wide range of reported values. For total solids the values range from 16 to 24%. The crude mean values are 19.3% total solids, 7.6% fat and 5.5% protein. These means are almost identical to the mean values for ewes cited by Jenness (1974) which were derived from several sources in temperate parts of the world, and which are shown in Table 9.5. Compared with other domestic species, ewe milk has a very high solid content. The fat content of ewe's milk is approximately twice that of European cows, and is similar to that of the

Table 9.4 Composition of milk from sheep in the tropics and sub-tropics

Breed	Location	Composition (%)			Source
		Total solids	Fat	Protein	
Awassi	Iraq	16.2	5.3	—	Eliya et al. 1972
Awassi	Lebanon	—	6.8	6.5	Bhattacharya and Harb 1973
Beni Ahsen	Morocco	22.3	9.9	5.5	Kabbali 1977
Blackhead Persian	Trinidad	17.3	5.9	5.6	Butterworth et al. 1968
Middle Atlas	Morocco	24.4	11.9	5.1	Kabbali 1977
Najdi	Saudi Arabia	17.0	5.7	5.5	Ramadan et al. 1977
Ossimi and Rahmani	Egypt	—	7.9	—	Sirry et al. 1950
Pelibuey	Mexico	—	7.0	5.6	Castellanos Ruelas and Zarazúa 1982
Shropshire	Philippines	—	7.8	4.7	Villegas and Cruz 1960
Tunisian Barbary	Tunisia	21.2	9.0	5.2	Sarson 1972
West African Dwarf	Nigeria	16.6	6.3	5.4	Adu and Ngere 1979
Crude mean		19.3	7.6	5.5	
Range		16–24	5–12	4.7–6.5	

Table 9.5 *Average composition of milk of domestic animals*

Species	Composition (%)				
	Total solids	Fat	Protein	Lactose	Ash
Sheep *(Ovis aries)*	19.3	7.4	5.5	4.8	1.0
Cattle *(Bos taurus)*	12.7	3.7	3.4	4.8	0.7
Cattle *(B. indicus)*	13.5	4.7	3.2	4.9	0.7
Water buffalo *(Bubalus bubalis)*	17.2	7.4	3.8	4.8	0.8
Goat *(Capra hirca)*	13.2	4.5	2.9	4.1	0.8
Horse *(Equus caballus)*	11.2	1.9	2.5	6.2	0.5
Donkey *(E. asinus)*	11.7	1.4	2.0	7.4	0.5
Bactrian camel *(Camelus bactrianus)*	15.0	5.4	3.9	5.1	0.7
Dromedary *(C. dromedarius)*	13.6	4.5	3.6	5.0	0.7

Source: After Jenness 1974.

buffalo cow. The protein content of ewe's milk is considerably greater than that of all other species, but the lactose content is similar to that of cow's milk.

The composition of milk is much affected by the stage of lactation. The milk produced during the few days after lambing, known as the colostrum, has a very high content of total solids. This is largely because colostrum has a very high protein content; Adu and Ngere (1979) reported that the colostrum of West African Dwarf ewes in Nigeria contained over 20% protein, four times as much as the milk produced later.

Even after the first week of lactation, milk composition changes. For Blackhead Persian ewes, the fat and protein contents of milk decline for the first 9 weeks of lactation, then rise slightly (Butterworth *et al.* 1968). Castellanos Ruelas and Zarazúa (1982) reported that after the seventh week of lactation, the fat content of milk from Pelibuey ewes rose significantly as lactation proceeded. Similarly, for Finnish × Scottish Blackface ewes, Peart *et al.* (1972) found that the fat and protein contents of milk increased while lactose content decreased slightly as lactation progressed. In their studies, the greatest change was observed for fat content which increased from 6.3% in week 4 to 7.6% in week 12 of lactation. During this period, milk yield fell substantially so that despite the increase in concentration of solids in the milk, the overall rate of production of milk solids fell steadily. Thus within lactation there appears to be a negative correlation between milk yield and solid content.

A similar negative relationship between yield and solid content can be postulated when comparing the milk composition of different breeds of ewe (Table 9.4). Those breeds which are known as high-yielding dairy breeds, such as the Awassi, tend to have lower solid contents than those breeds which are not specialised for dairy production. Thus it appears that selection of ewes for high milk yield has been partly at the expense of solid content.

Castellanos and Zarazúa (1982) reported that the fat content of ewe's milk was higher (by about two percentage units) in the afternoon than in the morning. The protein and lactose contents were unaffected by time of day. These authors postulated that the consumption of grass during the morning increased the acetic acid concentration in the rumen which in turn enhanced the production of milk fat.

Because of the higher solid content of ewe's milk, the price it realises in countries with significant milk production from ewes is usually greater than that for cow's milk (Gall 1975).

Milk contains almost all the minerals required by lambs and humans: calcium, phosphorus, potassium, sodium, chlorine and trace minerals. Only about one-third of the minerals in milk are in true solution, and the rest are associated with the milk solids. Thus when milk is made into cheese or other products it retains its value as a mineral source. The contents of several minerals in ewe's milk has been reported to be lower than in goat's milk (Abou-Dawood et al. 1980).

Milk contains substantial quantities of vitamin A and the B vitamins (especially riboflavin and thiamin), but relatively little of the other vitamins. Vitamin A is fat-soluble so is found in milk fat, whereas vitamin B is water-soluble. When milk is preserved, the quantities of vitamins are greatly reduced as they are destroyed by heating or storage.

Milk preservation and processing

Liquid milk can be preserved by controlled heating and cooling. The three heat treatments most commonly used in the preservation of milk are pasteurisation (72 °C for at least 15 seconds), sterilisation (108 °C for 25 minutes) and ultra-high temperature treatment known as 'UHT' (140 °C for a few seconds). These methods are described by Ashton (1977). However, heat treatment of milk so that it can be stored is not widespread in the tropics or sub-tropics, even for cow's milk. An alternative method of preservation is to add a chemical such as hydrogen peroxide. By adding thiocyanate as well as a small quantity of hydrogen peroxide the enzyme lactoperoxidase in the milk is activated to produce an antibacterial compound (Korhonen 1980), so that milk can be stored for several days without deterioration.

Milk is conserved in the form of yoghurt, cheese or other products by curdling and fermentation. Fermentation converts the lactose in milk to lactic acid, and thus creates a pH unfavourable for the growth of pathogenic and food-spoiling micro-organisms. In the production of fermented milk products, milk is partially pasteurised and concentrated by heating, is cooled and then inoculated with some of the product from the previous batch. The culture used usually comprises *Lactobacillus* spp. and *Streptococcus thermophilus*. Yoghurt is produced after a short incubation period, and is consumed within a few days of manufacture. Pressed yoghurt, which has a low water content, can be stored in oil for several months.

Cheese is usually made by adding a bacterial culture to milk and after incubation, adding a coagulant so that it separates into a curd containing protein and fat, and a liquid known as whey. Rennet (obtained from the abomasum of a pre-ruminant) is the most commonly used coagulant, but many other proteinases have a similar effect (Green 1977). The curd is cut, strained and finally pressed during the ripening period. Approximately 1.8 kg of hard cheese can be made from 10 kg ewe's milk (considerably more than the 1 kg from 10 kg of cow's milk). Many combinations of bacterial fermentation, temperature of incubation, method of curdling and pressing and duration of ripening give the variety of cheeses found throughout the world. One of the most famous cheeses made from ewe's milk is the French Roquefort which is inoculated with *Penicillium roqueforti* before ripening. Cheese made from ewe's milk has a distinct flavour and may be less acceptable than cheese made from cow's milk, particularly if there is no tradition of eating ewe's milk cheese as is the case in Mexico (Castellanos Ruelas and Zarazúa 1982).

Butter made from ewe's milk is said to be pale and unsatisfactory (Fraser and Stamp 1968). In hot parts of the world, butter-fat is conserved as ghee, made by boiling cream to remove the water. Heating is then continued until the non-fatty solids are golden brown and settle to the bottom. Ghee is used for cooking, and can be kept in sealed containers without deterioration for over a year.

In the Near East, where milk products play a large part in the economy of the sheep industry, approximately two-thirds of ewe's milk is used for home consumption and one-third is sold to merchants for processing (Kolding and Koford 1970). The most important milk products in the Near East are *laban* (yoghurt), *labaneh* (pressed yoghurt), *ambrise* (pressed yoghurt made from skim milk), *shanklish* (ambrise containing mould), *samne* (ghee) and cheeses with a variety of flavours (Kolding and Koford 1970; Gordin 1980). In the Sudan, milk is preserved as cheese, *zabadi* (semi-solid fermented milk) and *robe* (liquid fermented milk) (Ibrahim 1970).

Small-scale operations to purchase and process milk from small ruminants are well established in many areas (Gall 1975). Either traders move around the nomadic camps collecting milk and taking it back to their processing units, or they install mobile units close to the nomad camps. The traders have firm relationships with the producers, sometimes contracting and paying in advance for the season's production.

The standard of hygiene of small-scale producers is very variable. It is not possible to improve the standard of hygiene by installing expensive modern plants because these cannot compete economically with the traditional plants (Gall 1975). Even with a relatively simple modern cheese-making unit it is difficult to make a profit (Westergaard 1972). Instead, the traditional production and processing systems can be improved by giving technical and financial assistance (Gall 1975).

Milk intolerance

People in many parts of the tropics and sub-tropics are said to be 'milk intolerant'. They suffer abdominal rumbling, cramp and diarrhoea after consuming milk. This phenomenon sometimes arises because the intestinal mucosa has been damaged, but more usually it is observed in healthy individuals. These people have a deficiency of lactase in the small intestine. Lactase is an enzyme which splits the disaccharide lactose into glucose and galactose. If there is insufficient lactase, the lactose ingested in milk remains in the intestine and, as a result of osmosis, water moves into the intestine and causes diarrhoea (Rosenweig 1969).

Babies are able to digest lactose, but lactose intolerance can develop if children do not regularly drink milk. Milk intolerance is thought to be partly genetic, but is closely related to the milk-drinking habits of ethnic groups. For instance, in Nigeria, the Fulani regularly drink milk, whereas other tribes who do not are milk intolerant (Kretchmer 1972).

Lactose-intolerant individuals can, however, consume milk products such as yoghurt in which the lactose has been fermented to form lactic acid. It is probable that because lactose accounts for only 25% of the total solids in ewe's milk compared with 37% in cow's milk, ewe's milk is less likely to cause digestive upset.

Skins

Skins are a very valuable by-product of the sheep industry. Sheepskins are much smaller and thinner than cattle and buffalo hides and are used to produce 'light' leather for the uppers of shoes, gloves, handbags, purses, etc. Most skins sold on the international market have had the hair or wool removed. Hair sheep give better-quality skins than wool sheep which give more spongy or 'loose' leather. Sheepskins produced in India weigh approximately 640–730 g dry salted (Bawa and Bhote 1966) whereas the wet weights of skins of indigenous sheep (two-tooth males) in Botswana were approximately 2.2 kg (Owen and Norman 1977). The weight of 1 m^2 of sheep leather is approximately 1.6 kg (FAO 1970), so that the area of individual Indian skins is about 0.4–0.5 m^2.

The processing of skins includes flaying, curing and tanning. Flaying is the removal of the skin from the carcass. It is generally achieved by making a small cut on the inside of one of the hind legs and blowing in air to separate the subcutaneous tissue holding the skin to the carcass (Aten *et al.* 1955). The inflated carcass is pummelled, suspended by the hind legs and then the skin is removed by pulling. If a knife is used, care must be taken to ensure that the skin is not damaged by cuts. For commercial production the skin is slit down the belly and the inside of the legs, but in the production of water containers, for instance, the skin is removed as a tube.

To prevent skins putrefying, they are cured in one of three ways: drying, salting or pickling. Methods of curing are described in detail by Aten *et al.* (1955) and Carrie and Woodroffe (1960). Once properly cured, skins can be stored and transported without deterioration (Fig. 9.4).

The art of converting fresh or cured skins into leather is known as tanning. A comprehensive booklet on rural tanning techniques is published by the Food and Agriculture Organization of the United Nations (FAO 1960). Tanning by traditional methods using vegetable tannins (from bark, wood, leaves, pods or tubers) or minerals has been practised for centuries. More recently, synthetic tanning methods have

Fig. 9.4 Pile of sheepskins in market, Ethiopia

been developed. Tanning does not require expensive tools or chemicals, but skill is needed consistently to obtain a good-quality product. The processes of tanning are soaking, liming, unhairing, tanning and finishing. Skins can be tanned with the hair or wool intact, and both the skin and the pelage may be dyed.

The value of skins is reduced if they are damaged. Damage is inflicted on the skin of a live sheep by branding, wounds, grass seeds, ticks and skin diseases. After slaughter the skin is damaged by bruising, cuts during flaying, delay before curing, wetting, insects and many other factors. Skins are graded for quality when sold, and this includes damage as well as size and shape (Mann 1969).

The processing of skins for export can be carried to a number of stages from curing to the manufacture of finished leather. It is in the interests of

Table 9.6 *Production of sheepskins, showing major producing countries in the tropics and sub-tropics*

	Thousand tonnes per year
World	1304
Asia	370
India	37
Pakistan	33
Saudi Arabia	18
Syria	17
Iraq	9
Kuwait	9
Oceania	267
Africa	262
Ethiopia	138
South Africa	26
Sudan	13
Libya	12
Algeria	10
Morocco	10
Nigeria	8
Tunisia	6
Mali	5
Kenya	5
Europe	174
USSR	125
S. America	77
Peru	7
Bolivia	6
Brazil	5
N. and C. America	28
Mexico	6

Source: From FAO 1983a.

the exporting country to process the skins as much as possible to create employment and increase the foreign exchange earned by the skins. However, the degree of processing in a developing country may be limited by the technology available and because tanners in the importing countries frequently prefer to import skins in a cured form so as to have a wide choice of tanning, dyeing and finishing processes. Some exporting countries, India for example, have well established leather industries using advanced technology. The export of untanned skins from India is virtually banned, but exports of finished skins are encouraged (FAO 1970). One of the major obstacles to the export of leather from developing to developed countries is the tariff structure of the importing countries, which favours raw and rough tanned skins rather than processed products. Sheep and goat leather appears to be subjected to fewer import controls than cattle leather.

Skins are by-products so that the supply is almost independent of changes in demand for leather. World production of sheepskins increased rapidly during the late 1950s because of the increase in world demand for sheep meat. Since then the production of skins by less developed countries has fluctuated. The production in the major producing countries in the tropics and sub-tropics is shown in Table 9.6. Ethiopia is by far the largest producer, and other important producers are India, Pakistan, Syria and several countries in northern and eastern Africa. More than 95% of the sheepskins entering the international trade are absorbed by developed countries (FAO 1970).

Leather is usually described by the animal from which it originated, the tanning process, finish, potential end use, stage of processing, weight or any combination of these. In Brazil for instance, skins are mainly classified according to weight (Bellaver 1980) or size (Nogueira Padilha 1980). The value of sheepskins has been estimated at $800 t^{-1} (Jahnke 1982), so that for a country exporting 10 000 t y^{-1}, the income from skins is about $8 million.

Pelts

Pelts of high-quality fur, known as astrakhan, are produced from lambs of the fat-tailed Karakul breed. Karakul sheep were introduced from Asia into Europe, and from there were taken by German traders to South West Africa in 1908 (Nel et al. 1960; Ryder and Stephenson 1968). There is now a thriving Karakul industry in Namibia and the northern part of Cape Province of South Africa. Most pelts are exported, and the USA, Canada and Europe (particularly West Germany) are the main importers. In the last 10 years Karakul sheep have been introduced into Rajasthan, India, from the USSR. Performance of the pure-bred Karakuls is said to be satisfactory (India, CSWRI 1983) and their crosses with local coarse-wooled breeds have shown excellent promise for producing acceptable-quality pelts.

Pelts are produced only from lambs under 5 days old. The best pelts are from lambs less than 24 hours old, or from lambs born prematurely. The possibilities of inducing parturition to shorten the gestation period and thus improve pelt quality are being investigated (Skinner *et al.* 1970). Immediately after slaughter of the young lamb, its skin must be treated to prevent putrefaction. The simplest method of preservation is dry salting. The coats of Karakul lambs have tight curls of hair, and are usually black, although selection for white, grey and other colours may be practised (South Africa, Department of Agricultural Technical Services 1976; Zyl 1976). Pelts are graded according to curl type and quality (Pfeifer 1953). Adult Karakul sheep have very poor-quality wool and it is not possible to select for pelt quality on the basis of the adult coat. Instead, matings are arranged by reference to pictures of lamb pelts.

The Karakul industry is very specialised. It thrives in dry areas where the vegetation is very poor. The majority of Karakul lambs are slaughtered soon after birth, thus releasing ewes from the demands of lamb rearing. Almost all male lambs are slaughtered, and those female lambs which are not required for flock replacements. In flocks with a reasonably high litter size, female twin lambs from selected ewes are fostered. However, for Karakul sheep in southern Africa, the lambing percentage may be only 104% (Carstens 1970). Large litters are associated with small (low-quality) pelts (Asamov and Stepanov 1979) so that attempts to increase the reproductive rate of Karakul sheep have focused on reducing the lambing interval (South West Africa, Department of Agricultural Technical Services 1974) rather than on increasing litter size.

Hair

Hair is obtained from hair sheep in tanneries and also by clipping live animals. In India and Pakistan the production of hair is an important industry, and hair is exported to developed countries where it is used in the production of carpets, fabrics, ropes, etc. Good-quality woven carpet can be made from a blend of hair and coconut fibres, and when blended with jute fibres, hair can be used to make felted insulating material. Longer hairs fetch a higher price and are used to make brushes.

Carcass by-products

Guts

In several parts of the tropics and sub-tropics, most of the digestive tract of sheep is eaten. The small intestines and stomachs are cooked in a similar manner to meat, usually by stewing. However, the small intestines of sheep which are about 20 m long can be used for other (more valuable) purposes. One of the main uses of intestines is as sausage skins. For this purpose the fat is scraped off, the intestines cleaned by washing and the mucus removed by hand. Intestines are graded (according to size and quality) and are stored either dry or in salt. Sheep intestines have traditionally been used as strings for sports racquets and musical instruments (Ryder 1983a). Modern uses of intestines include surgical sutures and collagen sheet for use as an artificial skin after burn damage.

Condemned meat

Condemned meat and inedible offal can be made into meat meal for pigs and poultry. It must first be sterilised, the fat removed (or it will become rancid) and then dried, before being ground. Alternatively, these products can be used as pet food.

Fat

Fat as a human food gives a concentrated supply of energy and fat-soluble vitamins and is regarded as a delicacy in many parts of the tropics, although not in westernised societies. Sheep fat is known as tallow. It is harder than the fat of other species, and is used for making candles. The tail or rump fat is softer than fat from the other parts of the carcass (Parvaneh 1972). Sheep fat which is not of a suitable quality for human consumption can be utilised in animal foods, pet foods, in soap manufacture or as a lubricant.

Blood

Blood is a valuable by-product of meat, yet rarely is it well utilised. Blood should be sterilised before use. It may be absorbed directly into cereals and dried to make a high-quality food. About 200 g of blood meal (for feeding to pigs and poultry) can be obtained from each sheep slaughtered.

Bones

Bones can be used to produce animal food, fertiliser, gelatin or ornaments (particularly buttons and combs) for the tourist industry. Bones yield an appreciable amount of fat which is removed by boiling. Bones for use as food or fertiliser are sterilised by burning or prolonged boiling at ambient pressure, or for a shorter time at high pressure. Bones used for gelatine manufacture are crushed then subjected to high-pressure steam.

Hooves and horns

Horns can be used for ornaments, but more generally hooves and horns are boiled under pressure then dried and ground to give a high-quality fertiliser known as hoof and horn meal.

Dung

Faeces and urine contain the undigested portion of the food and the products excreted by the sheep. Even though they are sometimes called waste products, they are valuable as fertiliser or fuel. The quantity of faeces and urine produced by sheep is enormous. The number of sheep in less developed countries is approximately 600 million (FAO 1983a), and assuming a daily production of faeces per sheep of 300 g DM, the total annual production is 60 million t. For sheep grazing at a density of 10 ha^{-1} the dung deposited each year will be about 1 t DM ha^{-1}.

Few data are available on the composition of sheep manure. Farmyard manure typically contains 2% nitrogen, 0.4% phosphorus and 1.7% potassium in the dry matter, but different batches may have very different composition depending on their origin and storage (Cooke 1967). For sheep fed sorghum and *Dolichos lablab* hay in the Sudan, the nitrogen content of air-dry dung was 1.5% (Jewitt and Barlow 1949). The mineral content of dung depends strongly on the composition of the diet fed to the animals, and where the diet contains little crude protein, the nitrogen content of the manure is low. The nitrogen content of farmyard manure is largely supplied by the urine, whereas other minerals and organic matter are in the faeces. It follows that failure to collect the urine as well as the faeces gives a product with a low nitrogen content. Farmyard manure also contains trace elements needed by plants for growth: boron, magnesium, cobalt, copper, zinc and molybdenum (Russell 1973).

The energy contained in the organic matter of sheep faeces varies from about 3 MJ kg^{-1} DM for a concentrate diet to about 8 MJ kg^{-1} DM for a roughage diet (McDonald *et al.* 1981). The energy content is important in the use of dung as fuel, whereas the mineral content is important when dung is used as a fertiliser.

Fertiliser

The value of manure as a fertiliser for crops is recognised by most tropical farmers and shepherds. Manure contains nitrogen, phosphorus, potassium and trace minerals essential for plant growth, and it contains organic matter which improves soil structure, reducing erosion and increasing the capacity of the soil to hold water. In some parts of the world (e.g. in the Sudan) the nitrogen component of manure is thought to be the most beneficial (Jewitt and Barlow 1949), while in others (e.g.

northern Nigeria) the remarkable responses to small dressings of farmyard manure have been attributed to its phosphorus content (Hartley 1937). Because a large proportion of the minerals in manure are combined with organic substances and are released only when these decay, manure releases minerals over a long period of time.

Manure applied at a rate of 25 t ha^{-1} supplies approximately 120 kg N, 25 kg P and 100 kg K ha^{-1}, and these make an important contribution to soil fertility and crop yield. The data given by Jahnke (1982) show that the extra grain given by the manure from one sheep is approximately 15 kg y^{-1}.

Sheep may be herded or penned on fields where manure is required, or the droppings may be collected and distributed. For sheep kept in houses the design of systems using bedding or slats to minimise labour

Fig. 9.5 Woman carrying dung from tethered animals to fields, Rajasthan, India

costs and reduce disease hazards are discussed by Robertson (1977). For sheep kept in small village flocks, either faeces are gathered up daily from sheep which are penned or tethered, or they are allowed to build up in deep litter or beneath slats and are cleared out after a period of months. The manure is transported to the fields and spread (Fig. 9.5). Because minerals are easily lost from manure by leaching, it is important that manure is protected from heavy rain and ground water. The collection of urine as well as faeces gives manure with a higher content of nitrogen, but, particularly in hot climates, nitrogen in manure is easily lost to the air (Cooke 1967).

In Rajasthan, India, the sale of manure accounts for about 3% of the total income from the sheep flock (see Table 10.2), and is a small but important part of the economy of sheep farming. In return for grazing sheep on fields after harvest, a flock owner is given grain and sometimes money (Bose *et al.* 1964).

Grazing sheep deposit faeces and urine on the pasture. This gives a cycling of nutrients through the vegetation, sheep and soil. Urine is concentrated in relatively small areas, and typically in a temperate climate, 25% of the nitrogen it contains is lost by leaching (Keeney and MacGregor 1978). Herbage contaminated by faeces is rejected by sheep, so that the grazing pressure over the pasture is uneven. Sheep are less selective against contamination by the faeces of other species than their own faeces, and vice versa, so that mixed stocking results in less rejection of herbage and better utilisation of pasture.

Fig. 9.6 Biogas plant, Andhra Pradesh, India

Fuel

Faeces may be used as a source of fuel for cooking, lighting or heating either by direct burning or by conversion to biogas. Ryder (1981) reported that sheep dung trodden down in pens can be dried and removed in sheets. However, sheep droppings are less popular for direct burning than faeces from large ruminants, because they are more difficult to mould into convenient blocks of fuel.

Production of biogas involves the airtight fermentation of faeces and water to produce a gas (largely methane) which is collected and piped to where it is needed. There are many types of biogas unit: that recommended in China is described by McGarry and Stainforth (1978). Figure 9.6 shows a biogas plant in India. Only the organic matter of dung is used in the production of gas, so that the residue retains the minerals which are the valuable components for use as fertiliser.

Chapter 10

Sociology and economics

Sociological and economic factors are very important in determining the system of sheep production and its reaction to any improvements. Too often these aspects are disregarded and attention is given only to technical aspects such as nutrition and reproduction.

Systems of production have evolved in response to sociological and economic as well as biological factors. Systems of management in the tropics were discussed in Chapter 2. Both in extensive pastoral systems in dry areas and intensive systems in irrigated or wetter areas, traditional sheep-production systems in the tropics are allocated only minimal inputs. Either the sheep are scavenging, they are fed on scraps and agro-industrial by-products or they feed on ground which is not suitable for crop production. Labour costs are usually low and there are few other inputs such as veterinary treatment.

Throughout the tropics there is little prestige attached to working with sheep. For instance, in northwestern India castes such as the Gujars and Jats are traditionally associated with commercial shepherding, and these are not normally influential members of the community (ICAR 1981b). On the other hand, in countries where the sheep industry is recognised as being an important part of the national economy, as in Australia or South America, sheep producers enjoy a social status above average (Coop and Devendra 1982). There appears to be a strong correlation between social status of sheep owners and their income.

Sheep are kept for their tangible products (meat, wool, milk, manure and other products as described in Chapters 7-9) which are either consumed by the family or sold. In systems in dry areas the sheep as a source of meat for family consumption becomes relatively more important in the dry season when the production of milk from sheep and other species is low (Field 1982). As a proportion of total farm income, livestock account for less than 10% in the humid areas of West Africa (ILCA 1979b), i.e. they are secondary to crops. They may, however, form an important part of the total farming system by providing manure or as a source of cash to buy inputs for the arable system. In West Africa, small ruminants are kept by women to supplement the wage earned by the head of the household (Burns 1981). In dry areas sheep account for a larger proportion of farm income. For instance in India, Taneja (1978) estimated that 40% of the rural population in dry areas depended directly on animal husbandry, particularly on sheep, for the majority of their income.

In addition to their use for sale and for home consumption, sheep have other functions: in particular they are a form of investment. Storage of wealth in the form of livestock is necessary in less developed countries in which land is owned communally and where there is no other form of investment. In a survey in the Kinangop area of Kenya, the UNDP/FAO (1976b) found that the main reason for keeping sheep was because they could be used as a quick source of cash. As a source of wealth, sheep are not subject to the high inflation rates found in the unstable economies in many less developed countries, and they can increase in number after a disaster (such as a drought or disease epidemic) much more rapidly than large ruminants.

In some parts of the tropics the transfer of livestock between families is an important social transaction. An animal is lent to a friend or relative either because he needs it or to demonstrate the bond between the lender and borrower (Dahl and Hjort 1976). If a crisis occurs the lender can ask this friend for help in the form of one or two animals. In the event of this particular friend not being able to help, the man in need contacts other friends until he finds someone who is in a position to be able to lend him animals. The net effect is a distribution of animals from those who have plenty to those who are in need, and each individual family has its investment geographically spread out, thus minimising the risk of total loss from drought or disease.

Animal ownership is complex in pastoral societies. A substantial proportion of the cattle belonging to a pastoralist may be dispersed among friends. It is not clear whether this is true for small as well as large ruminants. In Africa, the exact rules of animal ownership depend on tribal traditions. For instance some animals 'belonging' to a pastoralist may belong to his wife and she has the exclusive rights to their fate. Some animals may be set aside for future marriages, and some belong to a relative who is unable to look after them at present. In the Afar tradition in northeastern Ethiopia a woman retains her own animals when she marries, but keeps them in her father's herd (Helland 1980). The group of animals managed by a family is therefore rarely the same as those actually owned by them.

In Tanzania, Mlambiti (1975) estimated that of small ruminants reared in traditional systems, 19% were used to pay the bride price of daughters getting married, and 13% were used as gifts. Although there is doubtless some prestige attached to the ownership of large flocks of sheep (Maro 1978), it is unlikely that prestige alone is an important reason for keeping sheep.

There is some evidence that more animals are sold from pastoral flocks and herds when the price is low than when the price is high. For instance, many sheep are sold in droughts even though the price is low. Similarly, in northern Nigeria, the peak sales of sheep and goats are at the beginning of the rainy season (Veen and Buntjer 1983) even though this is the beginning of the best growing season. This so-called perverse reaction of traditional stockowners who do not appear to respond logically to price changes can be understood when it is appreciated that the income generated by livestock is not their sole role (Baker 1976). A sheep owner sells animals when he needs money for food or diverse purposes such as a wedding or funeral, or to buy inputs such as seed, fertiliser and insecticide for crop production. It is now realised by international agencies such as the World Bank that far from being tradition-bound peasants, farmers in less developed countries will respond to market incentives, and adapt to technology which presents opportunities for improvement (Peters 1983), provided that these incentives are matched to the farmers' own objectives.

In contrast to cattle which are owned by only a small proportion of families – fewer than 20% of households in Africa own cattle (ILCA 1980) — small ruminants are more evenly distributed through the human population. In agro-pastoral systems in Africa, typically more than 50% of families own sheep and goats. Similarly, in Syria (ICARDA 1982) and other parts of southern and western Asia, the producers of sheep tend to be the less privileged of the farming community. Therefore, if the aim of a development project is to raise the living standard of the poorer sectors of the community it is much more likely to do so if it concentrates on production from small ruminants rather than cattle.

There are disadvantages associated with keeping sheep and goats. Crop damage is one major disadvantage. Anteneh (1982) reported that in the derived savannah zone of Nigeria the potential damage of crops was the main reason why families who did not keep sheep gave for not doing so. Especially in extensive systems, sheep and goats are more susceptible than large ruminants to theft.

Assessing productivity

Traditional sheep-production systems are frequently said to be unproductive because they produce few commercial outputs. This may, on the other hand, merely be because they are poorly documented. Biological productivity indices can be used to give some measure of the productivity of sheep-production systems. Meat and wool production are relatively easy to estimate as they are harvested in discrete quantities. Continuous outputs are more difficult to estimate. Thus milk, blood and manure production from sheep are rarely measured under field conditions. The inputs into a sheep-production system (especially food or even land) are also difficult to estimate. For this reason it is easier to express a productivity index in terms of 'per breeding ewe' or 'per unit weight of breeding ewe' than 'per unit input'. This approach is justified because the quantity of real inputs needed by a flock is more or less proportional to the number of breeding ewes in the flock or to their total weight.

Wilson (1980) used an index to describe the productivity of tropical flocks. His index is the weight of meat produced by weaned lambs (at 6 months of age) expressed per unit body weight of ewes in the flock. It is calculated as

Wilson's Index = lambing rate
  × survival rate of lambs to 6 months
  × mean weight of lambs at 6 months
  × dressing % (expressed as a fraction)
  ÷ mean weight of ewes of mature age   [10.1]

For flocks maintained under traditional management in semi-arid areas of Africa he obtained values of between 0.21 and 0.24.

Fall *et al.* (1982) used similar indices developed by ILCA in their study of West African Dwarf sheep in Senegal. Index 1, the total weight of weaned lamb per ewe per year had an average value of 11.5 kg y^{-1}; Index 2, the total weight of lamb weaned per kg of body weight of ewe per year was 0.47 y^{-1}; and Index 3, the weight of lamb weaned per unit metabolic body weight of ewe per year was 1.1 $kg^{0.27}$ y^{-1}. Wilson's data for flocks in semi-arid Africa, expressed in terms of the ILCA Index 2 are shown in Table 10.1. They range from 0.47 to 0.60 y^{-1}, similar to that for the West African Dwarf sheep. Of these three ILCA indices, Index 2 is the most useful:

ILCA Index 2 = total weight of weaned lamb
  ÷ lambing interval
  ÷ ewe weight   [1610.2]

Fitzhugh and Bradford (1983b) used a flock productivity index (FPI) based on the birth weight of lambs. It is calculated from

FPI = litter size
  × lamb survival
  × birth weight
  ÷ lambing interval   [10.3]

Values of FPI for American Hair breeds and West African Dwarf sheep range from 1.92 kg y^{-1} for West African Dwarf to 5.70 kg y^{-1} for Barbados Blackbelly (Table 10.1). This index is less useful than indices based on weaning weight. The FPI can be converted to the ILCA Index 2 by multiplying by weaning weight (which was assumed to be 10 kg) and dividing by average ewe weight and lamb birth weight. The values of

Table 10.1 *Productivity indices of sheep flocks. FPI and ILCA Index 2 are defined in eqns [10.3] and [10.2] respectively*

Breed	Location	FPI ($kg\ y^{-1}$)	ILCA Index 2 (y^{-1})	Source
West African Dwarf	Senegal	—	0.47	Fall et al. 1982
Afar [Abyssinian]	Ethiopia	—	0.51	Wilson 1980
Baggara	Sudan	—	0.60	Wilson 1980
Bambara [Bornu]	Mali	—	0.56	Wilson 1980
Masai	Kenya	—	0.47	Wilson 1980
Barbados Blackbelly	C. and S. America	5.70	0.53	Fitzhugh and Bradford 1983b
Blackhead Persian	Brazil	2.48	0.38	Fitzhugh and Bradford 1983b
Pelibuey and West African	C. and S. America	3.65	0.43	Fitzhugh and Bradford 1983b
Virgin Island	Virgin Islands	4.99	0.53	Fitzhugh and Bradford 1983b
West African Dwarf	W. Africa	1.92	0.42	Fitzhugh and Bradford 1983b

Index 2 calculated in this way are shown in Table 10.1. They are comparable to those of the flocks studied by Wilson (1980) and Fall *et al.* (1982).

Offtake

'Offtake' is a term used by scientists studying pastoralism and wildlife management. It refers to the proportion of animals which are removed from the herd or flock each year. It may also refer to the removal of useful products from the flock (such as milk) while the animal itself remains. Sometimes the term 'offtake' is used to include losses of animals through disease, predators, etc. (Konczacki 1978). It is difficult to differentiate between such losses and animals removed for productive purposes because sick animals are frequently slaughtered and consumed. The most useful definition of offtake is the number of animals which are removed from the flock for meat production and other purposes, but excluding those that are lost from the flock.

Offtake is calculated from the difference between the reproductive and mortality rates, taking into account the changes in flock size which normally fluctuates with a cycle of several years. The offtake of a flock is related to its productivity index. For a flock in which the ILCA Index 2 is $0.50 \, y^{-1}$, the proportion of the flock which are breeding ewes is 0.57, the weight of lambs at weaning as a percentage of mature weight is 0.4, and the proportion of weaned lambs which form the offtake (i.e. almost all the males) is 0.45, the offtake would be 0.32, or approximately 30%. This value is similar to or slightly greater than values of flock offtake found in the literature (Dahl and Hjort 1976; Sandford 1983) but considerably greater than the 10% offtake observed for cattle.

Economic assessment of production

The performance of flocks can also be expressed in terms of economics. Although to a family it is the total income from sheep production that is important, for analytical purposes it is more helpful to consider the productivity of systems in terms of productivity per unit input. In developed countries it is convenient to express biological and economic performance per hectare (e.g. Maxwell 1979). In the tropics the area of land allocated to a flock is rarely well defined so that it is easier to consider the economic parameters expressed per breeding ewe.

Few analyses of the economics of systems of sheep production in the tropics or sub-tropics have been conducted. Nevertheless, economic budgets have been reported for traditional production in India, commercial production in Ghana and several production systems in Kenya. The environments of these production units represent, respectively, a dry area, a humid area and in Kenya a range of environments from arid to temperate.

The analysis of sheep production in Rajasthan by ICAR (1981a) is given in Table 10.2. The analysis has been divided into the ewe flock which produces new-born lambs, and the rearing and finishing of these lambs (although in practice lamb production and rearing take place in the same flock). The cost of maintaining ewes was so low that, even though the output of the ewe flock was only 0.6 lamb per ewe per year, the annual recurrent cost of producing each new-born lamb was only Rs 14, or approximately US$2.00. Lamb rearing and finishing using concentrates is much more expensive, and the total cost of feeding the lamb was Rs 86. The total return per lamb (price obtained for the sale of

Table 10.2 *Economic analysis of sheep farming in Rajasthan, India. Lambs are weaned at 90 days weighing 22 kg. All values in Indian rupees per 100 animals*

	Rs
Ewes	
Costs	
Labour for grazing	1 800
Veterinary care	200
Mortality losses (7% at Rs 100.00)	700
Total	2 700
Returns	
820 g wool at Rs 15.00 kg^{-1}	1 230
100 kg manure at Rs 0.05 kg^{-1}	500
Total	1 730
Net costs	970
Cost of 100 lambs at birth if lambing rate is 0.60	1 350
Lamb rearing and finishing	
Costs	
Lambs at birth	1 350
11.25 kg pre-weaning concentrate at Rs 1.11 kg^{-1}	1 250
77 kg post-weaning concentrate at Rs 0.95 kg^{-1}	7 315
Veterinary care	200
Mortality losses	135
Total	10 250
Returns	
11 kg meat at Rs 10.00 Kg^{-1}	11 000
Skin	1 300
Wool	1 125
Head, hooves and offal	500
Manure	200
Total	14 125
Net return per 100 lambs	3 875

Source: After ICAR 1981a.

the lamb at slaughter plus its manure) was Rs 141, so that the next return was Rs 39. Meat accounted for almost 80% of the total sales from the lamb-finishing enterprise, with wool and skin accounting for another 9 and 8% respectively.

An economic analysis of sheep production in the humid tropics of West Africa was reported by ILCA (1979b). Table 10.3 shows the budget for a flock of 100 sheep on a commercial farm in Ghana. Total investment costs were high (US$176 per sheep) of which almost half was investment in animals. Total recurrent costs amounted to US$37 per sheep. The only return was the sale of animals, and with a weaning rate of 1.0, the total annual return was US$70 per sheep. Thus the annual net return was US$33 per sheep.

ILCA (1979b) also discussed the economics of traditional production systems in Ghana. In traditional systems the only investment cost is the initial purchase of stock. Breeding stock were valued at US$30–70, so that each ewe cost approximately 20% of the farmer's entire annual cash income. Annual recurrent costs are very low in the traditional system in

Table 10.3 *Economic analysis of a commercial sheep farm in the humid zone of Ghana. The farm has 100 sheep, 3.5 ha maize and 7 ha improved pasture*

	US$
Investment costs	
Building and fencing	4 200
Crop and pasture development	3 800
Stock purchases	8 000
Contingencies	1 600
Total	17 600
Total investment costs per sheep	176.00
Annual recurrent costs	
Salaries and wages	710
Maintenance of buildings and equipment	726
Pasture maintenance	50
Animal management	960
Others	1 280
Total	3 726
Annual recurrent costs per sheep	37.26
Annual gross returns	
Sale of sheep (100 animals at US$70)	7 000
Annual returns per sheep	70.00
Annual net return per sheep	32.74

Source: After ILCA 1979b.

West Africa as the small ruminants graze freely in the dry season and are tethered in the wet season, and the only supplementary feeding they receive is household waste. The returns from traditional West African flocks were estimated as US$30 per ewe per year, so that the income from traditional flocks was substantial considering the minimal input. However, the risk associated with keeping a small flock of sheep in the humid tropics is considerable and an epidemic can wipe out the entire flock.

Tables 10.4–10.6 show the economic budgets for three sheep flocks in Kenya (UNDP/FAO 1976b). For a flock of 100 Masai ewes in an arid area (Table 10.4), the reported annual costs are substantial because of the cost of dipping and drenching the animals. Table 10.4 also includes an annual labour cost of K.Sh. 12 per ewe. The weaning rate of 0.8 gave an income from the sale of surplus weaned lambs and culled adults of K.Sh. 44 per ewe. Thus the annual net return per ewe was almost K.Sh. 10.

Table 10.4 *Economic analysis of annual costs and returns of a Masai flock in Kenya with a weaning rate of 0.80*

Average flock structure	Head
Ewes	100
Current crop lambs	72
Second-year lambs, surplus to requirements	53
Working rams	4

Annual direct costs	K.Sh.
Drenching (104 sheep × 6, 80 lambs × 3, 53 weaners × 2; at 40 cents)	388.00
Dipping (30 times at 50 cents per ewe unit)	1500.00
Young ram (2/3 ram at K.Sh. 70.00)	46.70
Labour	1200.00
Miscellaneous	300.00
Total	3434.70

Annual gross returns	K.Sh.
Surplus lambs (45 at K.Sh. 70.00)	3150.00
Culled ewes (17 at K.Sh. 70.00)	1190.00
Culled ram (2/3 at K.Sh. 70.00)	46.70
Total	4386.70

Annual net return	952.00
Annual net return per ewe	9.52

Source: From UNDP/FAO 1976b.

Table 10.5 *Economic analysis of annual costs and returns of a Masai × Dorper flock in Kenya with a weaning rate of 0.80*

Average flock structure	Head
Ewes	100
Current crop lambs	75
Working rams	3

Annual direct costs	K.Sh.
Drenching (104 sheep × 6, 75 lambs × 3; at 40 cents)	339.60
Dipping (20 times at 50 cents per ewe unit)	1000.00
Young ram (3/4 ram, at K.Sh. 300.00)	225.00
Young replacement ewes (22 at K.Sh. 70.00)	1540.00
Labour	1200.00
Miscellaneous	300.00
Total	4604.60

Annual gross returns	K.Sh.
Lambs (75 at K.Sh. 100.00)	7500.00
Culled ewes (17 at K.Sh. 70.00)	1190.00
Culled ram (3/4 at K.Sh. 150.00)	112.50
Total	8802.50

Annual net return	4197.90
Annual net return per ewe	41.98

Source: From UNDP/FAO 1976b.

The budget for a Masai flock in which the ewes were mated with Dorper rams is shown in Table 10.5. This type of enterprise is not possible in the most harsh areas of Kenya. Total direct costs amounted to K.Sh. 46 per ewe, and with a weaning rate of 0.8, the total return was K.Sh. 88 per ewe and the net return was K.Sh. 42 per ewe.

In areas of Kenya at higher elevation and with higher rainfall, wooled sheep are found. Table 10.6 shows the budget for a flock of Corriedale ewes mated with Romney Marsh rams. Annual costs were slightly higher (at K.Sh. 57 per ewe) than in the Masai flocks, but gross returns were substantially higher (at K.Sh. 153 per ewe). Thus the annual net return per ewe was K.Sh. 96 per ewe. However, capital costs (which were not reported by UNDP/FAO) were probably substantially higher than in the Masai flocks.

From these few analyses of the economics of sheep production in the tropics some important points emerge. The greatest capital investment in all the above systems was the cost of the breeding animals; in traditional flocks it was the only investment, and in commercial flocks it was the largest single capital investment. In another study in Rajasthan

Table 10.6 *Economic analysis of annual costs and returns of a Corriedale × Romney Marsh flock in Kenya with a weaning rate of 0.86. Lambs sold as weaners*

Average flock structure	Head
Ewes (3 die before shearing)	100
Current crop lambs	86
Working rams	3
Young replacement ram	1

Annual direct costs	K.Sh.
Shearing (97 ewes and 3 rams; at 50 cents)	50.00
Wool classing, handling and transport (400 kg at 10 cents)	40.00
Drenching (104 adult sheep 5 times at 40 cents)	208.00
Dipping (140 adult sheep 5 times, 86 weaners 1 time; at 50 cents)	303.00
Young ram (1 at K.Sh. 200.00)	200.00
Transport to KMC (107 sheep at K.Sh. 4.00)	428.00
Labour (1 man per 100 ewes at K.Sh. 120 per month)	1440.00
Replacement ewe weaners (23 at K.Sh. 120.00)	2760.00
Miscellaneous	300.00
Total	5729.00

Annual gross returns	K.Sh.
Wool (100 adult sheep × 4 kg, at K.Sh. 4.00 kg^{-1})	1600.00
Weaner lambs (86 at K.Sh. 135.00)	11610.00
Culled ewes (20 at K.Sh. 100.00)	2000.00
Culled ram (1 at K.Sh. 100.00)	100.00
Total	15310.00
Annual net return	9581.00
Annual net return per ewe	95.80

Source: From UNDP/FAO 1976b.

by Bhati and Mruthyunjaya (1983) of sheep on fenced pasture, the cost of fencing was even greater than the cost of animals.

In all the flocks analysed, labour accounted for a large proportion of the recurrent costs. The single most important output was always the sale of lambs, and in the Kenyan flocks the sale of culled adults was also a substantial output. Even in the flocks of wool sheep (Bikaneri and Corriedale × Romney Marsh), wool accounted for only a small proportion of the total return. This suggests that a simple criterion for the initial comparison of flocks would be the value of animals sold divided by the labour requirement. Values of this ratio are shown in Table 10.7, and they range from 3 to 10.

In sheep flocks in temperate countries there is a large gap in performance between the best and worst flocks under similar conditions. For lowland flocks in the UK, the top third of the flock

Table 10.7 *Value of animals sold per unit labour requirement (S/L). S was calculated as the value of lambs and ewes sold per year, and L as the annual expenditure on labour*

Flock	Location	S	L	S/L
Bikaneri	India	Rs 128	Rs 30	4.3
West African Dwarf	Ghana	$7000	$710	9.9
Masai	Kenya	Sh. 4340	Sh. 1200	3.6
Masai × Dorper	Kenya	Sh. 8690	Sh. 1200	7.2
Corriedale × Romney Marsh	Kenya	Sh. 13610	Sh. 1440	9.5

Source: Data from Tables 10.2–10.6.

were 50% better than the average in terms of profit per hectare (MLC 1982). There is likely to be as least as great a difference between the best and worst flocks in the tropics. Examination of the factors associated with successful performance in individual flocks can identify aspects of production which respond to improvement.

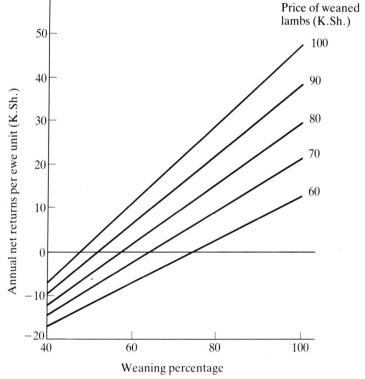

Fig. 10.1 Effect of weaning percentage and price of lambs on the net return of a Masai flock in Kenya. The lamb price is assumed to equal the price of culled ewes and rams (From UNDP/FAO 1976b)

Economic analyses are very sensitive both to production level and to the prices of inputs and outputs. For this reason the UNDP/FAO (1976b), in addition to presenting base budgets for each flock in Kenya, gave graphs to show how the budget is affected by changes in biological or economic factors. For instance, Fig. 10.1 shows the response of the net return of the Masai flock to changes in weaning rate and lamb price. For a weaning rate of 0.7, a lamb price of K.Sh. 65 is necessary to break even, and if the lamb price is K.Sh. 100 the net return is K.Sh. 20 per ewe. Similar graphs can be constructed for other enterprises, so that biological improvements or price incentives can be evaluated in economic terms.

To make the economic analyses more meaningful, the returns from sheep enterprises can be compared with the returns from other enterprises. In Table 10.8 the net returns per ewe for the five systems analysed are compared with the average monthly income of an agricultural labourer in each country. The next return from one ewe is equivalent to between 4 and 53% of 1 month's earnings. The lowest two values were obtained for the Masai flocks in Kenya where the typical flock size is large, and where other species are owned. The remaining three values obtained in Ghana, Kenya and India are 37, 41 and 53% respectively. In these systems a flock of only six ewes gives a net return equivalent to about 20% of the total annual earnings. These figures emphasise the importance of sheep in the economy of the poorer sectors in the tropics and sub-tropics.

Return on capital investment

The net annual return divided by the total capital investment provides a simple measure by which systems can be compared. The rates of return on capital invested for the six systems of sheep production are shown in Table 10.9. The Kenyan and Indian analyses do not give the total cost of investment, so that without making any further assumptions it is possible to calculate only the return per unit invested in ewes. These values show a wide range from 14 to 80%. The lower values are for the

Table 10.8 *Comparison of net return per ewe, R with the average monthly income, I, of a agricultural worker*

Flock	Location	R*	I†	R/I
Bikaneri	India	Rs 65	Rs 122	0.53
West African Dwarf	Ghana	$33	$88	0.37
Masai	Kenya	Sh. 10	Sh. 234	0.04
Masai × Dorper	Kenya	Sh. 42	Sh. 234	0.18
Corriedale × Romney Marsh	Kenya	Sh. 96	Sh. 234	0.41

Sources: * Tables 10.2–10.6.
† ILO 1982.

Table 10.9 *Rates of return on capital investment in six systems of sheep production*

Flock	Location	Return on investment		Source
		In ewes	Total	
Bikaneri	India	0.21	0.21	Table 10.2
West African Dwarf				
Commercial	Ghana	0.41	0.19	Table 10.3
Traditional	Ghana	0.50*	0.50*	ILCA 1979b
Masai	Kenya	0.14	0.14†	Table 10.4
Masai × Dorper	Kenya	0.60	0.60†	Table 10.5
Corriedale × Romney Marsh	Kenya	0.80	0.40‡	Table 10.6

* Does not include a charge for labour.
† Assumed that animals are the only capital investment.
‡ Assumed that total capital investment is twice the investment in ewes.

Masai and Bikaneri flocks in arid areas, whereas the high values are for more productive flocks in better environments.

It is possible to calculate the annual return per unit total investment by including the capital investment other than in sheep (or by estimating this if the data are not available). The rates of return per unit total investment range from 14 to 60% per annum (Table 10.9), with a mean of 34%. These rates of return are very high compared with the values of up to 4% obtained for sheep production in New Zealand and Australia (Coop and Devendra 1982). However, in New Zealand and Australia (as in most Western countries) there is very high investment in land, whereas in the tropics ownership of extensive land is communal and its value has not been included in the analyses.

It is difficult to make realistic comparisons between the analyses obtained by different authors because of the sensitivity of the analyses to current economic factors. Nevertheless, comparing the analyses within countries, some general points can be made. In Kenya, the highest net return was obtained for the Corriedale × Romney flock. However, this flock did not give the greatest return on capital invested because the investment costs were substantial. In the Ghana flocks, the return on capital invested was substantially higher for the traditional than the commercial flock because the capital investment in the traditional system was much lower than in the commercial system. On the other hand, the risk (of losses due to disease, etc.) is much greater in traditional systems in the humid tropics than in commercial systems (ILCA 1979b) so that the effect of capital investment is to reduce the risk of loss.

Estimates of rates of return for extensive livestock systems, especially in the tropics, are critically dependent on the costs imputed to the activity as well as on projected returns, which are subject to wide variation. The estimates cited above can therefore be taken only as a general indication that sheep in the tropics have a potentially high rate of return.

Marketing of sheep and sheep products

Unless the products of sheep are to be consumed by the family, some form of marketing is required to transfer the product from the producer to the consumer. In some cases this may be a simple process: the farmer may sell directly to a butcher who slaughters the sheep and sells it in the local village. At the other extreme, a sheep may be sold several times, from farmer to local trader, to a sequence of other traders who take it towards an urban area, and the animal may finally be sold to a business man who exports it by sea to another country where it is slaughtered and consumed.

Apart from retail butchers selling to consumers, there is little marketing of meat in less developed countries. Freezing and chilling facilities are available only in large towns and for meat for export. Most meat for these outlets is almost always produced in modern abattoirs. Skins are an important cacasss by-product, and local marketing systems have developed to transport these to the rural or urban tanneries.

Wool may be sold (Fig. 10.2) and used locally in traditional industries, but the majority of wool in the tropics and sub-tropics enters a complex marketing system. For instance, in India (Taneja 1978) wool traders visit flocks, examine the sheep and offer a price before shearing. These traders buy wool from several sheep owners and transport it (Fig. 10.3). Probably after being sold several times, the wool arrives at collecting and grading centres where it is graded and auctioned. In this system, the producers receive only a fraction of the selling price in the big markets.

The marketing of sheep and their products in tropical and sub tropical countries is not well documented. Nevertheless, some general points can be made regarding marketing of agricultural products. Several tasks and responsibilities are involved in marketing. These include: finding a buyer and transferring ownership; assembly and storage; sorting, packing and processing; providing the finance for marketing and taking risks; and assortment and presentation to consumers (Abbott and Makeham 1979). Each time the commodity is sold, its price increases so that the final price is several times greater than that received by the farmer. Traders are popularly criticised for making exorbitant profits, but if there is no legal or social limit to the number of middle men competition is likely to drive margins down towards the actual costs involved in the business of marketing. It is, therefore unlikely that the farmer could market his sheep more cheaply on his own (Anteneh 1982). The trader allows the farmer to sell his sheep easily for cash at irregular intervals and takes the risk that the animals lose

Fig. 10.2 Locally produced wool for sale in Nepal

condition or die before they are butchered. Sandford (1983) reported that in Africa sheep traders receive profit amounting to 6–13% of the value of the sheep; transport charges and international taxes account for a much larger proportion of the selling price.

Private traders may develop very large businesses. Abbott and Makeham (1979) cite an example of an African who built up a trading business extending from Nigeria to Khartoum. In some countries there are cultural traditions against marketing, and where this is so, immigrant peoples tend to set up as traders. Examples include the Asians in East Africa, and the Chinese in Southeast Asia.

Fig. 10.3 Wool in large bags for transportation, Rahasthan, India

As an alternative to private traders, marketing may be done by co-operatives or state enterprises. A co-operative is owned and run by those who use its services, whereas state marketing boards are run by the government. In theory, both co-operatives and state enterprises can achieve economies of scale, but the advantages of this in livestock marketing are probably small. Because co-operatives and state enterprises are run by individuals who may not have a strong interest in their success, they tend to be less efficient than private enterprises (e.g Stewart and Matlock 1980). Government officials must obey all regulations such as quarantine, which private traders usually manage to avoid. State marketing systems often attempt to control supplies and prices. In some cases in an attempt to maintain producers' margins and keep retail prices low, livestock boards are not allowed to charge economic slaughter fees (Disney et al. 1981). Improvement of slaughter facilities is impossible without subsidisation so that the marketing system does not function properly and prices paid to producers remain low and retail prices remain high. In effect, price controls may result in a decline in meat supply. Disney et al. (1981) concluded that despite laudable motives government intervention in meat marketing has led to more failures than successes.

In a study of marketing in Asian countries, Mittendorf (1981) classified problems into three categories: lack of marketing facilities inadequate marketing organisation and methods; and inadequate government policies and marketing-facilitating services (advisory training and applied marketing research).

Primary markets in which farmers and traders congregate to sell sheep to butchers have developed in Asia and Africa (Fig. 10.4). More sophisticated systems in which wholesale butchers place orders directly with producers are possible only where butchers and producers operate on a sufficiently large scale. International trade needs high standards of meat hygiene and inspection to minimise the transfer of disease.

Market information is informal in most less developed countries. No prices are displayed in markets and information is obtained only by talking to buyers or sellers who have conducted transactions. Inadequate market information results in producers bringing in products to sell when there is a glut, or in consumers buying inefficiently. One way of disseminating market information is by bulletins on the radio. Grading of sheep products according to quality is discussed in previous chapters. In general, little grading is practised in less developed countries. Meat, milk and wool are sold by weight or volume, and there is no quantitative assessment of quality. However, there is no evidence that formalisation of quality assessment would improve the efficiency of marketing.

Fig. 10.4 Market at Debre Birhan, Ethiopia

Table 10.10 *Export of live sheep and goats, 1982. Data given by continent and for major exporting countries in the tropics and sub-tropics*

	Thousand animals per year
Oceania	6097
Europe	4613
Africa	4218
Somalia	1680
Mauritania	650
Sudan	500
Niger	350
Mali	322
Namibia	250
Chad	150
Upper Volta	120
Asia	3079
Jordan	120
N. and C. America	292
S. America	134
USSR	0

Source: After FAO 1983b.

Compared with many other agricultural products such as cereals or vegetables, sheep have the advantage that they can walk. While being trekked to market, sheep may lose weight, but can often regain weight rapidly once they have reached their destination. However, trekking has its disadvantages. Compared with motorised transport, trekking is slow, sheep are more likely to pick up diseases on route and there are scarcities of food and water on trekking routes during droughts (at the time when pastoralists are most keen to sell their animals).

Alternative forms of transport include road, rail, boat and air transport. These are usually more expensive than trekking (Sandford 1983) depending on the distance and the relative costs of labour and fuel. Trucks are the most common form of road transport, but on rough roads a truck may last only 1 year (Abbott 1962). Bicycles, handcarts and animal transport are cheaper methods of moving small quantities of sheep products (milk, milk products, skins or wool) over short distances. Air transport can be justified only where the product has a high value. Abbott (1962) cites an example of fresh meat being transported by air to the coast of West Africa from the inland producing areas.

As production increases, more pressure is put on marketing services. Many complex factors interact to produce a suitable marketing system, and unless these are taken into account, the productivity of sheep systems will be seriously impaired.

In summary, sheep owners sell products in response to economic incentives. The economic conditions are related to the biology of the production system, but modified by many factors, particularly the demand for products and the marketing system. Economic and social aspects of sheep production therefore deserve careful attention when attempts are made to improve productivity.

International trade in sheep and sheep meat

International trade in sheep and goats, both live and as carcasses is shown in Tables 10.10–10.13. (Data for the species individually are not available.) The major exporters of live sheep and goats are in Oceania (Australia and New Zealand), Europe, Africa and Asia. In Africa and Asia, the major exporting countries are in the semi-arid areas. Somalia exports almost 1.7 million head per year. Mauritania and the Sudan are also major exporters. There is little export of sheep and goats from the humid tropics (Southeast Asia, the wetter parts of Africa and tropical America). Export of sheep and goat meat (Table 10.11) from tropical countries is very small in comparison with the export of live animals. Only Brazil exports substantially more sheep and goat meat than live animals, and India and Kuwait export similar quantities of both carcasses and live animals.

The four countries which import the most live sheep and goats are Saudi Arabia, Libya, Kuwait and Italy (Table 10.12). In addition to Saudi Arabia, many other countries in the Near East are prominent importers. In Africa, South Africa and the countries on the coast of West Africa (Ivory Coast, Nigeria and Senegal) import large numbers of

Table 10.11 *Export of sheep and goat meat (fresh, chilled or frozen), 1982. Data given by continent and for major exporting countries in the tropics and sub-tropics*

	Thousand tonnes per year
Oceania	603
Europe	127
Asia	78
Kuwait	2
India	2
S. America	35
Brazil	3
N. and C. America	1
Africa	0
USSR	0

Source: After FAO 1983b.

Table 10.12 *Import of live sheep and goats, 1982. Data given by continent and for major importing countries*

	Thousand animals per year
Asia	11383
Saudi Arabia	6251
Kuwait	1550
Iran	700
Jordan	538
Qatar	409
Lebanon	320
Iraq	300
UA Emirates	300
Bahrain	242
Yemen Arab Rep.	180
Yemen, Dem.	178
Syria	169
Africa	3654
Libya	2000
Ivory Coast	600
Nigeria	370
South Africa	250
Senegal	210
Europe	2376
Italy	1183
USSR	295
N. and C. America	310
Mexico	250
S. America	1
Oceania	1

Source: After FAO 1983b.

live sheep and goats. In tropical America, only Mexico is a substantial importer.

Table 10.13 shows the major importing countries of sheep and goat meat. The major importing areas are Europe, the Near East, Japan and Canada. In Africa, only Egypt imports $10\,000\,t\,y^{-1}$, and there is no significant importation of sheep and goat meat into tropical America. In general the proportion of sheep meat which is imported on the hoof is higher in less developed countries than developed countries. Thus, although Japan imports large amounts of sheep meat, she imports few sheep, but in Saudi Arabia, Kuwait and other parts of southern Asia, the reverse is true. The reasons for this are probably a combination of the close proximity of producing areas to consuming areas in less developed countries, the lack of refrigeration facilities and the importance of live animals at religious and festive occasions.

Table 10.13 *Import of sheep and goat meat (fresh, chilled or frozen), 1982. Data given by continent and for major importing countries*

	Thousand tonnes per year
Europe	343
UK	222
France	47
W. Germany	23
Italy	15
Belgium	12
Greece	10
Asia	339
Iran	105
Japan	85
Saudi Arabia	32
Iraq	20
Jordan	16
UA Emirates	16
Kuwait	14
Oman	10
USSR	120
N. and C. America	28
Canada	10
Africa	24
Egypt	10
Oceania	15
S. America	1

Source: After FAO 1983b.

The overall pattern is therefore a significant trade in sheep and goats from producing areas in the drier parts of Africa to the Near East, Libya and the coast of West Africa. There is only limited international trade in sheep either to or from tropical America or Southeast Asia.

Chapter 11

Improvement

The increasing human population in tropical countries, together with their demand for a better standard of living result in the need for increased food supplies. In agriculturally dependent countries the most logical route to development is via agricultural development (Peters 1983). Although the demand for more food can be met partly by an increase in primary food products (vegetables and cereals for human consumption) livestock products also have a part to play. Livestock, and sheep in particular, can utilise land that cannot otherwise be used to produce food, and they give products which are in demand from the wealthier sectors of the community. Traditional systems of livestock production have evolved in response to local resources, constraints and human demands. However, present systems may be no longer stable in the face of new pressures, particularly increased human population, more cropping of land in marginal areas and increased livestock numbers resulting from successful control of animal diseases. Thus, even if it was desired to maintain the status quo, this would not be possible. The present systems of production are bound to change.

These changes are often known as development. Development was defined by UNDP/FAO (1976a) as a self-sustaining transformation of the economic order, in which the well-being of all elements of the population, measured in terms of both levels of consumption of goods and services and levels of perceived access to security in established consumption patterns, is systematically increased. Development may be a natural consequence of the forces acting on and within present systems or it may be channelled in a particular direction by providing incentives and restrictions. Before these incentives or limitations can be imposed by either governments or development agencies, it is important to understand the direction in which the present system should be orientated, and the effect of the proposed incentives.

'It is always possible to create small ultra-modern islands in a pre-industrial society. But such islands will then have to be defended like fortresses, and provisioned, or they will be flooded by the surrounding sea' (Schumacher 1974). Appropriate development may be more difficult to achieve rapidly, but it is the route to long-term success.

What is improvement?

Improvement in productivity is usually thought of as (i) **more** produced in a given period, i.e. a greater rate of production of meat, wool, milk or any other product. However, improvement can also mean (ii) an **improved quality** of product, for instance the production of finer wool, (iii) a **different product,** such as the introduction of milking in a flock previously kept for meat, or (iv) the use of **less input** or a lower quality input, e.g. feeding sheep on an agro-industrial by-product which would not otherwise be used. The type of improvement appropriate for a particular area depends on the system of production and on the constraints acting on it. It is very important to define the objectives of improvement before they are advocated in the farming community. In India a great deal of effort is being put into breeding sheep which produce finer wool, yet wool accounts for less than 30% of the income from sheep flocks in the dry parts of tropical India, and in the southern parts of India hair sheep predominate (Acharya and Patnayak 1972). The productivity of both wool and hair flocks would be increased much more if, for instance, lamb survival could be improved, yet very little research has been conducted on this important topic.

The appropriate development in any given set of circumstances may comprise improvements of more than one of the four types listed above. For instance, improvement may consist of improving the growth-rate of lambs (type i) and in changing their diet from a cereal-based concentrate to one based on molasses (type iv).

Process of improvement

The process of improvement follows a sequence of stages (Gatenby 1982). In systems which evolve these stages are not identified, they simply happen, but when improvement is organised they are distinct:
1. Identify constraints limiting the present system – 'research'.
2. Solve identified problems – 'research' and 'development'.
3. Put research findings into practice – 'extension'.

Information 'discovered' by research scientists is rarely suitable for use directly, but must pass through a development phase in which it is put into a suitable form. The introduction of the new technology to farmers is then carried out by extension workers. Provided that adequate information is available on the sheep-production system, its response to external and internal changes, and its interaction with other production systems, it is now possible to use computer models to predict the best course of development (Spharim and Seligman 1983).

Before a farmer will adopt new practices he must be convinced that they are safe and reliable, and will give him long-term benefit. He is particularly likely to be wary if sheep form a large part of his livelihood. Sheep farmers are aware that the research scientists and extension workers who advocate the changes rarely share any economic risks, and almost never stake their lives on the success of the innovations in the field. The process of adoption is slow even in developed societies. It can be speeded up (provided the improvement really is beneficial) if there is a good relationship between extension workers and farmers and if the farmers are able to see for themselves the new methods demonstrated, preferably on a local farm but otherwise on an experimental station under realistic management.

Methods of agricultural extension in less developed oountries are described by Bradfield (1966), Benor and Harrison (1977) and Adams (1982). The level of training provided by extension workers must be appropriate to the requirements of the farmers who are often illiterate and find it difficult to absorb material presented in the form of lectures. Extension workers form a vital part of the development process, yet they are often treated with contempt by research workers. A good extension worker must understand the requirements of farmers and be able to communicate useful knowledge to them. To do this successfully, he or she does not need a university degree or diploma, but some form of training is useful. In India, for instance, extension workers known as village 'stockmen' receive a few months' training.

For a successful research and development programme, research workers must be in close collaboration with development and extension workers to find out how successfully the research results are being applied, and to identify new areas needing research attention. Ames (1983) reported that in Upper Volta there is very little feedback from extension workers to the higher levels of the University or Ministry of Rural Development. This is also true in many other countries. Too often research, development and extension workers are organised under separate administrative structures with the result that research findings are not put into practice and inappropriate research is conducted.

Identification of topics requiring research attention

It is of little use to continue research on topics which have been studied for years without significant achievement. Instead it is necessary to assess carefully the constraints on sheep-production systems, making sure that all aspects, including for instance, economic and social aspects, are properly considered. Once the limiting constraints are identified, effort can be put into easing the worst problems and thus better use can be made of the other resources available as inputs into the sheep-production system. Definite research objectives must be set if definite results are to be obtained.

Sometimes there is one obvious constraint to the improvement of a sheep-production system, but more often several constraints operate

together, and the key constraints must be identified (De Boer *et al.* 1982). Constraints can be grouped into:
1. Biological factors which affect the performance of sheep – particularly nutrition, health, genotype and management.
2. Economic and social factors including the purchasing power of consumers, government intervention into trade and prices, the availability and price of purchased inputs such as concentrate foods and prophylactics, the credit available for new enterprises, the attitudes of farmers and the priority they give to sheep, the confidence farmers have in responding to policies advised by the extension services, land tenure, marketing and transport facilities.

Biological factors

Nutrition

Nutritional constraints of inadequate food and its low quality are very serious limiting factors, particularly in the arid and semi-arid tropics. The nutritional requirements of sheep were discussed in Chapters 4 and 5. Alleviation of specific deficiencies, for example mineral deficiencies, by feeding supplements is relatively simple. In contrast the problem of a general shortage of good-quality food can be solved only by growing better fodder or by making better use of what is available. Practices such as better grazing management, the introduction of legumes, fertilisation, growing improved species of grass and growing crops specifically for fodder are all aimed at producing more, better-quality food. However, the resources which must be put into these practices are great and can be justified only for sheep kept in intensive systems. Only rarely are such improvements economically justifiable – for finishing units for lambs, to provide food for ewes in late pregnancy and early lactation, and possibly to provide a continual supply of good-quality food for modern ranching systems.

The alternative approach of better utilising what food is available is much more practicable. In many parts of the tropics there are agro-industrial by-products which are at present under-utilised or not utilised at all. Their cost is minimal and, provided that suitable methods of incorporating them in diets can be developed, this is an appropriate way of increasing productivity. For example, utilisation of straw has much potential. The use of chemical methods of straw treatment is limited by practical problems including the non-availability of chemicals and biological treatments may be more suitable.

Alleviation of nutritional constraints by the importation of foodstuffs other than supplements is practised by some oil-rich countries in the tropics, but is unlikely to offer an appropriate long-term solution.

Lack of drinking-water and overgrazing of pastures limit the productivity of sheep in the arid tropics, resulting in poor utilisation of what fodder would otherwise be available. Immediate solutions to these problems are, however, difficult to implement. Methods of reducing the stocking rate are discussed on p. 260.

Health

Diseases lower the efficiency of sheep-production systems. The effects of disease are closely related to nutrition and management. Although some diseases (e.g. enterotoxaemia) are effectively controlled by vaccination, many others are best prevented by management practices such as restricted movement of animals or sensible housing design.

Mack (1982) described the increase in productivity of small ruminants resulting from veterinary inputs in the humid zone of Nigeria. He reported a substantial response to treatment for PPR and pneumonia, but it was difficult to distinguish between the direct effects of veterinary treatment and the effects of improved management as a consequence of the stimulation of interest in small ruminants.

Veterinary inputs into less developed regions are a subject of controversy. While no one would doubt that there would be no production if no animals survived, it is also realised that one of the main problems facing animal production in the tropics is that there are too many animals competing for too few resources, particularly food. As a consequence many animals receive little more than their maintenance requirements, so that the vast majority of available resources are diverted towards maintenance not production. Following this reasoning, it is thought that veterinary inputs merely keep more animals alive and thus aggravate the problem of scarce resources. Veterinarians, on the other hand, argue that veterinary treatment minimises the wastage from animal-production systems, and thus enhances output. Where the argument is allowed to run to completion, it is usually agreed that veterinary inputs in the tropics are worth while (a) in a humid climate where diseases cause significant losses and where there are resources which can be sensibly utilised by animals, and (b) in improved systems which rely on substantial inputs of food, manpower, etc. There is, however, much less justification for substantial veterinary input in the dry tropics where any improvement in survival rate merely aggravates the problem of overgrazing and desertification. In the dry tropics, too, there is very little response to veterinary inputs unless accompanied by other inputs.

Genotype

It is often thought that the productivity of sheep-production systems in the tropics and sub-tropics is low because the indigenous breeds are inherently unproductive. Indigenous breeds have evolved in response to natural selection for the ability to survive environmental stresses (such as nutritional shortage in dry areas and disease challenge in humid areas), and have been subject to little deliberate selection by man.

Natural selection includes selection for an appropriate reproductive rate, so that the reproductive potential of indigenous breeds is usually reasonably good. On the other hand, the average productivity (growth-rate, etc.) of animals is usually low compared with the levels obtained in temperate countries. Even on a good diet the mean growth-rate of unimproved indigenous lambs is rarely more than 150 g d^{-1}. There is however, a large variation within the population in the response of productive traits to improved management.

There are two main approaches to genetic improvement: (1) crossbreeding, and (2) selection within the local breed. Crossbreeding may be between the local breed and either another indigenous breed or an imported breed. Even where the main approach is crossbreeding, selection will be practised as well.

Crossbreeding

Compared with selection, crossbreeding has certain advantages. It can give more rapid genetic change, heterosis can be exploited in the F_1 and subsequent crosses, and new characteristics can be introduced – for instance hair sheep can be crossed with a wooled breed. However, great care must be taken before crossbreeding is advocated as a method of increasing productivity. Many problems are encountered when exotic breeds of sheep are introduced into the tropics (Mason and Buvanendran 1982): in particular the mortality of crossbred and exotic sheep is higher than that of exotic animals. The use of large rams on small ewes can also give problems with dystocia. Osuagwuh *et al.* (1980) reported a 10% incidence of dystocia and 14% perinatal mortality when West African Dwarf ewes were mated with Permer, Uda and Yankasa rams, but no dystocia or perinatal deaths when the West African Dwarf ewes were sired by West African Dwarf rams.

Unless the aim is to keep the imported animals in a protected environment, crossbreeding to give sheep with a high percentage of exotic genes can be successful only when the introduced breed comes from an environment similar to that in which it is to be exploited. For instance Chios sheep from the Mediterranean were imported to Oman to improve the growth-rate and wool yield of the native Omani sheep (Steele 1983a,b). Chios ewes and Chios × Omani lambs performed satisfactorily under good management, but it has not yet been established what benefit the Chios would give in Oman under a low standard of management.

There are, however, many situations where the ideal sheep is some intermediate between a local adapted and an exotic breed. There are several ways in which this intermediate type of sheep can be produced.

A new breed can be formed by mating indigenous ewes with exotic rams. If desired the F_1 female progeny can be mated with indigenous rams to give 25% exotic genes in the F_2 generation. Further matings are arranged to give the required percentage of exotic genes. Once this has been established, the population is mated *inter se,* with selection, for several generations. Examples of new breeds created in this way are the Avikalin in India and the Dorper in South Africa. In producing a new breed the initial crossings must be done on a large enough scale (using at least ten rams) to avoid the deleterious effects of inbreeding. The major disadvantage of this method is that it takes several generations to establish the new breed.

Systematic crossing is an alternative way of using exotic genotypes. Sometimes it may be satisfactory merely to use exotic rams as terminal sires, but if the benefits of heterosis on reproductive performance are to be utilised there must be a more sophisticated system of crossing. Where the different types of genotypes are suited to different geographical regions, systematic crossbreeding leads to a stratified sheep industry.

Rotational crossing is a form of crossbreeding in which rams with different proportions of exotic blood are used in successive generations (Mason and Buvanendran 1982). Where rams with 0 and 100% exotic blood are used alternatively in rotational crossing this finally gives progeny with 33 and 67% exotic blood in alternate generations. For rotational crossing it is necessary to keep pure-bred flocks of both exotic and local sheep to supply rams. Exotic flocks are notoriously difficult to maintain in harsh tropical areas and AI using semen from rams in a more favourable area may be one solution. However, as discussed in Chapter 6, AI itself is not without problems.

Governments and international agencies may set up crossbreeding programmes, but flock owners may follow simple crossbreeding programmes themselves without external encouragement. One example of this which is recorded in scientific literature is the introduction of Blackhead Persian and Dorper rams into Masai flocks in East Africa (King *et al.* 1984).

No work appears to have been done on the use of prolific indigenous breeds to increase the reproductive rates of ewes in a nearby area. For instance, the prolific West African Dwarf sheep may have potential for increasing the reproductive rate of larger, less prolific breeds in West Africa. The possibilities of using prolific tropical breeds for crossing are discussed by Mason (1980a).

Where crossbreeding is used extensively, there may be a danger that some indigenous genotypes could become extinct. Unless genetic resources are conserved, genotypes which have desirable properties may not be available when they are required in the future. There appears to be less danger of genotype extinction with tropical sheep than with many other classes of animal because sheep generally live in unmodified harsh environments and large-scale genotype replacement is very slow.

Selection

The second general approach to genetic improvement, selection within the indigenous breed, has the advantage that all the sheep are adapted to the environment in which they will be used. Selection schemes are not easy to carry out in less developed countries. Testing may take the form of either performance or progeny testing. Performance testing is suitable for traits which are highly heritable and which can be measured in all individuals (e.g. growth and wool traits), whereas progeny testing is used for traits which can be measured in only one sex (e.g. milk traits). Thus rams for dairy flocks would be assessed by performance of their daughters. For reproductive traits, selection should be based on a ewe's performance at the first two or three lambings.

Selection may take place either in commercial flocks or on institutional farms. Commercial selection avoids the problem of trying to assess the genotype – environment interaction which arises if selection takes place under unrealistic management. An individual which is superior on the institutional farm may not be superior under the stresses of a commercial production system. However, it is very difficult to select within commercial flocks. Most shepherds in less developed countries are illiterate. In large extensive flocks recording is a problem, and in small intensive flocks there are too few animals for meaningful comparisons to be made.

Attempts to improve the genotype of sheep usually concentrate on either reproductive rate, growth and meat production, or wool production. There are no strong negative correlations between reproductive rate, meat production and wool production, so that it is possible to select simultaneously for all three traits (Turner 1972). Although simultaneous selection will give the best overall rate of genetic improvement, the rate of improvement of any individual trait will be lowered if the selection criteria are broadened.

Mason and Buvanendran (1982) recommend that, in all but dairy flocks, selection should concentrate on increasing reproductive rate. A strong argument can also be put against this recommendation. Indigenous breeds adapted to their environments have evolved a reproductive rate optimum for the environment. This has two important consequences. Firstly, unless the environment is improved, there will be no long-term benefit in selecting for fecund individuals. Secondly, there is little genotypic variation of reproductive rate within the population and this, together with the low heritability of reproductive traits, means that the rate of improvement of reproductive rate by selection within breeds is bound to be slow. Turner (1969) concluded that for sheep in Australia the best way to improve reproduction rate by selection is to select for litter size alone. This approach may not be the best in other tropical areas where breeding is aseasonal and the mortality of twins and triplets is generally much higher than that of singles. If it is decided to select for increased reproductive rate, it may be better to select ewes which have consistent short lambing intervals. However, selection for an increased reproductive rate should not be attempted unless the environment is also improved to some extent. Mason and Buvanendran

(1982) state that the simplest method of selection is to select ram lambs from lambs born in multiple births on the basis of criteria such as growth-rate or wool production.

Growth and other meat production characteristics of lambs are closely related to non-genetic factors – particularly nutrition and health. Within a breed, there is more scope for the improvement of growth traits than of reproductive traits because of the higher heritability and phenotypic variance of growth traits. If lambs are selected for weaning weights, these must be adjusted for the effects of age of dam, season of birth, litter size, etc. but selection of older animals is not substantially better if these adjustments are made.

Improvement of wool production is discussed in Chapter 8. The quantity and quality of wool produced can readily be improved by crossbreeding with imported genotypes but, as wool accounts for only a small proportion of the output from tropical flocks, care must be taken to ensure that an increase in wool production can be justified in terms of the extra inputs required, and is not accompanied by a fall in overall productivity.

In commercial dairy systems, genetic improvement of milk production follows a procedure similar to that recommended for dairy cows by selecting rams on the basis of the milk yield of their daughters. Selection under smallholder systems is very difficult.

Management

Analysis of production systems shows that management has a large effect on the productivity of flocks. Thus, whatever resources are available, a good shepherd is much more likely to have a successful flock than an incompetent or indifferent shepherd. Good management is difficult to define, but includes the following practices: care of the flock during grazing, and selection of good pasture; supplementation of grazing with loppings from trees or food from other sources; sensible use of any food supplements, particularly the supplementary feeding of ewes in late pregnancy and early lactation; weaning of lambs at an age to give the best compromise between a short lambing interval and rapid lamb growth; judicial marketing of lambs and other animals; care of ewes at lambing and young lambs to reduce peri- and postnatal losses; action to avoid diseases, and treatment of diseases at an early stage; and maintenance of a flock structure with a high proportion of reproductive ewes, by disposing of surplus males and non-productive females.

The success of a shepherd depends largely on his character and personal motivation. These qualities can be improved by experience or training to increase general and specific skills. Although in most tropical countries, agricultural and veterinary education is available at universities and colleges, there is little training for the farmer or shepherd who is classified as uneducated and illiterate. Extension services organised by Ministries of Agriculture provide occasional training in villages, usually concentrating on specific topics (such as care of the new-born) and concentrating on the species (cattle, buffaloes, pigs or poultry) which are thought to be of most economic importance.

Under communal grazing the shepherd may exercise only limited grazing management, but where pastures are improved this must be accompanied by more sophisticated management otherwise the technological advances will give little benefit. Sensible grazing management strikes a balance between the immediate food requirements of the sheep and the long-term productivity of the pasture, as well as considering other factors such as the control of helminths.

Good management is paramount in large-scale sheep enterprises. For instance, the Libyan government set up the Kufra sheep project in the desert 1000 km south of Benghazi (FAO 1972a). The aim was to establish a flock of 50 000 breeding ewes of the local fat-tailed breed and feed them on crops grown under irrigation. At the time of the report, the number of sheep had reached 11 000. Mortality was very low, and the technical success of the project was attributed to the high standard of management.

In intensive systems, heat stress may be a factor limiting productivity, and it may be of economic benefit to take measures to reduce the effects of heat stress. Several approaches can be adopted: to build houses or shelters which provide a cooler environment; to install coolers such as showers (Ramadan 1974); to feed at night when the sheep are not heat stressed and they are more willing to eat; to feed diets containing concentrates rather than roughages; to cool drinking-water for lactating ewes; or to feed thyroproteins which overcome the animal's natural reaction to slow down its metabolism in response to heat (Richards 1981).

Housing has traditionally been part of sheep-production systems in some areas, particularly in the humid tropics, and housing is often introduced in other systems as they are intensified. Sheep may be housed to give them a better environment, but more often housing is provided for management reasons – control, ease of feeding or collection of excreta. In all cases where sheep are housed, it is important that the house provides an appropriate environment, particularly where the sheep are confined inside the house and they are not able to escape to a more favourable environment. Besides its effect on thermoregulation, the house also affects the disease challenge, and both these aspects should be considered in the design of houses. In a confined space diseases are transferred easily between animals, especially if the relative humidity is high. Salmonellosis, respiratory diseases and coccidiosis are associated with poor housing design.

Some aspects of the design of houses for animals have been reviewed by Clark (1981), Linklater and Watson (1983), Wathes *et al.* (1983) and

Gatenby (1984). The design of a house can be divided into the design of its components – walls, roof and floor.

Most houses for sheep in hot climates have open sides to maximise ventilation which removes both sensible heat and water vapour which evaporates from the excreta and from the animal itself. In some cases it may be desirable to have some protection during certain seasons of the year, and this can be provided by one solid wall, a temporary wall (of straw bales, etc.) or a curtain of sacking.

The roof provides shade from direct sunlight and shelter from rain. It must be big enough to give protection in all weather conditions. A good roof is well insulated (thick), has large overhangs, is sloping so that it does not leak, and may have an open ridge to enhance ventilation. Stability in strong winds, fire resistance and durability are additional practical requirements.

The floor has little effect on the thermal environment, but a substantial effect on disease. It is essential that wet excreta are either removed regularly, separated from the sheep (e.g. by slats) or absorbed into deep litter.

In most countries some subsidised services for animal production are provided by the state. It is one characteristic of a good manager that he makes good use of appropriate services. For instance, where vaccinations against enterotoxaemia, PPR, etc. are provided, it would be foolish not to take them. It is sometimes thought that good management necessarily involves the use of modern techniques such as hormonal treatment to give large litters, growth promoters for rapid finishing etc.,While there may be a place for these technologies in some circumstances, more progress in improving sheep production is likely to result from the use of technology which is simple and effective, rather than that which is dramatic and expensive.

Economic and social factors

An understanding of the sociological and economic aspects of sheep production is necessary before changes in production systems are planned. Technological developments cannot be isolated from their economic and social effects (Chambers 1980). As discussed in Chapter 10, sheep farmers do not respond directly to biological factors, but to the economic and social implications of biological factors. It is therefore necessary to assess all potential improvements in terms of both the short-term and long-term benefits. For instance, the effect of supplementary feeding with ground cottonseed was analysed by McIntosh et al. (1976) and similar analyses can be undertaken for other potential improvements.

The economic environment in which a producer operates determines his level of operation and response to biological improvements. The price structure depends on many factors including the ability of consumers to purchase sheep products, the efficiency of the marketing system, and the ease of transport of sheep from producing areas to the market. The price structure can be manipulated by means of subsidies and taxes. In Saudi Arabia there is a very high demand for sheep meat, and many sheep and goats are at present imported from Africa. It might be thought sensible to set up breeding and fattening units in Saudi Arabia to satisfy the demand for meat. However, a feasibility study by Brown *et al.* (1976) showed that at current prices the proposed scheme was not feasible, largely because of the extremely high cost of labour. However, if the appropriate subsidy were forthcoming from the government (SR20 in 1976) these breeding and fattening units would have been economically viable.

The economic and social factors affecting sheep farmers are, however, complex and not properly understood nor easily described by simple economic analyses. There are many factors other than monetary gain which influence a farmer's decisions.

Sheep are only one part, and in all but arid areas only a small part, of the total farm system. Their advantage to the farmer is that sheep integrate well with the total system and the value of the inputs needed for sheep production is very low. Improvement of the sheep enterprise must not be at the expense of other enterprises which are seen by the farmer as being more important. For instance sheep production could possibly be improved by using some good land to grow fodder crops. But this land could also be used to grow crops for family consumption or for sale with a value far greater than the improvement in sheep productivity. Similarly, devoting more labour to the sheep may reduce lamb mortality, but is not likely to be undertaken if it means that cereals are not harvested at the right time.

Because producers respond to economic incentives, they thus will take up an innovation only if it gives them personally a substantial benefit and provided that it does not increase the risk associated with production. Therefore innovations which benefit certain sectors of the community but not the sheep owner, and those which may give higher productivity but which are not reliable, will not be adopted by farmers. Producers require not only a currently suitable economic environment but the confidence that the future will be good for sheep production.

At present small-scale producers sell sheep or goats when they need cash or when they require an animal for festive purposes. In the past this has been regarded as a primitive and undesirable use of small ruminants, and attempts have been made to change the attitudes of producers to operate purely for monetary gain. The value of small ruminants as a store of readily realisable capital is one of their attributes and should not be discouraged unless some alternative form of storing wealth, such as land-ownership or banking is provided. Encouraging a cash economy to increase the income of families may therefore be undesirable because it destroys the method of investment of the family.

It is clear that there are many production objectives. Scoville and Sarhan (1978) categorised these objectives into personal objectives (profit maximisation, risk avoidance, reserve of food or wealth) and government objectives (food supply, employment, public revenue maximisation, foreign exchange maximisation, egalitarianism). Because of the diversity of the socio-economic conditions in less developed countries, farmers show a diversity of production objectives. In Africa, Scoville and Sarhan (1978) found that for pastoralists the primary production objective was to increase their reserves, in Asia to produce more food, and in Latin America to make a larger profit.

One of the main biological factors limiting the productivity of sheep flocks is inadequate nutrition, but it is very difficult to improve the quality of pasture. In most parts of the world land is owned communally and farmers do not feel individual responsibility for the good of the communal grazing land (e.g. Maro 1977). The benefits to an individual of improving the quality of communal grazing are minimal (and an individual may not even be allowed to implement changes). On the other hand, if a farmer increases the number of animals in his flock this gives him an immediate personal advantage, but such a move is probably not in the long-term interests of either himself or the community as the cumulative effect of many farmers taking this action is that the resource base for production i.e. the pasture, is reduced in value as it becomes overgrazed (Campbell 1980). However, if a farmer refrained from adding to his flock when he had the chance he would be even worse off.

A change of the method of ownership from communal to individual would be an important step, but appears to be a very difficult one to achieve. A parallel step was, however, achieved over a period of many decades during the agrarian revolution in the UK, and allowed agriculture to be revolutionised (Ernle 1941). An agrarian revolution was launched in Algeria in 1971, but its effect on grazing land has been slow (Almeyra 1979).

The alternative approach to improved management is to assign an area of land to a group of people. Control of grazing pressure on these lands is possible only where there is a body with authority over the management of the land and the number of animals allowed on it. For example in the *hema* system in Saudi Arabia the grazing rights on communal pasture are controlled by an influential person or a group of people, so that in this case there is potential for the limitation of stocking rate (Eighmy and Ghanem 1982). In Botswana in order to stop overgrazing and degradation of the veld, some communal grazing areas are being made into communal farming areas. Ranch development with fencing and piped water will be encouraged. Groups and individuals will be given exclusive rights to grazing land (Botswana 1975). However, in Masai group ranches in Kenya, ranch committees have failed to limit livestock numbers (King et al. 1984). In India, trees belong to people, although land is communally owned. Thus there may be more potential for improving fodder supplies from trees than from pasture.

A number of authors (e.g. Morag 1972; Matthewman 1980) have commented on the possibility of developing a stratified sheep industry in certain areas of the tropics. In this way, harsh areas (those with a semi-arid climate) would be used to maintain breeding flocks from which lambs would be weaned and moved into a nearby area where they would be fed a superior diet and finished for slaughter. In northwestern Syria, for instance, there are large-scale finishing units which take in wethers from the locality and from rearing areas to the north in Turkey. These finishing units use locally grown barley, and the fattened sheep are mostly exported to other consuming countries in the Near East. The organisation of this type of enterprise requires co-ordination on a large scale.

Even with a price structure which is favourable to small-scale production of sheep, potential producers fail to start up sheep-production enterprises if they have inadequate capital (Gefu and Adu 1982; Adeyemo 1983). If this is the case, schemes to provide credit may be the most appropriate form of development. Credit is needed for investment primarily in animals, but in some circumstances also in infrastructure such as fencing. Unlike farmers in developed countries, sheep owners in less developed countries do not own land and so have no security against which to borrow capital. In the past, small farmers have been obliged to borrow from money-lenders at very high rates of interest. Since 1977, the FAO has operated a scheme for agricultural credit development which fosters credit associations in Asia and Africa (Ceres 1980). Problems of providing credit to small farmers include their lack of security, and in Nigeria, previous experience shows that there is a high rate of failure to repay loans. The system for allocating loans must be fair and not open to favouritism or bribery.

Sheep enterprises have the advantage that they require less capital than large livestock. Small farmers can increase the size of their flocks gradually by expansion and simple management measures. Substantial changes such as the development of a large commercial sheep unit from a smallholder system are unlikely to be undertaken by a peasant farmer.

Theft and predation are serious for most tropical flocks. The direct cost in terms of animal losses is often overshadowed by losses in productivity that result from management practices necessary to prevent predation, such as night-time confinement (Fitzhugh and Bradford 1983b).

In most tropical countries, the perceived value of sheep is low so that producers prefer other work or leisure rather than working with sheep. As the economic importance of sheep production is appreciated by producers, this attitude should change. Greater attention from extension services to small-scale sheep production should help to bring about this change. Associated with the higher status of sheep should come greater attention from extension services, and better training facilities for shepherds and flock owners. Agricultural education in schools (Boateng 1982) may also help to raise the perceived status of agricultural work.

Most attitudes of farmers are the logical response to their environment, even though these attitudes are often thought to be irrational by the outsider who has an incomplete picture of the farmer's environment. Attitudes of farmers can, however, slow down the rate of development. For instance, in southern India, when a ewe has twins, the farmer thinks this is a bad omen so he sells the ewe. While this may be a logical response to his unimproved environment which cannot support the growth of twins, once the environment is improved sufficiently for twins to grow satisfactorily, then it may still be difficult to persuade the farmer than an enhanced lambing rate is beneficial.

Overview of successful improvements

Rather than list attempted improvements in sheep production and categorise them into successful and unsuccessful (which would give a very long list of failures and a somewhat shorter list of successes), I will attempt to identify some general points concerning improvement programmes. All successful improvements have taken note of what is required, and have fully considered both biological and socio-economic aspects of production (even though this consideration may not have been consciously formulated). On the other hand, unsuccessful attempts have not considered these points. Successful improvements, therefore, are an integrated attempt at improvement. This does not mean that they are necessarily complex. On the contrary, the simpler the improvement the more successful it is likely to be. Rarely will a route of improvement which is initially formulated be precisely correct. More often it will be modified as it is tested.

Even traditional producers will respond to appropriate price incentives. For example, pastoralists will sell surplus livestock if given appropriate incentives in the form of grain or vegetables, and provided that there is an alternative means of storing wealth apart from animals (Sandford 1983). It is possible to create an environment in which the producers have economic incentive to sell livestock and their products (if this is not already created by the rising demand of the population for sheep products) by government intervention in the form of subsidies and taxes. However, care must be taken to ensure that these interventions not only enhance production in the short term but also lead to reliable production in the long term.

Consider a successful programme to encourage sheep production among the poorer sectors in India. It is realised that the rural poor have a very low standard of living and many are underemployed. Few of these people own anything, yet they have access to common grazing land. They are unable to obtain credit to start small businesses because they have no security and few management skills. The banks in India have devised a scheme in which loans are given out to landless peasants to allow them to purchase animals – two cows, about eighteen sheep, or some ducks – from which they can earn a living. These animals are insured and given veterinary treatment by a veterinarian employed by the bank, and care is taken to prevent fraud. This scheme takes into account the needs and resources of the people, and appears to be operating successfully in many parts of India.

Organisation and evaluation of research programmes

It is important for each nation to develop an agricultural research capacity suited to its own resource and institutional endowments (Ruttan 1982). The first essential for successful research is the identification of appropriate avenues of research. The resources available for research in many tropical countries are severely limited, and this has given rise to the feeling that less developed countries should concentrate on applied research to solve the immediate problems they face and that fundamental research should be left to countries which have greater resources for research. Thus, it seems more sensible for a less developed country to study the best way of including rice straw in the diet of finishing sheep, rather than to investigate the finer details of digestive physiology. However, appropriate applied research should identify topics which require further fundamental research. If this is not being carried out in a developed country, then less developed countries may feel the need to undertake this fundamental research themselves. Nevertheless, many tropical countries follow research programmes which are apparently unrelated to their needs. This arises partly because fundamental research is thought to carry greater prestige than applied research, and partly because nationals receive postgraduate training in some specific discipline in a developed country and when they return home, rather than use their research training to identify and solve local problems, they continue to study their specialised discipline.

Research within individual countries may be organised either as a national research programme or as part of an international programme, usually as development aid. There is little privately funded research in less developed countries (Pinstrup-Anderson 1982). As already discussed, development and extension as well as research itself are essential components of a research process. Education is usually included too, both to provide suitably trained personnel to staff the research programme, and also because it is recognised that education is one of the greatest resources that a country can have.

Within a national research programme the three activities: (i) research (ii) education and (iii) development and extension can be roughly said to take place, respectively, in (i) research institutes, (ii) universities and (iii) development and advisory services. There is, of course, overlap. For instance, most university staff undertake research, and many research institutes have an extension department. The categorisation of the research processes has in many countries evolved slowly, perhaps from a structure set up in colonial times. Particularly in large countries the integration of the research effort in different parts of the country is important. For instance, both India and Brazil have well-established networks for this organisation – the Indian Council of Agricultural Research (ICAR), and the Brazilian Public Corporation for Agricultural Research (EMBRAPA).

National research programmes are supplemented by international programmes which can be categorised into unilateral (funded by only one country) and multilateral (funded by more than one country). Projects funded by the FAO of the United Nations, and the World Bank (World Bank 1981) are examples of multilateral aid. Traditionally, international aid has taken the form of the posting of a specialist (formerly called 'expert') in a less developed country. In a few years this specialist was expected to conduct research, development and extension, as well as training up a local counterpart to succeed him. More recently, appreciating that there is little hope of successfully improving only one aspect of rural life, rural development projects have been undertaken with specialists concentrating on several aspects – perhaps crop production, human health, fodder production and animal health as well as animal husbandry. There is a tendency for such rural development projects to become isolated from national advisory services, and transfer of the project to the national system is difficult. Integration of rural development projects into national systems also has problems. National systems are often bureaucratic so that foreign specialists become frustrated at the hurdles raised to prevent the logical progression of the research process, and frequently the lack of finance.

The research institutes run by the Consultative Group on International Agricultural Research (CGIAR) are playing an increasing role in research on sheep husbandry. Initially the institutes under the CGIAR were concerned only with crop production, but there are now three institutes which are partly involved with sheep production: ILCA (International Livestock Centre for Africa) in Addis Ababa, Ethiopia; ILRAD (International Laboratory for Research on Animal Diseases) in Nairobi, Kenya; and ICARDA (International Centre for Agricultural Research in Dry Areas) in Aleppo, Syria. As its name suggests, ILRAD concentrates on diseases – particularly trypanosomiasis and East Coast fever. A substantial proportion of the research conducted by ILCA and

ICARDA is classified as 'pastoral systems research'. This is an approach in which the whole of a farming system is examined (ILCA 1983).

Initially the CGIAR institutes were isolated from national research programmes because there were no easy channels for communication, and local scientists were frequently jealous of the high salaries and splendid facilities available for the international scientists. Effort is now being put into breaking down these barriers, and it is realised that co-operation between the two groups is beneficial to both. International scientists are helped by local scientists to identify problems requiring attention. Conversely, local scientists can benefit by using facilities (particularly library and computing facilities) at the international institutes.

There are many other forms of international aid. For example, exchanges of scientists between countries can provide moral support and training necessary for the successful conduct of research in research institutes and universities in less developed countries. However, few of the scientists from developed countries who are involved in these exchanges become involved in the longterm research in less developed countries.

Whatever the organisation of research it can be successful only if the individual scientists and technicians are motivated. Workers need adequate facilities, but more important is to create an environment in which they are encouraged to follow scientifically the approaches they feel are most suitable. The promotion structure within research services is frequently poor with research scientists given tenured jobs for life and little incentive for stretching their ingenuity to the extreme.

It is easy for a research institute or department to slide into stagnation. In an attempt to ensure that resources are properly utilised, regular evaluation of research programmes is carried out. Methods of evaluating the benefits of agricultural research programmes were discussed by Schuh and Tollini (1979). Agricultural research can contribute to social and economic development in a number of ways, and the measurement of the direct benefits of research and its side-effects is not easy.

Future of sheep production in the tropics

Sheep cannot be considered in isolation from other farming enterprises. Compared with pigs or poultry, sheep are inefficient converters of grain into meat. Thus, as a country develops and its systems of animal production become more intensive, historically there has been a fall in the number of sheep. There are, however, three important niches which sheep can fill:
1. Sheep utilise rough pasture and farm wastes;
2. Sheep produce wool;

3. Sheep form a store of readily realisable capital for smallholders.

The most important sheep-rearing areas will continue to be areas of marginal land. As the cost of oil increases, the use of artificial fertiliser and other chemicals is not likely to increase, so that improvements based on improved management and better utilisation of resources are more feasible than improvements based on more inputs. Intensive methods (such as those which employ hormones) to achieve very high outputs are unlikely to be economical, even if they are technically possible.

Sheep-production systems can be improved considerably without radical changes in method. The potential of the present systems can best be exploited if there is effective marketing, particularly to reach the growing urban and export markets. Appropriate research, good extension services, and training for farmers and shepherds are needed if improvement is to be achieved.

Appendix I.

Glossary of technical terms

Barren description of ewe which fails to give birth.
Broken-mouthed description of a sheep which has lost some or all of its teeth.
Browse to consume vegetation from trees or shrubs; food which is browsed.
By-product an output from a system, but not a primary product (e.g. hooves are a by-product of meat production, and straw is a by-product of cereal production).
Case mortality the number of animals which die divided by the number of animals which are sick.
Carrying capacity the number of animals that can be supported on unit area of land.
Concentrate food with a high concentration of energy and little fibre (e.g. cereal).
Conception rate the number of females which conceive divided by the number of females mated.
Crossbred genotype derived by crossing one breed with another (often an indigenous breed crossed with an imported breed).
Crude protein nitrogen content of food × 6.25.
Culling removal of a ewe or ram from the flock, usually because of some defect.
Dagging removal of soiled wool from a sheep.
Docking removal of tail.
Dressing percentage ratio of dressed carcass weight to liveweight, expressed as a percentage.
Dystocia difficulty in giving birth.
Ewe female sheep which has given birth.
Exotic breed breed from another locality (usually from a temperate country).
Extension dissemination of technological developments to farmers.
Fecundity fertility × prolificacy.
Fertility number of ewes lambing divided by number of ewes mated.
Finishing preparing lambs for slaughter by feeding a good-quality diet.
Flushing giving ewes or gimmers extra food before mating, to enhance lambing rate.
FCE food conversion efficiency. Weight of product per unit weight of food input.
Gimmer female sheep aged 6 months (or from weaning) to first parturition.
Gimmer lamb female lamb.
Grading up increasing the proportion of exotic genotype in a flock by back-crossing ewes to exotic rams for several generations.
Growth promoter substance used in small quantities to increase growth-rate, usually during finishing.
h^2 heritability. The proportion of the phenotypic variance which can be attributed to additive genetic effects.
Heterosis the superior performance of crossbreds compared with the mean of the parents.
Hogg, hogget sheep aged 6 months to 2 years (or, in the UK and New Zealand, from weaning to first shearing).
ILCA Index 2 total weight of lamb weaned per unit body weight of ewe, per year.

Appendix I

Incidence (of disease) the number of new cases of the disease appearing each year, expressed as a proportion of the total sheep population.
Indigenous breed breed native to the locality.
Joining putting ewes or gimmers to the ram for mating.
Lamb young sheep, usually under 9 months (or before weaning); meat from young sheep.
Lambing interval period between successive parturitions (of one ewe).
Lambing rate number of lambs born divided by number of ewes joined.
Litter size number of lambs born divided by number of ewes lambing.
Local breed breed native to the locality (= indigenous breed).
M/D energy density. The concentration of ME in the diet, usually expressed as MJ kg^{-1} dry matter.
ME metabolisable energy. Digestible energy of food minus energy lost in methane and urine. Usually expressed in MJ.
Metabolic liveweight liveweight$^{0.75}$ (previously defined as liveweight$^{0.73}$). Usually in units of kg$^{0.75}$.
Morbidity rate proportion of animals showing signs of the disease.
Mortality rate proportion of the population which dies from the disease (*see also* **Case mortality**).
Mutton meat from (adult) sheep.
Oestrus period of sexual receptivity of ewe.
Offtake number of animals removed from the flock (for sale or consumption), divided by the number of animals in the flock.
Parturition lambing.
Prevalence (of disease) proportion of animals which show signs of the disease at any time.
Prolificacy mean litter size.
Productivity index measure of production of flock (*see, e.g.* **ILCA Index 2**).
Ram adult male sheep, usually used for breeding.
Rate of return on capital invested net annual return divided by total capital investment.
Relative humidity the absolute humidity divided by the absolute humidity at saturation at the same temperature.
Repeatability correlation between repeated measurements of the same trait on animals.
RDP rumen degradable protein. Crude protein which is degraded in the rumen.
Reproductive rate number of lambs weaned per ewe of reproductive age per year.
Roughage food with a low concentration of energy and much fibre (e.g. straw).
Stocking rate number of sheep per unit area of land.
Store lamb lamb which has been growing slowly, and which is ready to enter a finishing unit.
Terminal sire ram used for mating to give progeny intended for meat production but not for breeding.
Tup ram (less formal name).
Tupping mating.
UDP undegradable protein. Crude protein which is not degraded in the rumen.
Weaning percentage number of lambs weaned divided by number of ewes joined, expressed as a percentage.
Wether castrated male sheep, older than 6 months (or weaned).

Appendix II.

Breeds of sheep in the tropics

Some descriptions include the weights of adult sheep; these weights are either the mean weights recorded or typical ranges. Weights should be regarded as very approximate, since some authors report weights of animals in commercial systems, while others give those in experimental flocks.
The Near East refers to southern Asia between 25 and 60 °E.

Absa (Syria)
 variety of Awassi
 Mukhamed (1973)

Abyssinian (Ethiopia)
 meat, milk
 East African Fat-tailed type
 usually brown
 hair with mane
 horned or polled
 fat-tailed
 Italian: Pecora Abissina
 includes Adal, Horro
 varieties with woolly undercoat: Akele Guzai, Arusi-Bale, Menz, Rashaidi, Tucur
 Mason (1969)

Acchele Guzai (Ethiopia) – see Akele Guzai

Adal (E. Ethiopia)
 meat
 Abyssinian type
 white or light brown
 hair
 weight 20–40 kg
 polled
 fat-tailed
 Mason (1969), Galal (1983)

African (C. and S. America) – see West African

African Long-fat-tailed type
 tail usually long and fat – broad with hanging tip, twisted, cylindrical, strap-shaped, funnel-shaped or carrot-shaped
 includes Africander, Madagascar, Malawi, Nguni, Zimbabwe, Tanzania Long-tailed, Tswana
 Mason (1969)

African Long-legged type
 hair
 male horned, female polled
 thin tail
 includes Sudanese, West African Long-legged, Zaïre Long-legged, Baluba
 Mason (1969)

Africana (Colombia and Venezuela) – see West African

Africander (South Africa)
 African Long-fat-tailed type
 various colours
 hair
 polled or horned
 long fat tail
 Afrikaans: Afrikaner
 syn. Cape Fat-tailed
 varieties: Damara, Namaqua, Ronderib
 Mason (1969)

Ait Barka (S.E. of Marrakesh, Morocco)
 black variety of Berber
 Mason (1969)

Ait Haddidou (Morocco)
 variety of Berber

white, often with black marks on head
Mason (1969)

Ait Mohad (Morocco)
large variety of Berber
white
polled
Mason (1969)

Akele Guzai (Eritrea, Ethiopia)
coarse wooled variety of Abyssinian
usually black
Italian: Acchele Guzai
syn. Shimenzana (Italian: Scimenzana)
Mason (1969)

Aknoul (Morocco)
small earless variety of Berber
black
Mason (1969)

Algerian – see Algerian Arab, also Tunisian Barbary and Beni Guil

Algerian Arab (Algeria)
meat, coarsewooled
male horned, female polled
varieties include Ouled Jellal
Mason (1969)

Amalé (Sudan) – see Northern Sudanese

American Hair type (C. and S. America, West Indies)
weight 45–60 kg male, 30–40 kg female
includes African, Bahama, Barbados Blackbelly, Brazilian Woolless, Morada Nova, Pelibuey, Virgin Island, West African
Mason (1980a), Bradford and Fitzhugh (1983)

American Merino (USA)
fine wooled
Mason (1969)

American Rambouillet (USA)
finewooled, meat
male horned or polled, female polled
Mason (1969)

American Tunis (USA)
meat
coloured face and legs
polled

(fat-tailed)
Mason (1969)

Angola Maned (Angola)
black or pied
hair
horned
syn. Coquo
cf. West African Dwarf
Mason (1969)

Aouasse (Near East) – see Awassi

Ara-ara (Niger) – see Tuareg

Arab (Egypt) – see Barki

Arab (Somalia) – see Somali Arab

Arab (Turkey) – see Awassi

Arab (W. Africa) – see Maure

Arabi (S.W. Iran, S. Iraq, N.E. Saudi Arabia)
meat, coarsewooled
Near East Fat-tailed type
black, pied or white with black head
male horned, female polled
cf. Shafali (Syria)
Demiruren (1961), Mason (1969)

Arabian Long-tailed (Saudi Arabia) – see Najdi

Arbi (Near East) – see Arabi

Arrit (Keren, Eritrea, Ethiopia)
milk, meat
usually white, also blond, red or pied
hair or coarsewooled
polled, male occasionally horned
fat at base of tail
Mason (1969)

Arusi-Bale (Ethiopia)
variety of Abyssinian with woolly undercoat
brown, grey, roan or dun
male horned, female polled or horned
Mason (1969)

Assaf (Israel)
milk, meat, coarsewooled
origin: Israeli Improved Awassi ×

East Friesian
Sagi and Morag (1974), Gootwine et al. (1980)

Astrakhan – see Karakul

Atlantic Coast type (Morocco)
French: Race côtière atlantique
includes Beni Ahsen, Doukkala, Zemmour
Mason (1969)

Ausimi (Egypt) – see Ossimi

Australian Merino (Australia and New Zealand)
finewooled
male horned (or polled), female polled (or horned)
Mason (1969)

Avikalin (Rajasthan, India)
coarsewooled, meat
origin: Malpura (or Jaisalmeri) × Rambouillet
Acharya et al. (1976; 1977), Kishore et al. (1982)

Avikouillet (India) – see Avivastra

Avivastra (Rajasthan, India)
finewooled, meat
origin: Chokla × Rambouillet
Acharya et al. (1976; 1977), Kishore et al. (1982)

Awassi (Syria, Israel, Lebanon, Jordan, C. and W. Iraq, and S. Turkey)
meat, milk, coarsewooled
Near East Fat-tailed type
white with brown head and legs, sometimes black, white, grey or spotted face, occasionally all brown or black
weight 60–90 kg male, 30–50 kg female
male horned, female usually polled
Turkish: İvesi
syn. Arab (Turkey); Baladi, Deiri, Gezirieh, Shami (Syria), Syrian
varieties: Israeli Improved Awassi, N'eimi (Iraq), Shafali (Iraq)
Mason (1969), Epstein (1982)

Awsemy (Egypt) – see Ossimi

Azrou (Morocco) – see Middle Atlas

Baggara (W. Sudan)
ecotype intermediate between Desert Sudanese and Nilotic
McLeroy (1961)

Bagri (Punjab and Haryana, India)
meat, coarsewooled
white, brown or black head
Mason (1969), Bhat et al. (1980)

Bahama (West Indies)
American Hair type
meat
white with brown or black spots
hair
weight 65 kg male, 37 kg female
male horned, female polled
long thin tail
syn. Long Island
Mason (1980a)

Bahawalpuri (Pakistan) – see Buchi

Bahu (Kibali and Ituri, N.E. Zaïre)
brown or pied
dwarf
horned
syn. Zaïre Dwarf
Mason (1969)

Baladi (Egypt) – see Fellahi

Baladi (Syria) – see Awassi

Balami (Nigeria) – see Bornu

Balandji, Balani (W. Africa) – see Bornu

Balangir (N.W. Orissa, India)
coarsewooled
white, light brown or black
weight 24 kg male, 18 kg female
small ears
male horned, female polled
Acharya (1982)

Bali-Bali (W. Africa) – see Uda

Balonndi (W. Africa) – see Bornu

Baluba (Katanga, Zaïre)
African Long-legged type
hair, maned
lop ears
short tail
Mason (1969)

Baluchi (Baluchistan and Sind, Pakistan and S. Afghanistan)
meat, milk, coarsewooled
black marks on head and legs
male horned, female polled
fat-tailed
syn. Baluchi dumba, Beluj
Mason (1969)

Baluchi dumba (Pakistan) – *see* Baluchi

Bandur (India) – *see* Mandya

Bangladesh (Bangladesh)
meat, coarsewooled
weight 15 kg male, 10 kg female
Falvey and Hengmichai (1979a), Mason (1980a)

Bannur (India) – *see* Mandya

Bantu (South Africa) – *see* Damara, Nguni, Tswana

Bapedi (N.E. Transvaal, South Africa)
variety of Nguni
syn. Pedi, Transvaal Kaffir
Mason (1969)

Baqqara (Sudan) – *see* Baggara

Baraka (S.W. Keren, Eritrea, Ethiopia)
smaller variety of Northern Sudanese with long hair
Italian: Barca
syn. Begghié Korboraca, Shukria (Italian: Sciucria)
Mason (1969)

Barbados Blackbelly (West Indies)
American Hair type
meat
brown with black underparts, badger face
weight 50–70 kg male, 32–43 kg female
polled
syn. Blackbelly
Devendra (1972), Patterson (1976), Mason (1980a)

Barbaresca, Barbarine (N. Africa) – *see* Barbary

Barbary type (N. Africa)
coarsewooled, meat
Near East Fat-tailed type
usually white, also pied, black or brown
male horned, female polled
French: Barbarine, Italian: Barbaresca
includes Barki, Libyan Barbary, Tunisian Barbary
Mason (1969), Dumas (1980)

Barbary Halfbred (Libya)
meat, coarsewooled
male usually polled, female polled
fat-tailed
origin: White Karaman × Libyan Barbary
Mason (1969)

Barca (Ethiopia) – *see* Baraka

Barki (N.W. Egypt)
coarsewooled, meat, milk
Barbary type
usually brown or black head
male horned, female polled
fat-tailed
syn. Arab, Bedouin, Dernawi
Fahmy et al. (1964), Mason (1969)

Bedouin (Egypt) – *see* Barki

Bedouin (Saudi Arabia) – *see* Najdi

Begghié Korboraca (Eritrea, Ethiopia) – *see* Baraka

Beja (E. Sudan)
variety of Northern Sudanese, similar to Butana
McLeroy (1961), Mason (1969)

Bekrit (Morocco) – *see* Middle Atlas

Bellani (W. Africa) – *see* Bornu

Bellary (Karnataka, India)
• meat, coarsewooled
black, grey, white with black face, or pied
weight 35 kg male, 27 female
male horned or polled, female polled
similar to Deccani
Mason (1969), Acharya (1982)

Beluj (Pakistan) – *see* Baluchi

Beni Ahsen (N.W. Morocco)
coarsewooled or medium wool, meat
Atlantic Coast type
white with coloured face

horned or polled
Mason (1969)

Beni Guil (plateaux of E. Morocco and W. Algeria)
meat, coarsewooled
white with coloured head and legs
horned or polled
French: Race des plateaux de l'Est
syn. Petit Oranais, Hamyan (Algeria)
varieties: Harcha (chief), Tounsint and Zoulay
Mason (1969)

Beni Hassen (Morocco) – see Beni Ahsen

Beni Meskine (Morocco)
variety of Tadla
smaller and with coarser wool
Sardi is spectacled subvariety
Mason (1969)

Berber (mountains of Morocco)
meat, coarsewooled
usually white, also black or white with black head
horned
syn. Chleuh, Kabyle
varieties: Ait Barka, Ait Haddidou, Ait Mohad, Aknoul (dwarf, earless), Marmoucha, Tounfite
origin: (with Tadla) of Middle Atlas, South Moroccan and Zaian
Mason (1969)

Berbera Blackhead (N. and E. Africa) – see Somali

Bergamasca (Italy)
meat, coarsewooled
polled
lop ears
Mason (1969)

Bikaneri (N. Rajasthan, India)
coarsewooled
face usually tan or brown
polled
syn. Jangli
includes Bagri, Buchi, Chokla, Jaisalmeri, Magra, Malpura, Nali, Pugal, Sonadi
Lall (1956), Mason (1969)

Black Maure (W. Africa)
black, long-haired variety of Maure
syn. Nar, Zaghawa (Sudan)
Mason (1969)

Black Merino (Tunisia) – see Thibar

Blackbelly (Trinidad) – see Barbados Blackbelly

Blackhead Persian (C. and S. America and West Indies)
meat
white with black or brown head
hair
fat-rumped
syn. Brazilian Somali (Portuguese: Somalis Brasileira)
Shelton and Figueiredo (1981), Figueiredo et al. (1983)

Blackhead Persian (South Africa)
meat
white with black head and neck
hair
polled
fat-rumped
name also used for Somali in E. Africa
Mason (1969)

Blackheaded Somali (N. and E. Africa) – see Somali

Bornu (N.E. Nigeria, N. Cameroon, and Chad)
white variety of Fulani
weight 50 kg male, 40 kg female
syn. Balami, Balandji, Balani, Balonndi, Bellani, Fellata (Chad), Weiladjo, White Bororo
Adu and Ngere (1979), Mason (1969)

Bororo (W. Africa) – see Uda

Brazilian Somali (Brazil) – see Blackhead Persian

Brazilian Woolless (N.E. Brazil)
meat, skin
American Hair type
includes Morada Nova
syn. Deslanado do Nordeste, Northeastern Woolless

Buchi (Bahawalpur, Pakistan and Rajasthan, India)

meat, coarsewooled
Bikaneri type
usually black spots on head and legs
horned (short)
syn. Bahawalpuri, Cholistani
Masud and Hussain (1961)

Buchni (India) – *see* Buchi

Bukhara (USSR) – *see* Karakul

Bulandshahri (India) – *see* Muzaffarnagari

Cameroons Dwarf (W. Africa) – *see* West African Dwarf

Camura (Colombia) – *see* African

Cape Fat-tailed (South Africa) – *see* Africander

Chaffal (Syria) – *see* Shafali

Chakri (India) – *see* Magra

Chanothar (India) – *see* Sonadi

Chapper (India) – *see* Chokla

Charotar, Charotari (India) – *see* Patanwadi, Sonadi

Chevali (Syria) – *see* Shafali

Chios (Chios, Greece and Izmir, Turkey)
milk, meat, coarsewooled or medium wool
white with black or brown spots on face, belly and legs, sometimes black face
male horned, female usually polled
long fat tail
Turkish: Sakiz
Mason (1969), Lysandrides (1981)

Chleuh (Morocco) – *see* Berber

Chokla (Rajasthan, India)
coarsewooled, meat
Bikaneri type
brown head
weight 35 kg male, 23 kg female
polled
syn. Chapper, Shekhawati
Arora *et al.* (1975b), Bhat *et al.* (1980), Acharya (1982)

Cholistani (Pakistan) – *see* Buchi

Chotanagpuri (Bihar, India)
coarsewooled
dirty white to light tan
weight 20 kg
small ears
polled
very short tail
Taneja (1978), Bhat *et al.* (1980), Acharya (1982)

Choufalié (Syria) – *see* Shafali

Churro (Spain) – *see* Spanish Churro

Coimbatore (W. Tamil Nadu, India)
meat, coarsewooled
white with pigmented patches on head
weight 25 kg male, 21 kg female
male horned or polled, female polled
syn. Kurumbai
Taneja (1978), Bhat *et al.* (1980), Acharya (1982)

Congo Dwarf (Zaïre) – *see* Bahu

Congo Long-legged (Zaïre) – *see* Zaïre Long-legged

Constantinoise (Tunisia) – *see* Tunisian Barbary

Coquo (Angola) – *see* Angola Maned

Corriedale (New Zealand)
medium wool, meat
Mason (1969)

Creole (C. and S. America) – *see* Criollo

Criollo (C. and S. America)
meat, coarsewooled
white with black spots
short ears
horned
long tail
origin: Spanish Churro and Spanish Merino
Mason (1980b), Martínez (1983), Stagnaro (1983)

Criollo (Colombia) – *see* African

Criollo (Cuba) – *see* Pelibuey

Crioula (Brazil) – *see* Criollo

Cuban Hairy (Cuba) – *see* Pelibuey

Cubano Roja (Cuba) – *see* Pelibuey

Curraleiro (Brazil) – *see* Criollo

Cyprus Fat-tailed (Cyprus)
milk, coarsewooled, meat
Near East Fat-tailed type, similar to Awassi
usually white with black (or brown) on face esp. eyes and nose, occasionally black, brown or white
male usually horned, female usually polled
Mason (1969)

Dahman (Sahara) – *see* Tuareg

Damani (Dera Ismail Khan, Pakistan)
milk, coarsewooled, meat
brown markings on head
polled
short-tailed
syn. Lama
Masud and Hussain (1961), Mason (1969)

Damara (N. Namibia)
variety of Africander
pied, white, black or brown
hair or coarsewooled
usually polled
long fat tail
Mason (1969)

Dazdawi (Syria) – *see* Herrik

Deccani (Maharashtra, Andhra Pradesh and Karnataka, India)
meat, coarsewooled
black, white, or pied
weight 38 kg male, 29 kg female
male usually horned, female polled
short-tailed
similar to Bellary
Lall (1956), Mason (1969), Acharya (1982)

Deiri, Deiry (Syria) – *see* Awassi

Delaimi, Delimi (Syria) – *see* Shafali

Derna, Dernawi (Egypt) – *see* Barki

Desert Sudanese (Sudan)
milk, meat, skins
variety of Northern Sudanese
usually brown, also pied
lop ears
usually polled
long fleshy tail
includes Kababish
McLeroy (1961), Mason (1969)

Desi (India) – *see* Malpura, Patanwadi

Deslanado branco, Deslanado do Nordeste, Deslanado vermelho (Brazil) – *see* Brazilian Woolless

Dhamani (Pakistan) – *see* Damani

Dilem, Dillène (Syria) – *see* Shafali

Dinka (S. Sudan) – *see* Nilotic

Djallonké (W. Africa) – *see* West African Dwarf

D'man (Ksar es Souk, Morocco)
meat, coarsewooled
black, brown, white or pied
weight 50 kg male, 40 kg female
male polled or scurs, female polled
lop ears
female often tassels
Mason (1980a)

Döhne Merino (E. Cape Province, South Africa)
finewooled, meat
polled
origin: South African Merino × German Mutton Merino
Mason (1969)

Donggala (Sulawesi, Indonesia)
similar to East Java Fat-tailed
hair
fat-tailed
Mason (1969)

Dongola (N. Sudan)
coarsewooled
white or black
polled
syn. North Riverain Wooled
Mason (1969)

Dormer (South Africa)
meat, wooled
origin: Dorset Horn × German Mutton Merino
Mason (1969)

Dorper (South Africa)
meat
white with black head, often black feet
hair or coarsewooled
usually polled
origin: Dorset Horn × Blackhead Persian
variety: White Dorper
Meyer (1951), Mason (1969)

Dorset Down (S. England)
shortwooled, meat
brown face and legs
polled
Mason (1969)

Dorset Horn (S. England)
shortwooled, meat
brown face and legs
horned
syn. Dorset
Mason (1969)

Dorsian, Dorsie (South Africa) – *see* White Dorper

Doukkala (S.W. Morocco)
coarsewooled, meat
Atlantic Coast type
usually white, head usually black or brown
male horned, female polled
French: Doukkalide
syn. Doukkala-Abda
Mason (1969)

Doukkala-Abda, Doukkalide (Morocco) – *see* Doukkala

Douleimi (Syria) – *see* Shafali

Drasciani, Drashiani (Sudan) – *see* Northern Sudanese

Drysdale (New Zealand)
coarsewooled
Mason (1969)

Dulaimi (Syria) – *see* Shafali

Dumba (Pakistan) – *see* Pakistan Fat-tailed

Dwarf – *see* Bahu (Zaïre Dwarf), West African Dwarf

East African Blackheaded (W. Uganda and N.W. Tanzania)
East African Fat-tailed type
pied – usually white with black or brown head
hair
usually polled
short fat tail
Mason (1969)

East African Fat-tailed type
hair
broad, usually S-shaped fat tail
includes Abyssinian, East African Blackheaded, and Masai
Mason (1969)

East Friesian (Germany)
milk
polled
woolless tail
Mason (1969)

East Java Fat-tailed (E. Java and Madura, Indonesia)
meat, coarsewooled
white
weight 45 kg male, 35 kg female
male polled (occasionally scurs or small horns), female polled
fat-tailed
Mason (1978), Mason (1980a)

Egyptian (Egypt)
Near East Fat-tailed type – *see* Ossimi, Barki, Fellahi, Rahmani
short fat tail or long fat tail – *see* Ibeidi
long thin tail – *see* Kurassi
Mason (1969)

El Awas (Near East) – *see* Awassi

El Hammam (Morocco) – *see* Middle Atlas

Fartass (N. Africa)
polled variety of Beni Guil
Mason (1969)

Fat-rumped (Africa, Yemen) – *see* Somali

Fellahi (Nile Delta, Egypt)
coarsewooled, meat
Near East Fat-tailed type
usually brown
male horned, female usually polled
fat-tailed
syn. Baladi
being replaced by Ossimi and Rahmani
Mason (1969)

Fellata (W. Africa) – *see* Bornu

Fezzanais (Africa) – *see* Barbary

Forest (Ghana) – *see* West African Dwarf

Foulbé (W. Africa) – *see* Fulani

Fouta, Fouta Djallon, Fouta Jallon (W. Africa) – *see* West African Dwarf

Fouta Toro (W. Africa) – *see* Toronké

Fulani (Senegal to Cameroon)
meat
West African Long-legged type
white, pied or coloured
horned
French: Peul (Peuhl, Peulh, or Peul-Peul)
syn. Foulbé
varieties: Bornu, Samburu, Toronké, Uda, Yankasa
Mason (1969), Dumas (1980)

Ganjam (S. Orissa, India)
meat
brown, white or pied
weight 24 kg
drooping ears
male horned, female polled
syn. Patna
Mason (1969), Acharya (1982)

Garut (Indonesia) – *see* Priangan

Gasc, Gash (Sudan) – *see* Northern Sudanese

German Mutton Merino (Germany)
meat, finewooled
polled
Mason (1969)

Gezira (riverain of N. Sudan)
variety of Northern Sudanese
usually pied
smaller than Desert Sudanese with shorter tail and ears
syn. Northern Riverain
Mason (1969)

Gezirieh, Gezrawieh (Syria) – *see* Awassi

Ghimi (Ubari, Libya)
coarsewooled or medium wool, meat
fat-tailed
Mason (1969)

Goitred (Angola) – *see* Zunu

Goundoun (Niger)
variety of Macina
Mason (1969)

Grey Shirazi (Fars, Iran)
fur
similar to Karakul
usually grey
fat-tailed
Mason (1969)

Guinea (W. Africa) – *see* West African

Guinea Long-legged (W. Africa) – *see* West African Long-legged

Gujarati (India) – *see* Patanwadi

Hadjazi (Saudi Arabia) – *see* Hijazi

Hallenjoo (Dera Ghazi Khan, Pakistan)
larger variety of Khijloo
grey with black patches
Mason (1969)

Hamalé (Sudan) – *see* Northern Sudanese

Hamdani (Erbil, Iraq)
coarsewooled, meat
Kurdi type
black head
polled
lop ears
syn. Hamdanya
Mason (1969)

Hamdanya (Iraq) – *see* Hamdani

Hammam (Morocco) – *see* Middle Atlas

Hampshire Down (England)
shortwooled, meat
Mason (1969)

Hamyan (Algeria) – *see* Beni Guil

Harcha (N. Africa)
main variety of Beni Guil
Mason (1969)

Harrick (Syria) – *see* Herrik

Hassan (Karnataka, India)
meat, coarsewooled
white, often with black, brown or grey patches
weight 26 kg male, 23 kg female
drooping ears
male horned or polled, female polled
syn. Kolar
Mason (1969), Acharya (1982)

Hausa (W. Africa) – *see* Yankasa

Hedjazi, Hejazi (Saudi Arabia) – *see* Hijazi

Herrik (Syria, from N. Iraq)
coarsewooled, meat, milk
black face
usually polled
fat-tailed
syn. Dazdawi, Mosuli
Mason (1969)

Hijazi (W. Saudi Arabia)
meat
usually white
hair
often earless
male polled or scurs, female polled
often tassels
short fat tail
syn. Khazi, Mecca
Mason (1969)

Hissardale (Haryana, India)
shortwooled
white, sometimes with dark patches
weight 57 kg male, 35 kg female
usually polled
originated from Bikaneri × Australian Merino
Mason (1969), Acharya (1982)

Horro (W. Ethiopia)
meat
Abyssinian type
light brown
hair, male often maned
weight 30–40 kg female
polled
fat-tailed
Galal (1983)

Houda (W. Africa) – *see* Uda

Hungarian Combing Wool Merino (Hungary)
meat, finewooled
Mason (1969)

Ibeidi (El Minya, Upper Egypt)
white, usually with brown head, occasionally black
male horned, female usually polled
short fat tail
Mason (1969)

Ideal (S. America) – *see* Polwarth

Ile-de-France (N. France)
meat, medium wool
Mason (1969)

Improved Awassi (Israel) – *see* Israeli Improved Awassi

Indian Fat-tailed – *see* Pakistan Fat-tailed breeds

Indonesian (Indonesia) – *see* Donggala, East Java Fat-tailed, Javanese Thin-tailed, Priangan

Ingessana (S. Sudan)
Desert × Nilotic intermediate
Mason (1969)

Iraqi (Iraq) – *see* Awassi, Arabi, Kurdi

Israeli Improved Awassi (Israel)
variety of Awassi with Herrik blood, selected for milk
Mason (1969)

Ivesi, Iwessi (Near East) – *see* Awassi

Jaffna (Sri Lanka)
manure, meat
similar to South India Hair type
black, tan, black and white, tan and white, or white with coloured patches
weight 23 kg male, 18 kg female
male horned or polled, female polled
Mason (1969), Buvanendran (1978), Ravindran *et al.* (1983)

Jaisalmeri (Rajasthan, India)
meat, coarsewooled
Bikaneri type
white, face dark brown or black
weight 38 kg male, 30 kg female
usually polled
long lop ears
Mason (1969), Bhat *et al.* (1980), Acharya (1982)

Jalauni (S.W. Uttar Pradesh, India)
coarsewooled, milk, meat
face may have coloured markings
weight 40 kg male, 30 kg female
usually polled
long lop ears
Mason (1969), Bhat *et al.* (1980), Acharya (1982)

Jangli (India) – *see* Bikaneri

Javanese (Java, Indonesia) – *see* East Java Fat-tailed

Javanese Thin-tailed (Java, Indonesia)
meat, coarsewooled
white with black patches around nose and occasionally elsewhere
weight 35–60 kg male, 20–35 kg female
short-tailed
male horned, female usually polled
Mason (1978), Mason (1980a)

Joria (India) – *see* Patanwadi

Junin (Peru)
medium wool, meat
FAO (1972b), Gamarra and Carpio (1982)

Kababish (N. Kordofan, Sudan)
tribal variety of Desert Sudanese
Mason (1969)

Kabyle (Morocco) – *see* Berber

Kajli (Sargodha, Gujrat and Jhelum, Pakistan)
meat, coarsewooled
black nose, eyes and ear tips
long lop ears
Roman nose
male horned or polled, female polled
Masud and Hussain (1961), Mason (1969)

Karadi (Kurdistan)
Kurdi type

Karakul (Uzbekistan, USSR)
fur, milk
male horned, female polled
fat-tailed
syn. Astrakhan, Bukhara
Mason (1969)

Karandhai (India) – *see* Vembur

Karuvai (India) – *see* Kilakarsal

Kashmir Valley (S.W. Kashmir, India)
coarsewooled
usually coloured
usually polled
Mason (1969)

Kathiawari (India) – *see* Marwari, Patanwadi

Keezhakkaraisal, Kelakarisal (India) – *see* Kilakarsal

Kelantan (Malaysia and Thailand)
meat, coarsewooled
white, or black, light brown or pied
ears small, pointed sideways
male horned, female usually polled
short-tailed
syn. Malay, Thai
Smith and Clarke (1972), Mason (1980a), Mukherjee (1980)

Kenguri (Karnataka, India)
meat
South India Hair type
dark brown, sometimes light brown or spotted
hair
weight 32 kg male, 27 kg female
male horned, female polled

syn. Tenguri
Acharya (1982)

Kent (England) – *see* Romney Marsh

Khazi (W. Saudi Arabia) – *see* Hijazi

Khijloo (Dera Ghazi Khan, Pakistan)
coarsewooled, meat
black marks on face and feet, or tan face and feet
short fat tail
variety: Hallenjoo (larger, grey with black patches)
Mason (1969)

Khios (E. Mediterranean) – *see* Chios

Kilakarsal (S.E. Tamil Nadu, India)
meat
South India Hair type
dark brown with white patches
weight 29 kg male, 22 kg female
tassels
male horned, female polled
syn. Keezhakkaraisal, Karuvai
Bhat *et al.* (1980), Acharya (1982)

Kipsigis (W. Kenya)
variety of Masai
syn. Lumbwa
Mason (1969)

Kirdi, Kirdimi (W. Africa) – *see* West African Dwarf

Kolar (India) – *see* Hassan

Kooka (Pakistan) – *see* Kuka

Kuka (Sind, Pakistan)
coarsewooled, milk
face usually black
male horned, female polled
long lop ears
short-tailed
Mason (1969)

Kurassi (Qena and Aswan, Upper Egypt)
coarsewooled, meat
usually black or brown, also fawn, pied or white
polled
long thin tail
cf. Dongola
Mason (1969)

Kurdi type (Kurdistan)
coarsewooled
black head
polled
fat-tailed
includes Hamdani, Karadi
Mason (1969)

Kuruba, Kurumba, Kurumbai (India) – *see* Coimbatore

Kushari (Syria) = nomadic sheep

Kutchi (India) – *see* Patanwadi

Lakka (Chad) – *see* West African Dwarf

Lama (Pakistan) – *see* Damani

Lamkanni (Bahawalpur, Pakistan)
smaller variety of Lohi with grey ears and no tassels
syn. Lamochar
Mason (1969)

Landim (Mozambique)
variety of Nguni
polled
Mason (1969)

Large Peul (W. Africa) – *see* Toronké

Lati (Salt Range, N.W. Punjab, Pakistan)
meat, coarsewooled
fat-tailed
Mason (1969)

Legagora (Ethiopia) – *see* Menz

Libyan Barbary (Libya)
meat, coarsewooled, milk
white with brown or black (occasionally pied) face and legs, sometimes pied or coloured
male horned (occasionally 4), female usually polled
fat-tailed
cf. Barki (Egypt)
Mason (1969)

Lincoln Longwool (E. England)
longwooled, meat
Mason (1969)

Lohi (S. Punjab, Pakistan and India)
coarsewooled, meat, milk

white, sometimes with dark spots,
  head black or brown
long lop ears usually with tassel
polled
short-tailed
variety: Lamkanni (Bahawalpur)
Lall (1956), Mason (1969), Bhat *et al.*
  (1980)

Long Island (West Indies) – *see* Bahama

Louda (W. Africa) – *see* Uda

Lumbwa (Kenya) – *see* Kipsigis

Luo (W. Kenya)
variety of Masai
Mason (1969)

Maasai (E. Africa) – *see* Masai

Macheri (India) – *see* Mecheri

Macina (C. delta of Niger, Mali)
coarsewooled
white, pied or black
male horned, female horned (small) or
  polled
long thin tail
variety: Goundoun (Niger)
Mason (1969), Wilson (1983)

Madagascar (Malagasy Republic)
meat
African Long-fat-tailed type
white or light brown, often pied with red,
  brown or black, often white with black
  head
hair
male horned or polled, female polled
long, fat, sickle-shaped tail
syn. Malagasy, Malgache
Mason (1969)

Madras Red (Tamil Nadu, India)
meat
South India Hair type
reddish brown
hair
weight 36 kg male, 23 kg female
drooping ears
male horned, female polled
similar to Mecheri
syn. South Madras
Bhat *et al.* (1980), Acharya (1982)

Madurese (Madura, Indonesia) – *see*
  East Java Fat-tailed

Maghreb thin-tailed breeds – *see* Algerian
  Arab, Atlantic Coast (Morocco), Beni
  Guil, Berber, Tadla

Magra (Rajasthan, India)
coarsewooled, meat
Bikaneri type
white face with brown patches round eyes
weight 27 kg male, 24 kg female
polled
syn. Chakri, Mogra (Jodhpur)
Bhat *et al.* (1980), Acharya (1982)

Maiylambadi (India) – *see* Mecheri

Malagasy (Malagasy Republic) – *see*
  Madagascar

Malawi (Malawi)
African Long-fat-tailed type
similar to Rhodesian
syn. Nyasa
Mason (1969)

Malay (Malaysia) – *see* Kelantan

Malgache (Malagasy Republic) – *see*
  Madagascar

Malpura (Rajasthan, India)
coarsewooled
Bikaneri type
light brown face
weight 42 kg male, 24 kg female
small ears
polled
syn. Desi
Arora *et al.* (1975a), Mason (1969), Bhat
  et al. (1980), Acharya (1982)

Mama (Namibia) – *see* Namaqua
  Africander

Mandya (Karnataka, India)
meat
South India Hair type
pale red-brown patches on anterior
weight 35 kg male, 24 kg female
polled
short-tailed
syn. Bandur, Bannur
Mason (1969), Acharya (1982)

Manze (Ethiopia) – *see* Menz

Marathwada (S. Maharashtra, India)
meat
South India Hair type
black, red or pied
male horned, female polled
similar to Nellore
Mason (1969)

Marmoucha (Morocco)
varieties of Berber
small white and larger blackheaded
male horned, female polled
Mason (1969)

Marwari (Rajasthan, India)
coarsewooled, meat
black face
weight 31 kg male, 26 kg female
small ears
polled
syn. Kathiawari
Mason (1969), Bhat et al. (1980), Acharya (1982)

Masai (Tanzania and Kenya, also Uganda)
meat
East African Fat-tailed type
red–brown, occasionally pied
hair
male horned or polled, female usually polled
short fat tail or fat-rumped
includes sheep of Kipsigis, Luo, Nandi and Samburu (all in Kenya)
syn. Tanzania short-tailed
Mason (1969)

Mashona (Zimbabwe) – *see* Zimbabwe

Massina (W. Africa) – *see* Macina

Maure (N. of West Africa)
West African Long-legged type, similar to Tuareg
male horned, female polled
syn. Arab, Mauritanian, Moor, Moorish
varieties: Black Maure (long-haired), and Tuabir (white or pied, short-haired)
Mason (1969), Dumas (1980)

Mauritanian (W. Africa) – *see* Maure

Mayo-Kebbi (S.W. Mayo-Kebbi, Chad)
meat

white with black eyes and nose
hair
lop ears
male usually horned, female polled
long thin tail
Dumas (1980)

Mecca (Saudi Arabia) – *see* Hijazi

Mecheri (Salem and Coimbatore, Tamil Nadu, India)
meat, skin
South India Hair type
lighth brown often with white patches
hair
weight 35 kg male, 22 kg female
polled
short tail
similar to Madras Red
syn. Macheri, Maiylambadi (Coimbatore), Thuvaramchambali (Coimbatore)
Acharya (1982)

Meidob (N.W. Darfur, Sudan)
Zaghawa or Desert Sudanese, or cross
Mason (1969)

Menz (C. Ethiopia)
meat, coarsewooled
variety of Abyssinian with woolly undercoat
black or dark brown with white on head, neck and legs
weight 20–30 kg female
male usually horned, female usually polled
fat-tailed
syn. Legagora
Mason (1969), Galal (1983)

Meraisi (Egypt) – *see* Ossimi

Merino
finewooled
male usually horned, female usually polled
see American Merino, Australian Merino, German Mutton Merino, South African Merino, Soviet Merino, Spanish Merino
Mason (1969)

Middle Atlas (Morocco)
meat, coarsewooled
mixed types with Berber and Tadler blood
horned

French: Moyen Atlas
includes Azrou, Bekrit, El Hammam,
  Timhadit
Mason (1969)

Mogra (India) – *see* Magra

Mongalla (Equatoria, S. Sudan)
variety of Southern Sudanese
Mason (1969)

Moor, Moorish (W. Africa) – *see* Maure

Morada Nova (Ceará, Brazil)
meat, skin
American Hair type
red with white tail tip, or white
  hair
weight 39 kg male, 31 kg female
male scurs or polled, female polled
improved variety of Brazilian Woolless
Mason (1980b), Shelton and Figueiredo
  (1981), Figueiredo *et al.* (1983)

Moroccan – *see* Beni Ahsen, Beni Guil,
  Berber, Doukkala, Middle Atlas, South
  Moroccan, Tadla, Zaian, Zemmour

Mosuli (Syria) – *see* Herrik

Moulouya (N. Africa) – *see* Tounsint and
  Zoulay varieties of Beni Guil

Moyen Atlas (Morocco) – *see* Middle Atlas

Murle (S.E. Sudan)
variety of Toposa with more
  Southern Sudanese blood
black-pied or brown-pied
male usually horned, female usually
  polled
short fat tail
Mason (1969)

Muzaffarnagari (Uttar Pradesh, India)
meat, coarsewooled, milk
white, sometimes with dark patches
weight 50 kg male, 40 kg female
polled
long drooping ears
long thin tail
syn. Bulandshahri
Bhat *et al.* (1980), Acharya (1982)

Na'ami, Naimi (Iraq) – *see* N'eimi

Najdi (C. Saudi Arabia)
coarsewooled
usually black with white head
male polled or scurs, female polled
long fat tail
syn. Arabian Long-tailed, Bedouin
Mason (1969)

Nali (Rajasthan, India)
coarsewooled, meat
Bikaneri type
light brown face
weight 35 kg male, 24 kg female
ears large
polled
Arora *et al.* (1975c), Bhat *et al.*
  (1980), Acharya (1982)

Namaqua Africander (N.W. Cape
  Province, South Africa and S. Namibia)
meat
variety of Africander
white, usually with black or brown head
  hair or coarsewooled
male horned, female usually polled
long fat tail
syn. Mama (Namibia)
Mason (1969)

Nami (Iraq) – *see* N'eimi

Nandi (W. Kenya)
variety of Masai
Mason (1969)

Nar (W. Africa) – *see* Black Maure

Navajo (S.W. USA)
coarsewooled
horned or polled
origin: Spanish Churro
Blunn (1943), Mason (1969),
  Sidwell *et al.* (1970)

Near East Fat-tailed type (N.E. Africa
  and S.W. Asia)
coarsewooled, milk, meat
white, white with coloured face,
  black, brown or pied
male usually horned, female usually
  polled
usually large S-shaped, bilobed, fat tail
includes Barbary (Tunisia to Egypt),
  Ossimi, Barki, Fellahi, Rahmani
  (Egypt), Cyprus Fat-tailed, Awassi

(Syria, Israel and Iraq), Arabi (Iraq)
syn. Semitic Fat-tailed
Mason (1969)

Nedjed (Saudi Arabia) – *see* Najdi

N'eimi (Dulaim, N.W. Iraq)
superior variety of Awassi
black or red face, sometimes red fleece
Mason (1969)

Nejdi (Saudi Arabia) – *see* Najdi

Nellore (S.E. Andhra Pradesh, India)
meat
South India Hair type
white, red–brown or white with light brown appendages
weight 36 kg male, 30 kg female
long drooping ears
tassels
male horned, female polled
short-tailed
Lall (1956), Mason (1969), Bhat et al. (1980), Acharya (1982)

Nguni (S.E. Africa)
African Long-fat-tailed type
black, brown or pied
hair
male usually horned, female usually polled
includes Bapedi (Transvaal), Landim (Mozambique), Swazi and Zulu
Mason (1969)

Nidjy (Saudi Arabia) – *see* Najdi

Nigerian Dwarf (W. Africa) – *see* West African Dwarf

Nilgiri (Ootacamund, Tamil Nadu, India)
shortwooled, meat
weight 30 kg male, 25 kg female
male scurs or polled, female polled
Mason (1969), Bhat et al. (1980), Acharya (1982)

Nilotic (Sudan)
variety of Southern Sudanese
syn. Dinka, Nuer, Shilluk
Mason (1969)

North Country Cheviot (Scotland).
Mason (1969)

North Nigerian Fulani (W. Africa) – *see* Uda

North Riverain Wooled (Sudan) – *see* Dongola

Northeastern Woolless (Brazil) – *see* Morada Nova

Northern Riverain (Sudan) – *see* Gezira

Northern Sudanese (Sudan N. of 12° N. latitude and W. Eritrea, Ethiopia)
milk, meat
African Long-legged type
hair
lop ears
long thin tail
syn. Amalé (Italian: Hamalé), Drashiani (Italian: Drasciani), Gash (Italian: Gasc)
varieties: Baraka, Beja, Gezira, Desert Sudanese, Watish
Mason (1969)

Nu'amieh, Nuamiyat (Iraq) – *see* N'eimi

Nuba Maned (S. Kordofan, Sudan)
variety of Southern Sudanese
all colours but black
male horned, female polled
Mason (1969)

Nuer (Sudan) – *see* Nilotic

Nungua Blackhead (Ghana)
white with black head and neck
hair
weight 39 kg male, 32 kg female
origin: Blackhead Persian × West African Dwarf
Ngere (1973)

Nyasa (Malawi) – *see* Malawi

Ogaden (N. and E. Africa) – *see* Somali

Omani (Oman)
meat, coarsewooled

Osemi (Egypt) – *see* Ossimi

Ossimi (Lower Egypt)
coarsewooled, meat

Near East Fat-tailed type
white with brown head, often brown neck and occasionally brown spots
male horned, female polled
fat-tailed
syn. Ausimi, Meraisi
Ghoneim et al. (1968), Mason (1969)

Ouda (W. Africa) – see Uda

Ouled Jellal (Algeria)
variety of Algerian Arab
French: Ouled Djellal
Mason (1969)

Ousimi (Egypt) – see Ausimi

Oussi (Near East) – see Awassi

Pahang (E. Malaysia)
wooled
similar to Kelantan
male usually horned
Mason (1969)

Pakistan Fat-tailed breeds
see Baluchi, Khijloo, Lati
syn. Dumba, Indian Fat-tailed, Puchia

Patanwadi (Gujarat, India)
coarsewooled, meat, milk
brown face and points
weight 33 kg male, 27 kg female
ears drooping
polled
short tail
syn. Buti, Charotari, Desi, Gujarati, Joria, Kathiawari, Kutchi, Vadhiyari
Lall (1956), Mason (1969), Bhat et al. (1980), Acharya (1982)

Patna, Patnai, Patnia (India) – see Ganjam

Pecora somala a testa nera (N. and E. Africa) – see Somali

Pedi (South Africa) – see Bapedi

Pelibuey (Cuba)
meat
American Hair type
tan, white or tan and white
hair
polled
similar to African, West African
syn. Criollo (Cuba), Cuban Hairy, Cubano Rojo, Peliguey, Tabasco (Mexico)
Mason (1980a), Bradford and Fitzhugh (1983)

Peliguey (Brazil) – see Pelibuey

Pelona (Colombia) – see African

Persian, Persian Blackhead – see Blackhead Persian

Petit Oranais (N. Africa) – see Beni Guil

Peuhl, Puel, Peulh, Peul-Peul (W. Africa) – see Fulani

Pied (W. Africa) – see Uda

Polwarth (W. Victoria, Australia)
medium wool
horned and polled varieties
syn. Ideal (S. America)
Mason (1969)

Poogal (India) – see Pugal

Porsha (Syria)
variety of Awassi
Mukhamed (1973)

Preanger (Indonesia) – see Priangan

Priangan (W. Java, Indonesia)
ram fighting, meat, coarsewooled
variety of colours
male maned
weight 60 kg male, 40 kg female
horned
short-tailed
includes Garut
Mason (1969), Mason (1978)

Puchia (Pakistan) – see Pakistan Fat-tailed

Pugal (Rajasthan, India)
coarsewooled, meat
Bikaneri type
black and brown face
weight 32 kg male, 27 kg female
short ears
polled

Mason (1969), Bhat *et al.* (1980), Acharya (1982)

Rabo Largo (Bahia, Brazil)
meat, coarsewooled
white, spotted, or white with coloured head
horned
long fat tail
origin: African × Criollo
Shelton and Figueiredo (1981), Figueiredo *et al.* (1983)

Race des Plateaux de l'Est (N. Africa) – *see* Beni Guil

Radmani (Yemen PDR and Saudi Arabia)
hair
white
polled
fat-tailed
syn. Sha'ra
Mason (1969)

Rahmani (Damanhur, Lower Egypt)
coarsewooled, meat
Near East, Fat-tailed type
brown, fading with age
often earless
male horned, female usually polled
fat-tailed
Dudgeon and Mohammad Askar (1916), Mason (1969)

Rambouillet (France)
finewooled, meat
male horned, female polled
Mason (1969)

Ramnad White (S. Tamil Nadu, India)
meat
South India Hair type
white, sometimes with dark patches
hair
weight 31 kg male, 22 kg female
male horned, female polled
Taneja (1978), Acharya (1982)

Rasciaida (Ethiopia) – *see* Rashaidi

Rashaidi (N.E. Eritrea, Ethiopia)
coarsewooled
brown, white, red or pied
polled
Italian: Rasciaida
Mason (1969)

Red African (Venezuela and Colombia) – *see* African

Red Woolless (Brazil) – *see* Brazilian Woolless

Rehamma-Srarhna (Morocco)
variety of South Moroccan
Mason (1969)

Rhodesian (Zimbabwe) – *see* Zimbabwe

Rojo Africana (Venezuela) – *see* African

Romney Marsh (S. England)
longwooled, meat
polled
syn. Kent
Mason (1969)

Ronderib Africander (N.C. Cape Province, South Africa)
meat
variety of Africander
white
hair or coarsewooled
male usually horned, female horned or polled
short fat tail
being displaced by Dorper and Merino
Mason (1969)

Sahel (W. Africa) – *see* West African Long-legged

Saidi (Asyut, Upper Egypt)
coarsewooled, meat
black or brown
polled
long fat tail
syn. Sohagi
variety: Sanabawi
Dudgeon and Mohammad Askar (1916), Mason (1969)

Sakiz (E. Mediterranean) – *see* Chios

Samburu (Mali)
variety of Fulani
maroon
long horizontal horns
French: Sambourou
Mason (1969)

Samburu (N. Kenya)
variety of Masai varying in type and colour
Mason (1969)

Sanabawi (Upper Egypt)
variety of Saidi with smaller tail
red with red, black or white head
sometimes horned
Mason (1969)

Santa Inês (Brazil)
meat
white, red or spotted
hair
weight up to 100 kg male, 45 kg female
polled
long lop ears
origin: Bergamasca × Morada Nova
Shelton and Figueiredo (1981),
Figueiredo *et al.* (1983)

Sarda (Italy) – *see* Sardinian

Sardi (Morocco)
variety of Beni Meskine with black eyes and nose
Mason (1969)

Sardinian (Sardinia, Italy)
milk, meat, coarsewooled
occasionally black
horned or polled
cf. Tunisian Milk Sheep
Mason (1969)

Sciucria (Ethiopia) – *see* Baraka

Scottish Blackface (Scotland and N. England)
coarsewooled, meat
Mason (1969)

Semitic Fat-tailed (Near East) – *see* Near East Fat-tailed

Semmeri (India)
variety of Madras Red

Setswana (S.E. Africa) – *see* Tswana

Shafali (Hai, Kut Liwa, Iraq)
large variety of Awassi
often reddish-brown, sometimes black
Mason (1969)

Shafali (Syria)
similar to Awassi but smaller and red (or black)
French: Chaffal, Chevali or Choufalié
syn. Delaimi, Delimi, Dilem, Dillène, Douleimi, Dulaimi
Mason (1969)

Shagra (Syria)
variety of Awassi
Mukhamed (1973)

Shahabadi (Bihar, India)
coarsewooled
white sometimes with dark spots on face
weight 38 kg male, 27 kg female
drooping ears
long thin tail
Taneja (1978), Bhat *et al.* (1980), Acharya (1982)

Shami (Syria) – *see* Awassi

Sha'ra (Saudi Arabia) – *see* Radmani

Shekhawati, Sherawati (India) – *see* Chokla

Shilluk (Sudan) – *see* Nilotic

Shimenzana (Eritrea, Ethiopia) – *see* Akele Guzai

Shropshire (England)
meat, shortwooled
black-brown face and legs
polled
Mason (1969)

Shukria (Eritrea, Ethiopia) – *see* Baraka

Sidi Tabet Cross (Tunisia)
medium wool
black
origin: Portuguese Black Merino × Thibar
Mason (1969)

Sohagi (Egypt) – *see* Saidi

Somali (Somalia; also E. Ethiopia and N. Kenya)
meat
white with black head
hair
weight 30 kg female
polled

fat-rumped
syn. Berbera Blackhead, Blackheaded
  Somali (Italian: Pecora somala a testa
  nera), Ogaden
varieties: Adali, Toposa
Mason (1969), Galal (1983)

Somali Arab (coast of Somalia)
coarsewooled or hair
white
polled
fat-tailed
Mason (1969)

Somali Blackhead (E. Africa) – *see* Somali

Somalis Brasileira (Brazil) – *see* Blackhead
  Persian

Sonadi (Rajasthan and Gujarat, India)
meat, coarsewooled, milk
Bikaneri type
white with light brown head and neck
weight 39 kg male, 21 kg female
very long drooping ears
polled
long tail
syn. Chanothar, Charotar (Gujarati)
Arora *et al* (1977), Bhat *et al.* (1980),
  Acharya (1982)

South African Merino (South Africa)
finewooled
Mason (1969)

South India Hair type (India)
  includes Kenguri, Kilakarsal, Madras
  Red, Mandya, Mecheri, Nellore,
  Ramnad White, Tiruchy Black, Vembur

South Madras (India) – *see* Madras Red

South Moroccan (Morocco)
horned
varieties: Rehamma-Srarhna and
  Zembrane
origin: Tadla × Berber
Mason (1969)

Southern (W. Africa) – *see* West African
  Dwarf

Southern Goat (Algeria) – *see* Tuareg

Southern Sudanese (Sudan S. of 11 ° N.
  latitude)
white usually with black or tan patches
hair, male usually with ruff
horned or polled
syn. Sudanese Maned
varieties: Mongalla, Nilotic, Nuba Maned
Mason (1969)

Soviet Merino (USSR)
finewooled
Mason (1969)

Spanish Churro (N. and W. Spain)
milk, meat, coarsewooled
black spots on face and feet
male usually horned, female usually
  polled
Mason (1969)

Spanish Merino (Spain)
finewooled
white, also black variety
male occasionally polled, female usually
  polled
Mason (1969)

St Croix (West Indies) – *see* Virgin Island

Sudan Desert (Sudan) – *see* Desert
  Sudanese

Sudanese (Sudan) – *see* Dongola,
  Northern Sudanese, Southern
  Sudanese, Zaghawa
Mason (1969)

Sudanese Desert (Sudan) – *see* Desert
  Sudanese

Sudanese Maned (Sudan) – *see* Southern
  Sudanese

Suffolk (England)
meat, shortwooled
black face and legs
Mason (1969)

Swazi (S.E. Africa) – *see* Nguni

Syrian (Near East) – *see* Awassi

Tabasco (Mexico) – *see* Pelibuey

Tadla (Morocco)
meat, coarsewooled
white with coloured legs
horned

French: Race des Plateaux de l'Ouest
variety: Beni Meskine
Mason (1969)

Tadmit (Algeria and Tunisia)
meat, medium wool
male horned, female polled
Mason (1969)

Tanganyika Long-tailed (E. Africa) – *see* Tanzania Long-tailed

Tanganyika Short-tailed (E. Africa) – *see* Masai

Tanzania Long-tailed (Tanzania)
meat
African Long-fat-tailed type
various colours
hair
often earless and sometimes tassels
male often horned, female usually polled
long fat tail, short fat tail or long thin tail
syn. Ugogo
Mason (1969)

Tanzania Short-tailed (E. Africa) – *see* Masai

Targhee (Idaho, USA)
mediumwooled, meat
polled
Mason (1969)

Targhi, Targi, Targui (W. Africa) – *see* Tuareg

Teng, Tenguri (India) – *see* Kenguri

Thai (Thailand) – *see* Kelantan

Thalli (Pakistan) – *see* Thal

Thibar (Tunisia)
medium wool
black, occasionally with white spot on head or tail
syn. Black Merino
Mason (1969)

Thuvaramchambali (India) – *see* Mecheri

Timhadit (N. Africa) – *see* Middle Atlas

Tiruchy Black (Tamil Nadu, India)
meat
South India Hair type
black, sometimes with white face
weight 26 kg male, 18 kg female
male horned, female polled
fat-tailed
syn. Trichi Black
Bhat *et al.* (1980), Acharya (1982)

Toposa (S.E. Sudan)
variety of Somali with some Nilotic blood
white or white with black or brown head
male usually horned, female sometimes horned
fat-rumped
cf. Murle
Mason (1969)

Toronké (Senegal to Mali)
variety of Fulani
white or spotted
syn. Fouta Toro. Large Peul
Mason (1969)

Touabire (W. Africa) – *see* Tuabir

Touareg (Sahara) – *see* Tuareg

Tounfite (Morocco)
variety of Berber
white, occasionally with black marks on head and body
horned
Mason (1969)

Tounsint (Morocco)
variety of Beni Guil in Moulouya Valley
Mason (1969)

Transvaal Kaffir (South Africa) – *see* Bapedi

Trichi Black (India) – *see* Tiruchy Black

Tswana (Botswana and S.W. Zimbabwe)
African Long-fat-tailed type
often white or black and white
syn. Setswana
Mason (1969)

Tuabir (W. Africa)
usually white (also pied), short-haired
variety of Maure
French: Touabire

syn. White Arab, White Maure
Mason (1969)

Tuareg (Sahara)
meat
West African Long-legged type, similar to Maure
white, pied or fawn
male horned, female polled
French: Touareg
syn. Ara-Ara (Niger), Dahman or Southern Goat (Algeria)
Mason (1969)

Tucur (Amhara, Ethiopia)
coarsewooled or hair
variety of Abyssinian with woolly undercoat
white, brown or pied
Mason (1969)

Tunisian Barbary (Tunisia)
meat, coarsewooled
white with black or red–brown head, occasionally black or white
male horned, female usually polled
fat-tailed
French: Barbarine
syn. Constantinoise, Tunisienne
Mason (1969)

Tunisian Milk Sheep (Tunisia)
milk, meat, coarsewooled
white, black, brown or grey
male horned, female usually polled
syn. Sardinian
Mason (1969)

Tunisienne (Tunisia) – *see* Tunisian Barbary

Twareg (Sahara) – *see* Tuareg

Uda (N. Nigeria, Niger, Chad, and N. Cameroon)
pied variety of Fulani, front part of body black, rear part white
weight 50 kg male, 40 kg female
syn. Bali-Bali (Niger), Bororo (Chad), Houda, Louda, North Nigerian Fulani, Ouda, Pied
Mason (1969), Adu and Ngere (1979), Dettmers (1983)

Ugogo (Tanzania) – *see* Tanzania Long-tailed

Ussy (Near East) – *see* Awassi

Vadhiyari (India) – *see* Patanwadi

Van Rooy (Orange Free State, South Africa)
meat
all white
hair
polled
small fat rump
Afrikaans: Van Rooy-Persie
syn. Van Rooy White Persian
origin: Blackhead Persian × (Rambouillet × Ronderib Africander)
Mason (1969)

Vandor (South Africa)
origin: Dorset Horn × Van Rooy
Mason (1969)

Vembur (Tamil Nadu, India)
meat
South India Hair type
white with many irregular red and fawn patches
weight 34 kg male, 28 kg female
drooping ears
male horned, female polled
syn. Karandhai
Acharya (1982)

Venezuelan Criollo (Venezuela)
variety of Criollo
Mason (1969)

Vera (Bangladesh) – *see* Bangladesh

Vicanere (India) – *see* Bikaneri

Virgin Island (West Indies)
American Hair type
meat
usually white
hair
weight 45 kg male, 35 female
polled
syn. St Croix, White Virgin Island
Mason (1980a), Bradford and Fitzhugh (1983), Hupp and Deller (1983)

Vogan (Togo)
meat
red pied, black pied, or brown and

black pied
hair
weight 45 kg male, 40 kg female
origin: West African Dwarf ×
 West African Long-legged
Amegee (1983a)

Watish (riverain of N. Sudan)
variety of Northern Sudanese similar to
 Gezira
Mason (1969)

Weiladjo (W. Africa) – see Bornu

Wensleydale (Yorkshire, England)
longwooled, meat
Mason (1969)

Wera (Bangladesh) – see Bangladesh

West African (W. Africa) – see Macina,
 West African Long-legged and
 West African Dwarf
syn. Guinea
Mason (1969)

West African (C. and S. America)
meat
American Hair type
tan, brown or red
hair, male with mane
weight 49 kg male, 34 kg female
small horizontal ears
polled
similar to Pelibuey
syn. Africana, Camura, Criollo
 (Colombia), Pelona, Red African,
 Rojo Africana
Mason (1980a), Bradford and Fitzhugh
 (1983), Pastrana et al. (1983)

West African Dwarf (S. of W. Africa)
white, usually with red or black spots,
 occasionally black
hair, male with mane
weight 30 kg male, 20 kg female
male horned, female polled or scurs
syn. Cameroons Dwarf, Djallonké,
 Forest (Ghana), Fouta Jallon, Kirdi,
 Kirdimi, Lakka (Chad), Nigerian
 Dwarf, Southern, West African Maned
Mason (1969), Adu and Ngere (1979),
 Fall et al. (1982), Bradford and
 Fitzhugh (1983)

West African Long-legged type (W. Africa)
hair
includes Fulani, Maure and Tuareg
syn. Guinea Long-legged, Sahel
Mason (1969)

West African Maned (W. Africa) – see
 West African Dwarf

West Indian (West Indies)
American Hair type
includes Bahama, Barbados Blackbelly,
 Pelibuey, Virgin Island, West African

White Arab (W. Africa) – see Tuabir

White Bororo (W. Africa) – see Bornu

White Dorper (Transvaal, South Africa)
meat
all white variety of Dorper
hair
usually polled
syn. Dorsian, Dorsie
Mason (1969)

White Fulani (W. Africa) – see Bornu,
 Yankasa

White Maure (W. Africa) – see Tuabir

White Persian (South Africa) – see
 Van Rooy

White Virgin Island (West Indies) – see
 Virgin Island

White Woolless (Brazil) – see Brazilian
 Woolless

Wiltiper (Zimbabwe)
meat
usually black, also white and brown
hair
horned or polled
origin: Wiltshire Horn × Blackhead
 Persian
Mason (1969)

Wooled Persian (South Africa)
meat, coarsewooled
usually brown
polled

fat-tailed
Mason (1969)

Woolless (Brazil) – *see* Brazilian Woolless

Yankasa (N.C. Nigeria)
variety of Fulani
white with black eyes and nose
weight 40 kg male, 30 kg female
syn. White Fulani
Mason (1969), Adu and Ngere (1979), Dettmers (1983)

Zaghawa (N. Darfur, Sudan)
black
hair
male horned, female polled
syn. Black Maure (Chad)
McLeroy (1961), Mason (1969)

Zaian (Khenifra, Morocco)
variety of Tadla with Berba blood
Mason (1969)

Zaïre Dwarf (Zaïre) – *see* Bahu

Zaïre Longlegged (N.E. Zaïre)
African Long-legged type
white or brown pied
hair
lop ears
male horned, female polled
short-tailed
syn. Congo Long-legged
Mason (1969)

Zembrane (Morocco)
variety of South Moroccan with more Berber blood than Rehamma-Srarhna
Mason (1969)

Zemmour (N.W. Morocco)
meat, coarsewooled
Atlantic Coast type
white with pale brown face
male horned, female polled
Mason (1969)

Zenu (Angola) – *see* Zunu

Zimbabwe (Zimbabwe)
African Long-fat-tailed type
often brown, black or pied
hair
weight 65 kg female
male often horned, female usually polled
includes Mashona
syn. Rhodesian
Ward (1959), Mason (1969)

Zoulay (Morocco)
variety of Beni Guil in Upper Moulouya Valley from crossing Tounsint with Berber
Mason (1969)

Zulu (S.E. Africa) – *see* Nguni

Zunu (Angola)
hair
long thin tail
syn. Goitred, Zenu
Mason (1969)

References

Abate, A., Kayongo-Male, H., Abate, A. N., Wachira, J. D. (1984) Chemical composition, digestibility, and intake of Kenyan feedstuffs by ruminants – a review, *Nutrition Abstracts and Reviews B* **54**: 1–13

Abbott, J. C. (1962) Marketing – its role in increasing productivity. *Freedom from Hunger Campaign, Basic Study No. 4.* Food and Agriculture Organization of the United Nations, Rome, Italy

Abbott, J. C., Makeham, J. P. (1979) *Agricultural Economics and Marketing in the Tropics.* Longman, London, UK

Abdel-Aziz, A. S., Iman, S. A., Ragab, M. T., Sharafeldin, M. A. (1978) Phenotypic and genetic parameters of the three body-weight traits and the first greasy fleece-weight in Fleisch Merino sheep in Egypt, *Egyptian Journal of Animal Production* **18**: 29–38

Abou-Dawood, A. E., Ghita, I. I., Taha, S. M. (1980) Major and minor components and trace elements of the Egyptian ewes' and goats' milk, *Egyptian Journal of Dairy Science* **8**; 109–15

Aboul-Naga, A. M. (1975) Reproductive performance of Merino, sub-tropical Egyptian sheep and their crosses, *Journal of Agricultural Science* (UK) **84**: 383–6

Aboul-Naga, A. M. (1976) Location effect on the reproductive performance of three indigenous breeds of sheep under the subtropical conditions of Egypt, *Indian Journal of Animal Science* **46**: 630–6

Aboul-Naga, A. M. (1977) Location effect on the lamb performance of three indigenous breeds of sheep under subtropical conditions of Egypt, *Indian Journal of Animal Sciences* **47**: 29–33

Aboul-Naga, A. M. (1978) Using Suffolk sheep for improving lamb production from sub-tropical Egyptian sheep. 1. Reproductive performance, *Journal of Agricultural Science* (UK) **90**: 125–30

Aboul-Naga, A. M., Afifi, E. A. (1977) Environment and genetic factors affecting wool production from sub-tropical coarse wool sheep, *Journal of Agricultural Science* (UK) **88**: 443–7

Aboul-Naga, A. M., El-Shobokshy, A. S., Afifi, E. A. (1980) Effect of length of suckling period on the performance of the local ewes bred three times per two years, *Egyptian Journal of Animal Production* **20**: 147–52

Aboul-Naga, A. M., El-Shobokshy, A. S., Marie, I. F., Moustafa, M.A. (1981a) Milk Production from subtropical non-dairy sheep. I. Ewe performance, *Journal of Agricultural Science* (UK) **97**: 297–301

Aboul-Naga, A. M., El-Shobokshy, A. S., Moustafa, M. A. (1981b) Milk production from subtropical non-dairy sheep. II. Method of measuring, *Journal of Agricultural Science* (UK) **97**: 303–8

Acharya, R. M. (1982) Sheep and goat breeds of India. *Animal Production and Health Paper 30.* Food and Agriculture Organization of the United Nations, Rome, Italy

Acharya, R. M., Kulkarni, V. G., Singh, M. (eds) (1980) *Canary Colouration of Indian Wool.* Central Sheep and Wool Research Institute, Avikanagar, Rajasthan, India

Acharya, R. M., Malhi, R. S., Arora, C. L. (1972) Studies on sampling procedure for assessing wool quality in Indian sheep: determination of body region for sampling – a preliminary note, *Indian Journal of Animal Sciences* **42**: 1043–4

Acharya, R. M., Malik, R. C., Gaur, D., Kishan, L., Sharma, M. M. (1977) Evaluation of the performance of new strains of sheep evolved at CSWRI Avikanagar and their further improvement through selection. In *Annual Report 1976,* pp. 7–11. Central sheep and Wool Research Institute, Avikanagar, Rajasthan, India

Acharya, R. M., Malik, R. C., Mehta, B. S., Gaur, D. (1976) Evaluation of the performance of new wool strains of sheep evolved at CSWRI Avikanagar. In *Annual Report 1975,* pp. 8–17. Central Sheep and Wool Research Institute, Avikanagar, Rajasthan, India

Acharya, R. M., Patnayak, B. C. (1972) Progress in sheep production research in India, *Indian Farming,* October 1972, 7pp.

Adalsteinsson, S. (1982) Inheritance of colours in sheep. In *Proceedings of the World Congress on Sheep and Beef Cattle Breeding, Volume I: Technical,* pp. 459–64. Dunsmore Press, Palmerston North, New Zealand

Adams, M. E. (1982) *Agricultural Extension in Developing Countries.* Longman, Harlow, UK

Adams, N. R. (1979) Clover disease in western Australia. In Tomes, G. J., Robertson, D. E., Lightfoot, R. J. (eds) *Sheep Breeding,* 2nd edn. pp. 373–8. Butterworths, London, UK

Adebowale, E. A. (1981) The maize replacement value of fermented cassava peels (*Manihot utilissma* Pohl) in rations for sheep, *Tropical Animal Production* **6**: 54–9

Adeleye, I. O. A. (1982) The effects of concentrate to grass hay ratio on the feedlot performance of West African Dwarf (WAD) rams, *Tropical Animal Production* **7**: 127–33

Adeleye, I. O. A., Oguntona, E. (1975) Effects of age and sex on liveweight and body composition of the West African Dwarf sheep, *Nigerian Journal of Animal Production* **2**: 264–9

Ademosun, A. A. (1970) Nutritive evaluation of Nigerian forages. 2. The effect of stage of maturity on the nutritive value of *Stylosanthes gracilis, Nigerian Agricultural Journal* **7**: 164–73

Adeyanju, S. A., Ogutuga, D. B. A., Ilori, J. O., Adegbola, A. A. (1975) Cocoa husk in maintenance rations for sheep and goats in the tropics, *Nutrition Reports International* **11**: 351–7

Adeyemo, R. (1983) Agricultural credit for small-scale farmers in Nigeria, *Tropical Agriculture* (Trinidad) **60**: 162–7

Adu, I. F., Ngere, L. O. (1979) The indigenous sheep of Nigeria, *World Review of Animal Production* **15**(3): 51–62

Agar, N. S., Evans, J. V., Roberts, J. (1972) Red blood cell potassium and haemoglobin polymorphism in sheep, A review, *Animal Breeding Abstracts* **40**: 407–36

Agar, N. S., Rawat, J. S., Roy, A. (1969) Blood potassium and haemoglobin polymorphism in Indian sheep, *Journal of Agricultural Science* (UK) **73**: 197–202

Agarwala, O. N., Nath, K, Mahadevan, V. (1971) Use of superphosphate as a phosphorus supplement for lambs – effect of calcination or supplementation with oral cobalt or parenteral vitamin B_{12}, *Journal of Agricultural Science* (UK) **77**: 467–71

Agarwala, O. N., Saha, G., Grover, K., Mahadevan, V. (1970) Nutritive value of gram (*Cicer arietinum*) straw for adult sheep, *Indian Veterinary Journal* **47**: 677–9

Aguilar Corona, J. O. (1980) Evaluation of the production and characters of wool from criollo ewes in Ajusco district, DF, *Veterinaria* (Mexico) **11**: 30

Ahmed, N., Tiwari, S. B., Sahni, K. L. (1975) Effect of castration on live weight growth and meat production in cross-bred lambs, *Indian Journal of Animal Sciences* **45**: 454–8

Ali, M. H., Ranjhan, S. K., Pathak, N. N. (1980) Note on the effect of supplement feeding on the performance of Muzaffarnagari lambs weaned at 60 and 80 days of age, *Indian Journal of Animal Sciences* **50**: 95–7

Allden, W. G. (1968) Undernutrition of the Merino sheep and its sequelae. I. The growth and development of lambs following prolonged periods of nutritional stress, *Australian Journal of Agricultural Research* **19**: 621–38

Allden, W. G. (1970) The effects of nutritional deprivation on the subsequent productivity of sheep and cattle, *Nutrition Abstracts and Reviews* **40**: 1167–84

Allden, W. G., Whittaker, I. A. M. (1970) The determinants of herbage intake by grazing sheep: the interrelationship of factors influencing herbage intake and availability, *Australian Journal of Agricultural Research* **21**: 755–66

Allison, A. J. (1975) Flock mating in sheep. 1. The effect of number of ewes joined per ram on mating behaviour and fertility, *New Zealand Journal of Agricultural Research* **18**: 1–18

Alliston, J. C., Hinks, C. E. (1981) A note on the use of the 'Danscanner' for prediction of the composition of a sample joint from beef cattle, *Animal Production* **32**: 345–7

Al-Mahmood, F. T., Younis, A. A., Farhan, S. M. A. (1976) Optimum age for fattening Awassi lambs, *Iraqi Journal of Agricultural Science* **11**: 66–75

Al-Mallah, M. Y., Mohammed, A. S., Darwash, K. N. (1979) Effects of breed and castration on performance of Iraqi lambs, *Mesopotamia Journal of Agriculture* **14**: 91–100

Almeyra, G. M. (1979) How real was Algeria's agrarian revolution? *Ceres* **12**(2): 31–5

Altaif, K. I., Dargie, J. D. (1978) Genetic resistance to helminths. The influence of breed and haemoglobin type on the response of sheep to primary infections with *Haemonchus contortus, Parasitology* **77**: 161–75

Al-Tawash, M. Y., Alwash, A. H. (1983) The effect of the level of raw and alkali-treated sugar cane bagasse on the digestibility of the rations and performance of Awassi lambs, *World Review of Animal Production* **19** (2): 25–9

Alwash, A. H., DePeters, E. J. (1982) The use of date stones for feeding and fattening ruminant animals, *World Review of Animal Production* **18**(3): 29–32

Alwash, A. H., Jumah, A. N., Hassan, S. A. (1983) Relative value of single cell proteins (SCP) and soyabean meal as a protein supplement in the ration of Awassi lambs, *World Review of Animal Production* **19**(3): 67–70

Amegee, Y. (1981) The Vogan sheep in Togo. II. Lamb carcass quality. Paper presented at the 12th Animal Science Symposium of the Ghana Animal Science Association, 26–28 August 1981. University of Ghana, Legon, Ghana.

Amegee, Y. (1983a) Le mouton de Vogan (croisé Djallonké × Sahélien) au Togo [Study on the Vogan sheep (Djallonké × Sahelian crossbred) in Togo], *Revue d'Élevage et de Médecine Vétérinaire des Pays Tropicaux* **36**: 79–84

Amegee, Y. (1983b) La prolificité du mouton Djallonké en milieu villageois au Togo [Study of the prolificacy of Djallonké sheep in a village environment in Togo], *Revue d'Élevage et de Médecine Vétérinaire des Pays Tropicaux* **36**: 85–90

Ames, G. C. W. (1983) Planning agricultural research and extension in the Sahel, *Livestock International* **11**(3): 60–3

Ammar-Khodja, F., Brudieux, R. (1982) Seasonal variations in the cyclic luteal ovarian activity in the Tadmit ewe in Algeria, *Journal of Reproduction and Fertility* **65**: 305–11

Anderson, J. (1972) The oestrus cycle in Merino, Masai and Somali sheep, *Zootecnica e Veterinaria* **27**: 77–94

Anderson, N. (1982) Internal parasites of sheep and goats. In Coop, I. E. (ed.) *Sheep and Goat Production,* pp. 175–91. Elsevier, Amsterdam, Netherlands

Anteneh, A. (1982) Production objectives and market forces. In Gatenby, R. M., Trail, J. C. M. (eds) *Small Ruminant Breed Productivity in Africa,* pp. 7–12. International Livestock Centre for Africa, Addis Ababa, Ethiopia

Anwar, C. M., Ahmad, S. L., Schneider, B. H. (1965) Experiments on fattening sheep. 6. The effect of substitution of guar meal for cottonseed cake and the implantation of diethylstilbestrol, *Agriculture* (Pakistan) **16**: 501–7

AOAC (1975) *Official Methods of Analysis of the Association of Official Analytical Chemists,* 12th edn. Association of Official Analytical Chemists, Washington, DC, USA

ARC (1980) *The Nutrient Requirements of Ruminant Livestock,* Commonwealth Agricultural Bureaux, Farnham Royal, Slough, UK

Arendt, J., Symons, A. M., Laud, C. (1981) Pineal function and photoperiod in the ewe. In *Photoperiodism and Reproduction in Vertebrates, International Colloquium, Nouzilly,*

France, September 1981, pp. 219–29. Institut National de la Recherche Agronomique, Versailles, France

Armstrong, D. T., Evans, G. (1983) Factors influencing success of embryo transfer in sheep and goats, *Theriogenology* **19**: 31–42

Arnold, G. W. (1981) Grazing behaviour. In Morley, F. H. W. (ed) *Grazing Animals*, pp. 79–104. Elsevier, Amsterdam, Netherlands

Arnold, G. W., Wallace, S. R., Rea, W. A. (1981) Associations between individuals and home-range behaviour in natural flocks of three breeds of domestic sheep, *Applied Animal Ethology* **7**: 239–57

Arora, A. L., Pant, K. P. (1979) Quantitative analysis of sperm dimensions in relation to fertility of Soviet Merino rams, *Indian Veterinary Journal* **56**: 478–81

Arora, C. L., Acharya, R. M. (1972) A note on haemoglobin and potassium types in Nali breed of Indian sheep and their relationship with body weights and wool yield, *Animal Production* **15**: 95–7

Arora, C. L., Acharya, R. M., Badashiya, B. S., Dass, N. C. (1975a) Characterization of Malpura breed in Rajasthan and future prospects for its improvement, *Indian Journal of Animal Sciences* **45**: 843–8

Arora, C. L., Acharya, R. M., Bhadashiya, B. S., Dass, N. C. (1975b) Characterization of Chokla breed in Rajasthan and future prospects of its improvement, *Indian Journal of Animal Sciences* **45**: 345–50

Arora, C. L., Acharya, R. M., Sangolli, B. V., Nagaraja, M. S. (1977) Characterization of Sonadi breed in Rajasthan and future prospects for its improvement, *Indian Journal of Animal Sciences* **47**; 120–5

Arora, C. L., Acharya, R. M., Singh, R., Buch, S. D. (1975c) Characterization of Nali breed in Rajasthan and future prospects for its improvement, *Indian Journal of Animal Sciences* **45**: 849–55

Arora, R. K., Patni, P. C., Gupta, N. P., Pokharna, A. K. (1978) Physical and mechanical characteristics of wools of some cross-bred sheep, *Indian Journal of Textile Research* **3**: 116–18

Arrowsmith, S. P., Steenkamp, J. D. G., Roux, P. le (1974) The influence of breed and plane of nutrition on the chronology of teeth eruption in sheep, *South African Journal of Animal Science* **4**: 127–30

Arthur, G. H., Noakes, D. E., Pearson, H. (1982) *Veterinary Reproduction and Obstetrics*, 5th edn. Baillière Tindall, London, UK

Asamoah-Amoah, F. (1977) Effect of flushing on the lambing percentage of local type forest ewes. In *Animal Research Institute Annual Report 1975/76*, pp. 41–3. Council for Scientific and Industrial Research, Achimota, Ghana

Asamov, S. A., Stepanov, B. M. (1979) Biologicheskie osobennosti razvitiya mnogoplodnykh yagnyat [Biological aspects of development of lambs from multiple births], *Ovtsevodstvo* **4**: 35–7

Ashmawy, G. M., Al-Azawi, W. A. (1982a) A comparative study of fleece characteristics in Iraqi sheep. I. Greasy fleece weight, shrinkage percentage and fibre type ratio, *Egyptian Journal of Animal Production* **22**: 53–61

Ashmawy, G. M., Al-Azawi, W. A. (1982b) A comparative study of fleece characteristics in Iraqi sheep. II. Fibre diameter, medullated fibre percentage and fibre length, *Egyptian Journal of Animal Production* **22**: 63–71

Ashton, T. R. (1977) Ultra-high-temperature treatment of milk and milk products, *World Animal Review* **23**: 37–42

Asiedu, F. H. K. (1978) Grazing behaviour of sheep under tree crops in Ghana, *Tropical Agriculture* (Trinidad) **55**: 181–8

Asiedu, F. H. K., Appiah, M. (1983) Note on the effects of body weight and lambing interval on the reproductive performance of West African sheep raised under tree crops, *Tropical Agriculture* (Trinidad) **60**: 61–3

Asiedu, F. H. K., Oppong, E. N. W., Opoku, A. A. (1978) Utilisation by sheep of herbage under tree crops in Ghana, *Tropical Animal Health and Production* **10**: 1–10

Asker, A. A., El-Khalisi, I. J. (1965) Some observations on commercial flocks of sheep in Iraq, *Annals of Agricultural Science* (Cairo) **10**(2): 17-28
Assoku, R. K. G. (1981) Studies of parasitic helminths of sheep and goats in Ghana, *Bulletin of Animal Health and Production in Africa* **29**: 1-10
Aswad, A., Abdou, M. S. S., Al-Bayaty, F., El-Sawaf, S. A. (1976) The validity of the 'Ultra-sonic Method' for pregnancy diagnosis in ewes and goats, *Zentralblatt für Veterinärmedizin* **A23**: 467-74
Aswad, A., Rebesko, B., Fathalla, R., Fahri, R. (1974) Sinhronizacija porođaja [Parturition synchronisation], *Veterinarski Glasnik* **28**: 609-14
Aten, A., Innes, R. F., Knew, E. (1955) *Flaying and Curing of Hides and Skins as a Rural Industry*. Food and Agriculture Organization of the United Nations, Rome, Italy
Atkins, K. D., McGuirk, B. J., Thornberry, K. J. (1980) Genetic improvement of resistance to body strike, *Proceedings of the Australian Society for Animal Production* **13**: 90-2
Atmadilaga, D. (1958) Study on the milk yield of Indonesian sheep with special reference to the Priangan breed (preliminary communication), *Hemera Zoa* **65**: 3-14
Australian Bureau of Animal Health (1981) *Sea Transport of Sheep*. Australian Government Publishing Service, Canberra, Australia
Ayalon, N., Shemesh, M. (1979) Use of milk progesterone measurement for determining pregnancy and ovarian activity in cattle and sheep, *Refuah Veterinarith* **36**: 79
Bachhil, V. N. (1980) Observations on carcass yield and their inter-relationship with live weights of some meat producing animals, *Indian Veterinary Medical Journal* **4**: 119-21
Badreldin, A. L. (1951) Growth and carcase percentage in Ossimi and Rahmani sheep, *Bulletin No. 3*. Department of Animal Breeding, Fouad I University, Cairo, Egypt
Baker, P. R. (1976) The social importance of cattle in Africa and the influence of social attitudes on beef production. In Smith, A. J. (ed.) *Beef Cattle Production in Developing Countries*, pp. 360-74. University of Edinburgh Centre for Tropical Veterinary Medicine, Edinburgh, UK
Barlow, R. M. (1982) Infectious diseases of sheep and goats. In Coop, I. E. (ed.) *Sheep and Goat Production*, pp. 151-74. Elsevier, Amsterdam, Netherlands
Baron, J. (1955) Étude des procédés utilisés par les Maures pour empêcher les jeunes animaux de téter [Methods used by the Moors to prevent young animals from suckling], *Revue d'Élevage et de Médecine Vétérinaire des Pays Tropicaux* **8**: 15-23
Basuthakur, A. K., Nivsarkar, A. E., Singh, R. N. (1980) Studies on some pre and post slaughter parameters in Malpura lambs. 1. Phenotypic biometry and dressing percentage, *Indian Veterinary Journal* **57**: 473-8
Bautista, O. R., Arango, T., Fajardo, R. (1976) Efecto de raza, castración e implante con lactona del ácido resorcílico (RAL) sobre el engorde de ovinos [Effect of breed, castration and resorcyclic acid lactone implants on fattening performance in sheep], *Memoria, Asociación Latinoamericana de Producción Animal* **11**: 30
Bautista, O. R., Gonzalez, S. A. (1976) Efecto de la raza, castración e implante con lactona del ácido resorcílico (RAL) sobre los rendimientos de carne y subproductos en ovinos [Effect of breed, castration and an implant of resorcyclic acid lactone (RAL) on the yield of meat and other products in sheep], *Memoria, Asociación Latinoamericana de Producción Animal* **11**: 76
Bawa, H. S., Bhote, R. A. (1966) *Livestock Products Production and Utilisation*. Indian Council of Agricultural Research, New Delhi, India
Beasley, P. (1967) Photosensitization in sheep, *Queensland Agricultural Journal* **93**: 350-2
Bellaver, C. (1980) As peles [Skins]. *Circular Tecnica No. 3*. Empresa Brasileira de Pesquisa Agropecuária, Bagé, Brazil
Bellaver, C., Arruda, F. de A. V., Simplicio, A. A. (1980a) Avaliação da produção leiteira das raças deslanadas Morada Nova e Santa Inês [Evaluation of milk yield in Morada Nova and Santa Inês ewes]. In *Pesquisa em ovinos no Brasil 1975-1979*, p. 110. Empresa Brasileira de Pesquisa Agropecuária, Bagé, Brazil
Bellaver, C., Oliveira, E. R. de, Portella, J. da S., Figueiredo, E. A. P. de (1980b) Avaliação de carcaças em ovelhas da raca Santa Inês [Evaluation of carcasses of Santa Inês ewes]. In *Pesquisa em ovinos no Brasil 1975-1979*, p. 40. Empresa Brasileira de Pesquisa Agropecuária, Bagé, Brazil

Belschner, H. G. (1959) *Sheep Management and Diseases*, 6th edn. Angus and Robertson, Sydney, Australia
Benjamin, Y., Kali, J., Barkai, D., Benjamin, R. W., Eyal, E. (1982) [Feeding cattle manure for maintenance of mature ewes], *Hassadeh* **63**: 565-9
Benor, D., Harrison, J. Q. (1977) *Agricultural Extension: the Training and Visit System.* World Bank, Washington, DC, USA
Berg, R. T., Butterfield, R. M. (1976) *New Concepts of Cattle Growth.* Sydney University Press, Sydney, Australia
Berg, R. T., Walters, L. E. (1983) The meat animal: changes and challenges, *Journal of Animal Science* **57** (suppl. 2): 133-46
Berger, Y., Ginisty, L. (1980) Bilan de 4 années d'étude de la race ovine Djallonké en Côte d'Ivoire [Results of a 4-year study of the Djallonké breed of sheep in the Ivory Coast], *Revue d'Élevage et de Médecine Vétérinaire des Pays Tropicaux* **33**: 71-8
Bessa, M. N., Teixeira, F. J. L., Freitas, J. P. de, Souza, A. A. de, Albuquerque, J. J. de L. (1980) Peso aos 180 e 240 dias de ovinos deslanados Morada Nova–variedade branca [Body weight at 180 and 240 days of white Morada Nova sheep]. In *Pesquisa em ovinos no Brasil 1975-1979,* p. 26. Empresa Brasileira de Pesquisa Agropecuária, Bagé, Brazil
Bhadula, S. K., Bhat, P. N. (1980) Note on growth curves in sheep, *Indian Journal of Animal Sciences* **50**: 1001-3
Bhaik, N. L., Kohli, I. S. (1980) Reproductive performance of Chokla and Magra ewes in the subtropical area of North-western arid zone of Rajasthan. 2. Parturition and involution of uterus, *Indian Veterinary Journal* **57**: 327-33
Bhargava, B., Ranjhan, S. K. (1974) Effect of feeding treated groundnut-cake with urea-molasses-impregnated wheat straw on growth rate and utilization of nutrients in Nali lambs, *Indian Journal of Animal Sciences* **44**: 464-73
Bhargava, B., Ranjhan, S. K. (1976) Comparative nutritional value and the effect of feeding of oat hay and wheat straw on growth, wool yield and utilization of nutrients in Nali lambs, *Indian Journal of Animal Sciences* **46**: 139-43
Bhat, P. N., Asker, A. A., Badwey, E. F., El-Maa'ly, A., Abid, M. A. (1978) Effect of early weaning on body weight of Awassi lambs, *Indian Journal of Animal Sciences* **48**: 98-102
Bhat, P. N., Bhat, P. P., Khan, B. U., Goswami, O. B., Singh, B. (1980) Animal genetic resources in India. In *Animal Genetic Resources in Asia and Oceania,* pp. 119-88. Tropical Agriculture Research Center, Ministry of Agriculture, Forestry and Fisheries, Tsukuba, Ibaraki, Japan
Bhati, G. N., Mruthyunjaya (1983) Economics of sheep farming on different pastures in arid land of western Rajasthan, *Indian Journal of Animal Sciences* **53**: 732-7
Bhatia, D. R., Mohan, M., Patnayak, B. C., Ratan, R. (1981) Note on growth of Malpura lambs with or without supplemental concentrates, *Indian Journal of Animal Sciences* **51**: 238-42
Bhatia, D. R., Mohan, M., Patnayak, B. C., Sheikh, Q. D. (1973) Nutritive value of the available pasture in semi-arid areas of Rajasthan, *Indian Journal of Animal Sciences* **43**: 838-43
Bhatnagar, V. S., Seth, O. N., Pandey, M. D. (1974) Study of wool follicles in Bikaneri (Magra) sheep, *Agra University Journal of Research, Science* **23**: 71-8
Bhattacharya, A. N., Harb, M. (1973) Sheep production on natural pastures by roaming Bedouins in Lebanon, *Journal of Range Management* **26**: 266-9
Bhuvanakumar, C. K., Subramanian, P. (1981) Effect of body weight on greasy fleece weight in Merino, Nilagiri and their crosses, *Cheiron* **10**: 40-2
Bickoff, E. M. (1968) Oestrogenic constituents of forage plants. *Review Series Commonwealth Bureau of Pastures and Field Crops No. 1.* Commonwealth Agricultural Bureaux, Farnham Royal, Slough, UK
Bidarkar, D. K. (1982) Dressing percentage of Mandya and Mandya cross breed lambs, *Livestock Adviser,* **7**(6): 55-6
Bidarkar, D. K., Mudaliar, A. S. R. (1981) The effect of feeding different levels of antibiotic on the digestibility of nutrients in lambs, *Indian Journal of Animal Research* **15**: 124-6

Bindon, B. M., Piper, L. R. (1979) Assessment of new and traditional techniques of selection for reproduction rate. In Tomes G. J., Robertson, D. E., Lightfoot, R. J. (eds) *Sheep Breeding,* 2nd edn. pp. 387–401. Butterworths, London, UK

Bircham, J. S., Crouchley, G. (1976) Free water intake of ewe lambs offered water from different sources, *New Zealand Journal of Experimental Agriculture* **4**: 41–4

Bishop, A. H. (1979) Recent developments in farm fence construction, *World Animal Review* **29**: 29–37

Blood, D. C., Henderson, J. A., Radostits, O. M. (1979) *Veterinary Medicine,* 5th edn. Baillière Tindall, London, UK

Blunn, C. T. (1943) Characteristics and production of old-type Navajo sheep, *Journal of Heredity* **34**: 141–52

Boateng, A. K. (1982) Vocational education for future farmers in developing countries: possible contribution of school agriculture, *Agricultural Progress* **57**: 80–3

Bodisco, V., Duque, C. M., Valle, S. A. (1973) Comportamiento productivo de ovinos tropicales en el período 1968–1972 [Performance of tropical sheep in 1968–1972], *Agronomía Tropical* (Venezuela) **23**: 517–40

Bogdan, A. V. (1977) *Tropical Pasture and Fodder Plants.* Longman, London, UK

Bohra, H. C., Ghosh, P. K. (1983) Nitrogen metabolism in water-restricted Marwari sheep of the Indian desert, *Journal of Agricultural Science* (UK) **101**: 735–9

Bohra, S. D. J., Nagarcenkar, R., Sharma, K. N. S. (1979) Factors affecting pre-weaning body weights in Malpura sheep, *Indian Veterinary Journal* **56**: 125–8

Bores, R., Romano, J. N., Castellanos, A. (1983) The use of henequen (sisal) pulp in maintenance rations for Pelibuey sheep, *Tropical Animal Production* **8**: 73–4

Bosc, M. J. (1972) The induction and synchronisation of lambing with the aid of dexamethasone, *Journal of Reproduction and Fertility* **28**: 347–57

Bose, A. B., Malhotra, S. P., Bharara, L. P. (1964) A socio-economic study of households raising sheep in central and lower Luni Basin, *Annals of Arid Zone* **3**: 44–53

Boshoff, D. A., Gouws, D. J., Burger, F. J. L. (1977) Studies on the sex physiology of breeds of sheep: use of teaser rams and flushing to advance the breeding season of Karakul ewes, 1972–1974. Final report. In *Agricultural Research 1975,* p. 58. Department of Agricultural Technical Services, Pretoria, South Africa

Botswana (1975) National policy on tribal grazing land. *Government Paper No. 2 of 1975.* Republic of Botswana, Gaberone, Botswana

Botswana, Animal Production Research Unit (1980) *Livestock and Range Research in Botswana, Annual Report 1979.* Ministry of Agriculture, Gaberone, Botswana

Botswana, Animal Production Research Unit (1981) Management with fire. In *Beef Production and Range Management in Botswana,* pp. 87–8. Ministry of Agriculture, Gaberone, Botswana

Bourke, M. E. (1967) A study of mating behaviour of Merino rams, *Australian Journal of Experimental Agriculture and Animal Husbandry* **7**: 203–5

Bradfield, D. J. (1966) *Guide to Extension Training.* Food and Agriculture Organization of the United Nations, Rome, Italy

Bradford, G. E., Fitzhugh, H. A. (1983) Hair sheep: a general description. In Fitzhugh, H. A., Bradford, G. E. (eds) *Hair Sheep of Western Africa and the Americas, a Genetic Resource for the Tropics,* pp. 3–22. Westview Press, Boulder, Colorado, USA

Brinckman, W. L. (1981) Concentrate feeding of sheep in Nigeria. In Smith, A. J., Gunn, R. G. (eds) Intensive animal production in developing countries, pp. 421–8. *Occasional Publication No. 4.* British Society of Animal Production, Thames Ditton, UK

Britt, J. H. (1979) Prospects for controlling reproductive processes in cattle, sheep and swine from recent findings in reproduction, *Journal of Dairy Science* **62**: 651–65

Brown, D. E., Cameron, C. W., Ngere, L. O. (1972) Synchronisation of oestrus and early post-partum rebreeding of Nungua Blackhead sheep in Ghana, *Ghana Journal of Agricultural Science* **5**: 183–7

Brown, W., Farnworth, J., Clarke, S. H. (1976) A feasibility study of three farming types in Saudi Arabia. 3. 1,000 ewe sheep farm. *Publication No. 73.* Joint Agricultural Research and Development Project, University College of North Wales, UK, and Ministry of Agriculture and Water, Saudi Arabia

Bunch, T. D., Foote, W. C., Spillett, J. J. (1976) Sheep–goat hybrid Karyotypes, *Theriogenology* **6**: 379–85
Burns, M. (1955) Observations on Merino × Herdwick hybrid sheep with special reference to the fleece, *Journal of Agricultural Science* (UK) **46**: 389–406
Burns, M. (1966) Merino birthcoat fibre types and their follicular origin, *Journal of Agricultural Science* (UK) **66**: 155–73
Burns, M. (1967a) The Katsina wool project. I. The coat and skin histology of some Northern Nigerian hair sheep and their Merino crosses, *Tropical Agriculture* (Trinidad) **44**: 173–92
Burns, M. (1967b) The Katsina wool project. II. Coat and skin data from 3/4 Merino and Wensleydale crosses, *Tropical Agriculture* (Trinidad) **44**: 253–74
Burns, M. (1968) Some observations on the adaptations of livestock to tropical environments, *Ghana Journal of Science* **9**: 41–9
Burns, M. (1972) Effects of ova transfer on the birthcoats of lambs, *Journal of Agricultural Science* (UK) **78**: 1–6
Burns, M. (1981) Intensive mutton and wool production in the tropics. In Smith, A. J., Gunn, R. G. (eds) Intensive Animal Production in Developing Countries, pp. 407–10. *Occasional Publication No. 4*. British Society of Animal Production, Thames Ditton, UK
Burns, M., Clarkson, H. (1949) Some observations on the dimensions of follicles and other structures in the skin of sheep, *Journal of Agricultural Science* (UK) **39**: 315–34
Burns, M., Ryder, M. L. (1974) Effect of egg transfer on the skin follicles and birthcoats of Finnish Landrace and Soay lambs, *Journal of Agricultural Science* (UK) **82**: 209–16
Burns, R. H., Chaudhary, B. N. (1965) A tentative standard for carpet wools based on dimensional measurements, Mimeograph
Bush, E. C. (1965) A comparison of the milking capacity of Corriedale and Corriedale × Border Leicester ewes, *East African Agricultural and Forestry Journal* **31**: 31–4
Butterworth, M. H. (1966) A note on the maintenance requirement of adult sheep under tropical conditions, *Animal Production* **8**: 155–6
Butterworth, M. H. (1967) The digestibility of tropical grasses, *Nutrition Abstracts and Reviews* **37**: 349–68
Butterworth, M. H., Blore, T. W. D. (1969) The lactation of Persian Blackhead ewes and the growth of lambs. The effect of three different nutritional regimes during gestation on subsequent growth, *Journal of Agricultural Science* (UK) **73**: 133–7
Butterworth, M. H., Houghton, T. R., Macartney, J. C., Prior, A. J., Middlemiss, C. P., Edmond, D. E. (1968) Some observations on the lactation of Blackhead ewes and the growth of lambs: the composition and yield of milk, *Journal of Agricultural Science* (UK) **70**: 203–7
Buvanendran, V. (1978) Sheep in Sri Lanka, *World Animal Review* **27**: 13–16
Buvanendran, V., Adu, I. F., Oyejola, B. A. (1981) Breed and environmental effects on lamb production in Nigeria, *Journal of Agricultural Science* (UK) **96**: 6–15
Bwangamoi, O., DeMartini, J. (1970) A survey of skin diseases of domestic animals and defects which down-grade hides and skins in East Africa. III. Sheep, *Bulletin of Epizootic Diseases in Africa* **18**: 243–6
Calder, M. W. (1982) Livestock and meat marketing and grading. In Coop, I. E. (ed.) *Sheep and Goat Production* pp. 243–58. Elsevier, Amsterdam, Netherlands
Campbell, C. (1980) USFGG sheep program in Africa, Mideast could up US feedgrain exports, *Foreign Agriculture* **18** (8): 9–11
Campbell, R. S. F. (1976) Veterinary training for the tropics with particular reference to Australian experience, *World Animal Review* **17**: 16–21
Carew, B. A. R., Mosi, A. K., Mba, A. U., Egbunike, G. N. (1980) The potential of browse plants in the nutrition of small ruminants in the humid forest and derived savanna zones of Nigeria. In Houérou, H. N. le (ed.) *Browse in Africa, the Current State of Knowledge*, pp. 307–11. International Livestock Centre for Africa, Addis Ababa, Ethiopia
Carpenter, R. H., Spitzer, J. C. (1981) Response of anoestrous ewes to norgestomet and PMSG, *Theriogenology* **15**: 389–93

Carrie, M. S., Woodroffe, F. W. (1960) *Felmongers Handbook.* Department of Scientific and Industrial Research, New Zealand
Carstens, N. W. (1970) An economic analysis of mutton sheep and Karakul farming in the Northern Cape, 1968/69. *Economic Series No. 70.* Department of Agricultural Economics and Marketing, Pretoria, South Africa
Carter, H. B. (1943) The use of the tanned sheepskin in the study of follicle population density. In *Bulletin No. 164,* pp. 23–8. Council of Scientific and Industrial Research, Australia
Castañon Canet, J. A. T. (1978) Análisis cuantitativo y cualitativo de la producción de lana en la raza Romney Marsh en el hato de Ajuchitlan, Querétaro [Quantitative and qualitative analysis of wool yield in the Romney Marsh breed in a flock at Ajuchitlan, Querétaro], *Veterinaria* (Mexico) **9**: 125
Castellanos Ruelas, A., Zarazúa, M. V. (1982) Quantitative and qualitative study of milk production of the Pelibuey sheep, *Tropical Animal Production* **7**: 232–40
Castillo Rojas, H., Hernández Ledezma, J. J., Berruecos, J. M., López Angeles, J. J. (1977) Comportamiento reproductivo del borrego Tabasco mantenido en clima tropical. III. Pubertad y duración del estro [Reproductive performance of Tabasco lambs maintained in a tropical climate. III. Puberty and duration of oestrus], *Técnica Pecuaria en México* **32**: 32–5
Castillo Rojas, H., Valencia, M. Z., Berruecos, D. J. M. (1972) Comportamiento reproductivo del borrego 'Tabasco' mantenido en clima tropical y subtropical. I. Indices de fertilidad [Reproductive performance of 'Tabasco' sheep maintained in a tropical and subtropical climate. I. Fertility indices], *Técnica Pecuaria en México* **20**: 52–6
Catchpole, V. R., Henzell, E. F. (1971) Silage and silage-making from tropical herbage species, *Herbage Abstracts* **41**: 213–21
Ceres (1980) New credit systems for smallholders begin to show promise, *Ceres* **13**(2): 3–4
Cerrón, J. G., Pumayalla, A. D. (1974) Medidas corporales y producción en el ovino tipo Junin [Body measurements and production of sheep of the Junin type], *Memoria, Asociación Latinoamericana de Producción Animal* **9**: 57–8
Chadhokar, P. A. (1982) *Gliricidia maculata,* a promising legume fodder plant, *World Animal Review* **44**: 36–43
Chakravarty, A. K., Subbayyan, R. (1969) Grazing studies in arid and semi-arid zones of Rajasthan. Role of reserve pasture for summer grazing, *Indian Journal of Animal Science* **39**: 519–25
Chambers, R. (1980) The small farmer is a professional, *Ceres* **13**(2): 19–23
Chatterjee, R., Arora, R. K. (1974) Scope and practical importance for eliminating highly medullated fibers in Indian fleeces, *Textile Research Journal* **44**: 325–6
Chawdhary, S. S. (1978) Evaluation of carcass traits in Nali, Lohi, Magra and Hissardale × Magra breeds of sheep, *Thesis Abstracts, Haryana Agricultural University* **4**: 185–6
Chenost, M., Geoffroy, F. (1975) Intérêt du son dans l'alimentation des ruminants aux Antilles françaises. Premiers résultats et perspectives d'utilisation [Importance of bran in the feeding of ruminants in the French Antilles. First results and prospects for its use], *Nouvelles Agronomiques des Antilles et de la Guyane* **1**: 37–45
Chicco, C. F., Carnevali, A. A., Shultz, T. A., Shultz, E., Ammerman, C. B. (1971) Yuca y melaza en la utilización de la urea en corderos [Cassava and molasses in the utilisation of urea by lambs], *Memoria, Asociación Latinoamericana de Producción Animal* **6**: 7–17
Chicco, C. F., Shultz, E., Shultz, T. A. (1972) Algunas observaciones sobre niveles de melaza en suplementos con úrea y biuret para ovinos [Some observations on level of molasses in supplements with urea and biuret for sheep], *Agronomía Tropical* (Venezuela) **22**: 271–9
Choque, H., Cardozo, A. (1974) Pesos vivos y caracteres reproductivos en ovinos de altura [Liveweights and reproductive characters in sheep at high altitudes], *Memoria, Asociación Latinoamericana de Producción Animal* **9**: 53–4
Chorey, P. A., Naithani, S. P., Roy, A. (1965) The effect of antibiotic feeding on growth of lambs and kids, *Indian Veterinary Journal* **42**: 433–8

Clark, J. A. (ed.) (1981) *Environmental Aspects of Housing for Animal Production.* Butterworths, London, UK
Cloete, C. W. P., Vosloo, L. P., Smith, W. A. (1975) Vleisproduksie uit die wolkudde [Meat production from the wool flock], *South African Journal of Animal Science* **5**: 209–11
Cloete, S. W. P., Villiers, T. T. de, Kritzinger, N. M. (1983) The effect of temperature on the ammoniation of wheat straw by urea, *South African Journal of Animal Science* **13**: 202–3
Coelho, J. F. S., Galbraith, H., Topps, J. H. (1978) The effect of a combination of trienbolone acetate and 17β-oestradiol on the performance, carcass composition and blood characteristics of castrated male lambs, *Animal Production* **26**: 360
Coetzee, C. G. (1973) Intensive mutton production from wool-bearing sheep: determination of the optimum slaughter mass of early-weaned sheep. In *Agricultural Research 1972*, pp. 59–61. Department of Agricultural Technical Services, Pretoria, South Africa
Cohen, R. D. H. (1980) Phosphorus in rangeland ruminant nutrition: a review, *Livestock Production Science* **7**: 25–37
Colas, G., Courot, M. (1979) Storage of ram semen. In Tomes, G. L., Robertson, D. E., Lightfoot, R. J. (eds) *Sheep Breeding*, 2nd edn. pp. 521–32. Butterworths, London, UK
Combellas, J., Centeno, A., Mazzani, B., Combellas, J. (1972a) Aprovechamiento de la parte aérea del maní. 2. Henificación, consumo y digestibilidad *in vivo* [Use of the aerial parts of groundnut. 2. Making into hay, intake and digestibility *in vivo*], *Agronomía Tropical* (Venezuela) **22**: 281–5
Combellas, J., González, J. E. (1972) Rendimiento y valor nutritivo de forrajes tropicales. 2. *Cenchrus ciliaris* L. cv. Biloela [Yield and nutritive value of tropical forages. 2. *Cenchrus ciliaris* L. cv. Biloela], *Agronomía Tropical* (Venezuela) **22**: 623–34
Combellas, J., González, J. E. (1973) Rendimiento y valor nutritivo de forrajes tropicales. 5. Pasto para (*Brachiaria mutica* Stapf) [Yield and nutritive value of tropical forages. 5. Para grass (*Brachiaria mutica* Stapf)], *Agronomía Tropical* (Venezuela) **23**: 277–86
Combellas, J., González, J. E., Trujillo, A. (1972b) Rendimiento y valor nutritivo de forrajes tropicales. 1. Bermuda cv. coastal (*Cynodon dactylon* (L) Pers.) [Yield and nutritive value of tropical forages. 1. Coastal Bermuda grass (*Cynodon dactylon* (L) pers.)], *Agronomía Tropical* (Venezuela) **22**: 231–8
Combellas, J. de (1978) Comparación de los pesos al nacer, pesos al destete y ganancias en peso de corderos West African y corderos mestizos de Dorset Horn [A comparison of body weight at birth and weaning and gains of West African and crossbred Dorset Horn lambs]. In *Informe Anual, Instituto de Producción Animal*, pp. 71–2. Universidad Central de Venezuela, Maracay, Venezuela
Combellas, J. de (1980) Determinación de la producción de leche en ovejas West African × Dorset Horn en su primera lactancia [Determination of milk yield of West African × Dorset Horn ewes in their first lactation]. In *Informe Anual 1979, Instituto de Producción Animal*, pp. 69–70. Universidad Central de Venezuela, Maracay, Venezuela
Combellas, J. de, Rondon, Z., Guillén, J. (1980) Sincronización del estro en ovejas [Synchronisation of oestrus in ewes]. In *Informe Anual 1979, Instituto de Producción Animal*, p. 71. Universidad Central de Venezuela, Maracay, Venezuela
Cooke, G. W. (1967) *The Control of Soil Fertility.* Crosby Lockwood, London, UK
Coop, I. E. (1966) Effect of flushing on reproductive performance of ewes, *Journal of Agricultural Science* (UK) **67**: 305–23
Coop, I. E., Clarke, V. R., Claro, D. (1972) Nutrition of the ewe in early lactation. 1. Lamb growth rate, *New Zealand Journal of Agricultural Research* **15**: 203–8
Coop, I. E., Devendra, C. (1982) Systems, biological and economic efficiencies. In Coop, I. E. (ed.) *Sheep and Goat Production*, pp. 297–307. Elsevier, Amsterdam, Netherlands
Courot, M., Ortavant, R. (1981) Endocrine control of spermatogenesis in the ram, *Journal of Reproduction and Fertility*, Suppl **30**: 47–60
Cowlishaw, S. J., Unsworth, E. F. (1976) Factors affecting the *in vitro* digestibility of tropical grasses, *Turrialba* **26**: 44–53
Crowder, L. V., Chheda, H. R. (1982) *Tropical Grassland Husbandry*, Longman, London, UK

Cuq, P., Rozier, J., Adomefa, K. (1978) Diagnose différentielle de l'espèce sur les carcasses et les abats des moutons et des chèvres de l'Afrique tropicale de l'Ouest [Species differentiation of sheep and goat carcasses and offal in tropical West Africa (Senegal)], *Revue d'Élevage et de Médecine Vétérinaire des Pays Tropicaux* **31**: 401–9

Cuthbertson, A., Kempster, A. J. (1979) Sheep carcass and eating quality. In *The Management and Diseases of Sheep*, pp. 377–99. Commonwealth Agricultural Bureaux, Farnham Royal, Slough, UK

Cyprus, Agricultural Research Institute (1981) The effect of machine milking on the milk yield of Chios sheep. In *Annual Report 1980*, p. 61. Ministry of Agriculture and Natural Resources, Nicosia, Cyprus

Daflapurkar, D. K., Kaushik, S. N., Katapal, B. G. (1979) Comparative studies on wool production and quality traits in Patanwadi and Deccani crosses, *Indian Veterinary Journal* **56**: 764–7

Dahl, G., Hjort, A. (1976) *Having Herds, Pastoral Growth and Household Economy*. Department of Social Anthropology, University of Stockholm, Stockholm, Sweden

Darmono, S. (1982) Persentase kejadian haemonchosis serta perbandingan jumlan cacing jantan dengan cacing betina *Haemonchus contortus* pada domba di rumah potong hewan kodya Bogor [Percentage incidence of haemonchosis and the ratio of male worms to female worms of *Haemonchus contortus* in sheep at the slaughter-house in Bogor], *Penyakit Hewan* **14**(24): 43–6

Darwish, M. T. H., El-Samman, S., Abou-Hussein, E. R. M. (1973) Meat production from Rahmani rams, *Egyptian Journal of Animal Production* **13**: 35–48

Dass, G. S., Acharya, R. M. (1970) Growth of Bikaneri sheep, *Journal of Animal Science* **31**: 1–4

Dattilo, M., Congiu, F. (1979) Utilizzazione energetica degli alimenti per la produzione del latte in ovini di razza Sarda [Utilisation of food energy for milk production in ewes of the Sardinian breed], *Rivista di Zootecnia e Veterinaria* **3**: 183–90

Dawson, T. J., Denny, M. J. S., Russell, E., Ellis, B. (1975) Water usage and diet preferences of free ranging kangaroos, sheep and feral goats in the Australian arid zone during summer, *Journal of Zoology* **177**: 1–23

De Boer, A. J. (1983a) Economic analysis of small ruminant production and marketing systems. In Blond, R. D. (ed.) *Partners in Research. A Five Year Report of the Small Ruminant Collaborative Research Support Program*. pp. 19–25. University of California, Davis, California, USA

De Boer, A. J. (1983b) Economic analysis of improved small ruminant production systems. In Blond, R. D. (ed.) *Partners in Research. A Five Year Report of the Small Ruminant Collaborative Research Support Program*. pp. 63–70. University of California, Davis, California, USA

De Boer, A. J., Job, M., Matthewman, R. W. (1982) Economic modelling of small farms in western Kenya. In *Proceedings of the Small Ruminant CRSP (SR-CRSP) Kenya Seminar, March 1982*, pp. 91–6. Small Ruminant Collaborative Research Support Program, Nairobi, Kenya

Demiruren, A. S. (1961) The sheep industry and types and breeds of sheep in Afghanistan, *National Wool Grower* **51** (July), 4pp.

Dettmers, A. (1983) Performance of hair sheep in Nigeria. In Fitzhugh, H. A., Bradford, G. E. (eds) *Hair Sheep of Western Africa and the Americas, a Genetic Resource for the Tropics*, pp. 201–18. Westview Press, Boulder, Colorado, USA

Devendra, C. (1972) Barbados Blackbelly sheep of the Caribbean, *Tropical Agriculture* (Trinidad) **49**: 23–9

Devendra, C. (1973) Effect of level of inclusion of citrus meal on the digestibility of a concentrate diet for sheep in Trinidad, *Tropical Agriculture* (Trinidad) **50**: 221–4

Devendra, C. (1975) The utilisation of rice straw by sheep. 1. Optimal level in the diet, *Malaysian Agricultural Journal* **50**: 169–86

Devendra, C. (1976) The utilisation of poultry excreta by sheep, *Malaysian Agricultural Journal* **50**: 513–22

Devendra, C. (1979) Goat and sheep production potential in the ASEAN region, *World Animal Review*, **32**: 33–41

Devendra, C. (1981) The energy requirement for maintenance of pen-fed sheep in Malaysia, *MARDI Research Bulletin* **9**: 233–40

Devendra, C., Burns, M. (1983) *Goat Production in the Tropics*, 2nd edn. Commonwealth Agricultural Bureaux, Farnham Royal, Slough, UK

Devendra, C., McLeroy, G. B. (1982) *Goat and Sheep Production in the Tropics*. Longman, Harlow, UK

Dhanda, O. P., Arora, K. L. (1977) A note on synchronisation of oestrus in Nali sheep, *Indian Journal of Animal Sciences* **47**: 369–70

Diez, J., Quijandria, B., Calle, R. (1974) Efectos genéticos y ambientales sobre peso y lana en Corriedales [Genetic and environmental effects on body weight and wool in Corriedales], *Memoria, Asociación Latinoamericana de Producción Animal* **9**: 56–7

Disney, J. G., Norman, G. A., Silverside, D. E., Silvey, D. (1981) Meat products, processing and marketing. In Smith, A. J., Gunn, R. G. (eds) Intensive animal production in developing countries, pp. 461–7. *Occasional Publication No. 4*. British Society of Animal Production, Thames Ditton, UK

Dolberg, F., Saadullah, M., Haque, M., Ahmed, R. (1981) Storage of urea treated straw using indigenous material, *World Animal Review* **38**: 37–42

Dolling, C. H. S., Moore, R. W. (1961) Efficiency of conversion of food to wool. 2. Comparison of the efficiency of the same Merino ewes on two different rations, *Australian Journal of Agricultural Research* **12**: 452–61

Doney, J. M. (1979) Nutrition and the reproductive function in female sheep. In *The Management and Diseases of Sheep*, pp. 152–60. Commonwealth Agricultural Bureaux, Farnham Royal, Slough, UK

Doney, J. M., Gunn, R. G., Horak, F. (1982) Reproduction. In Coop, I. E. (ed.) *Sheep and Goat Production*, pp. 57–80. Elsevier, Amsterdam, Netherlands

Doney, J. M., Peart, J. N., Smith, W. F., Sim, D. A. (1983) Lactation performance, herbage intake and lamb growth of Scottish Blackface and East Friesland × Scottish Blackface ewes grazing hill or improved pasture. *Animal Production* **37**: 283–92

Dry, F. W. (1955) Multifactorial inheritance of halo hair abundance in New Zealand Romney sheep, *Australian Journal of Agricultural Research* **6**: 608–23

Dry, F. W. (1975) *The Architecture of Lambs' Coats*. Massey University, Palmerston North, New Zealand

Dudgeon, G. C., Mohammad Askar, E. (1916) *The Sheep in Egypt*. Government Press, Cairo, Egypt

Duerden, J. E. (1932) A down pelage in the ovidae, *Nature* (UK) **130**: 736–7

Duerden, J. E., Spencer, M. R. (1930) The coat of the Angora goat, *Bulletin No. 83*. Department of Agriculture, South Africa

Dumas, R. (1977) *Étude sur l'élevage des petits ruminants du Tchad* [Study of the production of the small ruminants of Chad]. Institut d'Élevage et de Médecine Vétérinaire des Pays Tropicaux, Maisons Alfort, France

Dumas, R. (1980) Contribution à l'étude des petits ruminants du Tchad [Contribution to the study of small ruminants of Chad], *Revue d'Élevage et de Médecine Vétérinaire des Pays Tropicaux* **33**: 215–33

Dutra, J., Mies Filho, A., Costa, I. A., Selaive, A. B. (1980) Congelação de sêmen ovino [Freezing of ram semen]. In *Pesquisa em ovinos no Brasil 1975–1979*, p. 69. Empresa Brasileira de Pesquisa Agropecuária, Bagé, Brazil

Dwivedi, I. S. D. (1977) Effect of shed temperature and grazing time on semen quality of exotic rams, *Indian Veterinary Journal* **54**: 732–4

Economides, S., Hadjipanayiotou, M., Georghiades, E. (1981) The nutritive value of straw, and barley and lucerne hay, and the effect of nitrogen supplementation on the nutritive value of straw to sheep. *Technical Bulletin No. 39*. Agricultural Research Institute, Ministry of Agriculture and Natural Resources, Nicosia, Cyprus

Edey, T. N. (1969) Prenatal mortality in sheep: a review, *Animal Breeding Abstracts* **37**: 173–90

Egan, J. K., Doyle, P. T. (1982) The effect of stage of maturity in sheep upon intake and digestion of roughage diets, *Australian Journal of Agricultural Research* **33**: 1099–105

Eighmy, J. L., Ghanem, Y. S. (1982) The hema system: prospects for traditional subsistence systems in the Arabian Peninsula, *Culture and Agriculture* **16**: 10–15

El-Hag, G. A. (1976) A comparative study between desert goat and sheep efficiency of feed utilisation, *World Review of Animal Production* **12**(3): 43–8

El-Hag, G. A., George, A. E. (1983) Potential of agroindustrial by-products as feed for ruminants in the Sudan. In Kiflewahid, B., Potts, G. R., Drysdale, R. M. (eds) *By-product Utilization for Animal Production,* pp. 16–22. International Development Research Centre, Ottawa, Canada

El-Hag, G. A., Mukhtar, A. M. S. (1978) Varying levels of concentrate in the rations of Sudan Desert sheep, *World Review of Animal Production* **14**(4): 73–9

El-Hag, M. G., El-Hag, G. A. (1983) Evaluation of acid insoluble ash (AIA) as a marker for predicting the dry matter digestibility of high roughage and high concentrate diets for Sudan Desert sheep and goats, *World Review of Animal Production* **19**(2): 7–12

El-Homosi, F. F., El-Hafiz, G. A. A. (1982) Reproductive performance of Ossimi and Saidi sheep under two pre-pubertal planes of nutrition, *Assiut Veterinary Medical Journal* **10**(19): 59–66

Eliya, J., Juma, K. H. (1970) Birth weight, weaning weight and milk production in Awassi sheep, *Tropical Agriculture* (Trinidad) **47**: 321–4

Eliya, J., Juma, K. H., Al-Shabibi, M. (1972) A note on the composition and properties of Awassi sheep milk, *Egyptian Journal of Animal Production* **12**: 51–5

El-Karim, A. I. A. (1980) Effect of docking on growth and carcass characteristics of Dubasi Desert sheep, *Tropical Animal Production* **5**: 15–17

El-Kouni, M. H., Karam, H. A., Galal, E. S. E., Afifi, E. A. (1974) Crossbreeding and the effect of certain environmental factors on body weight of Barki and German Merino sheep, *Journal of Agricultural Science* (UK) **82**: 349–52

Elliott, R. C., Mills, W. R., Reed, W. D. C. (1966) Survival feeding of Africander cows, *Rhodesia Zambia and Malawi Journal of Agricultural Research* **4**: 69–75

Elliott, R. C., O'Donovan, W. M. (1969) Compensatory growth in Dorper sheep. In Oliver, J. (ed.) *Proceedings of the Second Symposium on Animal Production,* pp. 41–51. Rhodesian Branch South African Society of Animal Production, Salisbury, Rhodesia

Ellis, P. (1974) The development of animal health services, *Agricultural Administration* **1**: 199–219

Ellis, P. R., James, A. D. (1979) The economics of animal health (1) Major disease control programmes, *Veterinary Record* **105**: 504–6

El-Oksh, H. A., Kadry, A. E. H., Ghanem, Y. S., Galal, E. S. E. (1979) Estimates of some physical wool characteristics of Barki sheep under semi-arid desert conditions, *Beiträge zur Tropischen Landwirtschaft und Veterinärmedizin* **17**: 391–5

El-Serafy, A. M., El-Ashry, M. A., Swidan, F. (1976) Comparison between carcass traits of two age groups of fat tailed sheep, *Egyptian Journal of Animal Production* **16**: 65–8

El-Shaffei, M. A., El-Shazly, K., Abou-Akkada, A. R., Borhami, B. E. A., Abaza, M. A. (1975) Early weaning of naturally suckled Fleish Merino lambs, *Alexandria Journal of Agricultural Research* **23**: 213–25

El-Sherbiny, A. A., El-Oksh, H. A., El-Sheikh, A. S., Khalil, M. H. (1978) Effect of light and temperature on wool growth, *Journal of Agricultural Science* (UK) **90**: 329–34

El-Sherbiny, A. A., Shesha, A., Farid, A. (1972) Effect of crossing Ossimi sheep with Merino on milk production and mothering quality, *Agricultural Research Review* **50**(3): 37–43

El-Sherif, M. M., Salem, A. A. (1978) Studies on metabolic profile test in sheep and its relation to infertility, *Assiut Veterinary Medical Journal* **5**(9/10): 77–87

El-Wishy, A. B. (1974) Some aspects of reproduction in fat-tailed sheep in subtropics. IV. Puberty and sexual maturity, *Zeitschrift für Tierzüchtung und Züchtungsbiologie* **91**: 311–16

El-Wishy, A. B., El-Sawaf, A., El-Mikkawi, F. (1971) Some aspects of reproduction in fat-tailed sheep in subtropics. I. Reproductive behaviour of local Ausimi and imported Awassi ewes, *Veterinary Medical Journal, United Arab Republic* **19**: 131–55

Entwistle, K. W. (1972) Early reproductive failure in ewes in a tropical environment, *Australian Veterinary Journal* **48**: 395–401

Entwistle, K. W., Knights, G. (1974) The use of urea-molasses supplements for sheep grazing semi-arid tropical pastures, *Australian Journal of Experimental Agriculture and Animal Husbandry* **14**: 17–22

Epstein, H. (1977) The Awassi sheep in Israel, *World Review of Animal Production* **13**(2): 19–26

Epstein, H. (1982) Awassi sheep, *World Animal Review* **44**: 9–18

Eriksen, P. J. (1978) Slaughterhouse and slaughterslab design and construction. *Animal Production and Health Paper 9*. Food and Agriculture Organization of the United Nations, Rome, Italy

Ernle, Lord (1941) *English Farming Past and Present*, new edn. Longmans, Green and Co., London, UK

Evans, J. V., Blunt, M. H., Southcott, W. H. (1963) The effects of infection with *Haemonchus contortus* on the sodium and potassium concentrations in the erythrocytes and plasma, in sheep of different haemoglobin types, *Australian Journal of Agricultural Research* **14**: 549–58

Everist, S. L. (1969) Use of fodder trees and shrubs. *Division of Plant Industry Advisory Leaflet No. 1024*. Queensland Department of Primary Industries, Brisbane, Australia

Everitt, G. C. (1962) On the assessment of body composition in live sheep and cattle, *Proceedings of the Australian Society of Animal Production* **4**: 79–89

Eyal, E. (1972) Biological and environmental factors affecting assessment of true production and the genetic potentials of dairy sheep. In *Symposium on Milk Recording Practices for Sheep and Goats in Memoriam of Dr M. Finci, Israel, 19–24 March 1972*. Ministry of Agriculture, Israel

Eyal, E., Folman, Y. (1978) The nutrition of dairy sheep in Israel. In Boyazoglu, J. G., Treacher, T. T. (eds) Milk production of the ewe, pp. 84–93. *Publication No. 23*. European Association for Animal Production, Brussels, Belgium

Eyal, E., Lawi, A., Folman, Y., Morag, M. (1978) Lamb and milk production of a flock of dairy ewes under an accelerated breeding regime, *Journal of Agricultural Science* (UK) **91**: 69–79

Fahmy, M. H., Ghanem, Y. S., El-Essawy, H. F. (1964) A study of some body measurements in a cross between Merino and Barki sheep living under desert conditions: 1-growth rates, *Bulletin de l'Institute du Désert d'Égypte* **34**: 43–61

Falconer, D. S. (1981) *Introduction to Quantitative Genetics*, 2nd edn. Longman, London, UK

Fall, A., Diop, M., Sandford, J., Wissocq, Y. J., Durkin, J., Trail, J. C. M. (1982) Evaluation of the productivities of Djallonké sheep and N'Dama cattle at the Centre de Recherches Zootecniques, Kolda, Senegal. *Research Report No. 3*. International Livestock Centre for Africa, Addis Ababa, Ethiopia

Falvey, J. L. (1979) A review of sheep in the highlands of North Thailand, *World Review of Animal Production* **15**(1):39–47

Falvey, L., Hengmichai, P. (1979a) A comparison of three sheep genotypes in the northern Thailand highlands, *Thai Journal of Agricultural Science* **12**: 201–16

Falvey, L., Hengmichai, P. (1979b) Carcass studies of small ruminants in the northern highlands, *Thai Journal of Agricultural Science* **12**: 301–8

FAO (1960) *Rural Tanning Techniques*. Food and Agriculture Organization of the United Nations, Rome, Italy

FAO (1966) *Report of the FAO/OIE International Conference on Sheep Diseases*. Food and Agriculture Organization of the United Nations, Rome, Italy

FAO (1970) *The World Hides, Skins, Leather and Footwear Economy*. Food and Agriculture Organization of the United Nations, Rome, Italy

FAO (1972a) The Kufra sheep project. *Report to the Government of the Libyan Arab Republic. No. TA3050*. Food and Agriculture Organization of the United Nations, Rome, Italy

FAO (1972b) *Veterinary Institute for Tropical and High Altitude Research, Lima, Peru, Report No. 21, July 1st–Dec 31st 1971*. Food and Agriculture Organization of the United Nations, Rome, Italy

FAO (1983a) *FAO Production Yearbook 1982*, vol. 36 Food and Agriculture Organization of the United Nations, Rome, Italy

FAO (1983b) *FAO Trade Yearbook 1982*, vol. 36. Food and Agriculture Organization of the United Nations, Rome, Italy

Farid, A., Makarechian, M., Sefidbakht, N. (1976) Crossbreeding of Iranian fat-tailed sheep. 1. Preweaning growth performance of Karakul, Mehraban, Naeini and their reciprocal crosses, *Iranian Journal of Agricultural Research* **4**: 69–79

Farid, M. F. A., El-Shennawy, M. M., Mehrez, A. Z., Salem, A. M. M. (1983) Protein requirements for maintenance of Barki desert sheep, *World Review of Animal Production* **19** (3): 31–6

Faure, A. S., Boshoff, D. A., Burger, F. J. L. (1980) Comparison between the intravaginal sponge and subcutaneous implant for synchronisation of oestrus in Karakul ewes, 1976–1977. In *Agricultural Research 1979*, pp. 83–5. Division of Agricultural Information, Pretoria, South Africa

Fayez, I., Marai, M., Taha, A. H. (1976) Wool follicle characteristics in the Awassi fat-tailed sheep, *Acta Anatomica* **96**: 55–69

Fehilly, C. B., Willadsen, S. M., Tucker, E. M. (1984) Interspecific chimaerism between sheep and goat, *Nature* (UK) **307**: 634–6

Field, C. R. (1982) The smallstock program of the UNESCO-integrated project in arid lands. In *Proceedings of the Small Ruminant CRSP (SP-CRSP) Kenya Seminar, March 1982*, pp. 151–67. Small Ruminant Collaborative Research Support Program, Nairobi, Kenya

Figueiredo, E. A. P. de, Oliveira, E. R. de, Bellaver, C., Simplicio, A. A. (1983) Hair sheep performance in Brazil. In Fitzhugh, H. A., Bradford, G. E. (eds) *Hair Sheep of Western Africa and the Americas, a Genetic Resource for the Tropics*, pp. 125–40. Westview Press, Boulder, Colorado, USA

Figueiredo, E. A. P., Simplicio, A. A., Pant, K. P. (1982) Evaluation of sheep breeds for early growth in tropical North-east Brazil, *Tropical Animal Health and Production* **14**: 219–23

Figueiró, P. R. P., Silva, O. L., Savian, J. F. (1980) Efeito do anabólico zeranol sobre o crescimento, produção de carne e lã em cordeiros [Effect of zeranol on growth, meat production and wool yield of lambs]. In *Pesquisa em ovinos no Brasil 1975–1979*, p. 113. Empresa Brasileira de Pesquisa Agropecuária, Bagé, Brazil

Finci, M. (1957) The improvement of the Awassi breed of sheep in Israel, *Bulletin of the Research Council of Israel* **B6**: 1–106

Fitzhugh, H. A., Bradford, G. E. (eds) (1983a) *Hair Sheep of Western Africa and the Americas, a Genetic Resource for the Tropics*. Westview Press, Boulder, Colorado, USA

Fitzhugh, H. A., Bradford, G. E. (1983b) Productivity of hair sheep and opportunities for improvement. In Fitzhugh, H. A., Bradford, G. E. (eds) *Hair Sheep of Western Africa and the Americas, a Genetic Resource for the Tropics*, pp. 23–52. Westview Press, Boulder, Colorado, USA

Flamant, J. C., Casu, S. (1978) Breed differences in milk production potential and genetic improvement of milk production. In Boyazoglu, J. G., Treacher, T. T. (eds) Milk production of the ewe, pp. 1–17. *Publication No. 23*. European Association for Animal Production, Brussels, Belgium

Flamant, J. C., Morand-Fehr, P. (1982) Milk production in sheep and goats. In Coop, I. E. (ed.) *Sheep and Goat Production*, pp. 275–95. Elsevier, Amsterdam, Netherlands

Fletcher, I. C., Lishman, A. W., Thring, B., Holmes, J. A. (1980) Plasma progesterone levels in lactating ewes after hormone-induced ovulation during the non-breeding season, *South African Journal of Animal Science* **10**: 151–7

Folman, Y., Volcani, R., Eyal, E. (1966) Mother – offspring relationships in Awassi sheep. I. The effect of different suckling regimes and time of weaning on the lactation curve and milk yield in dairy flocks, *Journal of Agricultural Science* (UK) **67**: 359–68

Forbes, J. M. (1982) Effects of lighting pattern on growth, lactation and food intake of sheep, cattle and deer, *Livestock Production Science* **9**: 361–74

Fox, C. W., Husnaoui, M. (1967) The effect of diethylstilbestrol on growth rates of young Awassi lambs. In *Proceedings of the Centennial Agricultural Symposium*, pp. 325-30. American University of Beirut, Beirut, Lebanon

France, IEMVT (1978) Comparison of growth rate and carcasses in entire male, castrated male and female lambs. In *Rapport d'activité Année 1977*, p. 76. Institut d'Élevage et de Médecine Vétérinaire des Pays Tropicaux, Maisons Alfort, France

Frandson, R. D. (1981) *Anatomy and Physiology of Farm Animals*, 2nd edn. Lea and Febiger, Philadelphia, USA

Franklin, M. C. (1951) The drought feeding of sheep, *Australian Veterinary Journal* **27**: 326-33

Fraser, A., Stamp, J. T. (1968) *Sheep Husbandry and Diseases*, 5th edn. Crosby Lockwood, London, UK

French, M. H. (1942) The growth rates of local and crossbred sheep, *East African Agricultural Journal* **8**: 24-5

French, M. H. (1945) Geophagia in animals, *East African Medical Journal* **22**: 103-10

Funes, F. (1975) Digestibility and nutritive value of pangola grass (*Digitaria decumbens* Stent) in relation to nitrogen fertilisation and harvesting time, *Cuban Journal of Agricultural Science* **9**: 369-78

Gaili, E. S. E. (1979) Effect of breed-type on carcass weight and composition in sheep, *Tropical Animal Health and Production* **11**: 191-8

Gaillard, Y. (1979) Caractéristiques de la reproduction de la brebis Oudah [Reproductive characteristics of the Uda ewe], *Revue d'Élevage et de Médecine Vétérinaire des Pays Tropicaux* **32**: 285-90

Galal, E. S. E. (1983) Sheep germ plasm in Ethiopia, *Animal Genetic Resources Information* **1**: 4-12

Galal, E. S. E., Ghanem, Y. S., Farid, M. A., Fahmy, M. H., Seoudy, A. E. M. (1975) Carcass traits and feed-lot performance of Barki, Merino and Awassi breeds of sheep and some of their crosses, *Egyptian Journal of Animal Production* **15**: 33-46

Galbraith, H., Topps, J. H. (1981) Effect of hormones on the growth and body composition of animals, *Nutrition Abstracts and Reviews* **B51**: 521-40

Gall, C. (1975) Milk production from sheep and goats, *World Animal Review* **13**: 1-8

Gamarra, M. A., Carpio, M. A. (1982) Sheep improvement in Peru. In *Proceedings of the World Congress on Sheep and Beef Cattle Breeding*, vol. II, pp. 139-43. Dunsmore Press, Palmerston North, New Zealand

Ganesakale, D. (1975) Effect of season of mating on the economic traits in sheep, *Cheiron* **4**: 40-4

García, C. E., Chicco, C. F., Carnevali, A. A. (1973) Una nota sobre el uso de la harina de hoja de plátano en la alimentación de rumiantes [Meal of banana leaves in the feeding of ruminants], *Agronomía Tropical* (Venezuela) **23**: 293-9

Gardiner, M. R, Craig, J. (1970) Factors affecting survival in the transportation of sheep by sea in the tropics and sub-tropics, *Australian Veterinary Journal* **46**: 65-9

Gartner, R. J. W., Anson, R. J. (1966) Vitamin A reserves of sheep maintained on mulga (*Acacia aneura*), *Australian Journal of Experimental Agriculture and Animal Husbandry* **6**: 321-5

Gartner, R. J. W., Johnston, D. C. (1969) Hepatic vitamin A concentrations in sheep in north-western Queensland, *Australian Journal of Experimental Agriculture and Animal Husbandry* **9**: 473-6

Gatenby, R. M. (1982) Research on small ruminants in sub-Saharan Africa. In Gatenby, R. M., Trail, J. C. M. (eds) *Small Ruminant Breed Productivity in Africa*, pp. 13-24. International Livestock Centre for Africa, Addis Ababa, Ethiopia

Gatenby, R. M. (1984) Shelter for animals in hot countries. In Grace, J. (ed.) *The Effects of Shelter on the Physiology of Plants and Animals*. Swets and Zeitlinger, Lisse, Netherlands

Gatenby, R. M., Monteith, J. L., Clark, J. A. (1983) Temperature and humidity gradients in a sheep's fleece. I. Gradients in the steady state, *Agricultural Meteorology* **29**: 1-10

Gaur, D., Chopra, S. C., Balaine, D. S., Chhikara, B. S. (1977) Pre-weaning and post-weaning body weights of Nali and Corriedale × Nali lambs, *Indian Journal of Animal Sciences*, **47**: 534-8

Gavora, J. S., Spencer, J. L. (1983) Breeding for immune responsiveness and disease resistance, *Animal Blood Groups and Biochemical Genetics* **14**: 159-80

Geenty, K. G., Sykes, A. R. (1981) Intake and growth performance of grazing lambs weaned at 4 and 12 weeks of age, *Proceedings of the New Zealand Society of Animal Production* **41**: 235-41

Gefu, J. O., Adu, I. F. (1982) Observations on the herd size of sheep and goats in Kano State, Nigeria, *World Review of Animal Production* **18**(2): 25-8

George, J. M. (1982) The mating potential of Indonesian sheep, *Animal Reproduction Science* **4**: 251-6

Ghanekar, V. M., Soman, B. V. (1972) Effect of season on body weight and wool yield, *Wool and Woollens of India,* July, 23-31

Ghanem, Y. S., Farid, M. F. A. (1982) Vitamin A deficiency and supplementation in Desert sheep. 1. Deficiency symptoms, plasma concentrations and body growth, *World Review of Animal Production* **18**(2): 69-74

Ghauri, R. H., Qazi, A. Q., Schneider, B. H. (1964) Experiments on fattening sheep in Pakistan. 5. Bagasse pulp as a roughage in sheep fattening rations, *Agriculture* (Pakistan) **15**: 223-9

Gherardi, S. G., Monzu, N., Sutherland, S. S., Johnson, K. G., Robertson, G. M. (1981) The association between body strike and dermatophilus of sheep under controlled conditions, *Australian Veterinary Journal* **57**: 268-71

Ghoneim, K. E., Aboul-Naga, A., Labban, F. (1968) Effect of crossing Merino with Ossimi sheep on growth and body weight, *Journal of Animal Production of the United Arab Republic* **8**(1/2): 45-56

Ghoneim, K. E., Kazzal, N. T., Abdallah, R. K., Thannoon, N., Alatroshy, A. H. (1973) *Some Wool Characteristics of Karadi Sheep in North of Iraq.* University of Mosul College of Agriculture and Forestry, Hamman Al-Alil, Iraq

Ghoneim K. E., Taha, A. H., Kazzal, N. T., Abdallah, R. K. (1974) Effects of non-genetic factors and estimation of genetic parameters on fleece weight of Awassi sheep in Iraq, *Tropical Agriculture* (Trinidad) **51**: 51-6

Gihad, E. A. (1976a) Value of dried poultry manure and urea as protein supplements for sheep consuming low quality tropical hay, *Journal of Animal Science* **42**: 706-9

Gihad, E. A. (1976b) Intake, digestibility and nitrogen utilization of tropical natural grass hay by goats and sheep, *Journal of Animal Science* **43**: 879-83

Gill, R. S., Negi, S. S. (1971) Requirement of maintenance digestible crude protein for adult sheep, *Indian Journal of Animal Sciences* **41**: 980-4

Göhl, B. (1981) *Tropical Feeds: Feed Information Summaries and Nutritive Values.* Food and Agriculture Organization of the United Nations, Rome, Italy

Gómez, R., Hernández, J. (1980) Response of growing Pelibuey sheep to varying levels of dietary energy, *Tropical Animal Production* **5**: 292

Gomide, J. A., Noller, C. H., Mott, G. O., Conrad, J. H., Hill, D. L. (1969) Effect of plant age and nitrogen fertilization on the chemical composition and *in vitro* cellulose digestibility of tropical grasses, *Agronomy Journal* **61**: 116-20

Gonzales Reyna, A., Alba, J. de (1978) Resultado economico de ovinos Peliguey en el tropico seco de Mexico [Economic performance of Pelibuey sheep in the dry tropics in Mexico], *Memoria. Asociación Latinoamericana de Producción Animal* **13**: 203-9

González-Stagnaro, C., Perozo-Gory, F., Goycochea-Llaque, J. (1980) Eficiencia reproductiva de ovejas 'West African' en exploitaciones comerciales de zonas aridas [Reproductive efficiency of West African sheep in commercial flocks in arid regions]. In *9th International Congress on Animal Reproduction and Artificial Insemination, 16-20 June 1980. III. Symposia (free communications),* p. 52. Editorial Garsi, Madrid, Spain

Gooden, J. M., Beach, A. D., Purchas, R. W. (1980) Measurement of subcutaneous backfat depth in live lambs with an ultrasonic probe, *New Zealand Journal of Agricultural Research* **23**: 161-5

Gooding, E. G. B. (1982) Effect of quality of cane on its value as livestock feed, *Tropical Animal Production* **7**: 72-91

Goot, H. (1972a) Effect of milking on fleece weight, cotts, hairiness and canary discoloration in Awassi dairy ewes, *Israel Journal of Agricultural Research* **22**: 221–4

Goot, H. (1972b) Milk yield and lactation length of single and twin-rearing dairy ewes. In *Symposium on Milk Recording Practices for sheep and Goats, in Memoriam of Dr M. Finci, Tel-Aviv, Israel, 19–24 March 1972.* Ministry of Agriculture, Israel

Goot, H., Foote, W. C., Eyal, E., Folman, Y. (1980) Crossbreeding to increase meat production of the native Awassi sheep. *Agricultural Research Organization Special Publication No, 175.* Institute of Animal Science, Israel

Gootwine, E., Alef, B., Gadeesh, S (1980) Udder conformation and its heritability in the Assaf (Awassi × East Friesan) cross of dairy sheep in Israel, *Annales de Génétique et de Sélection Animale* **12**: 9–13

Gordin, S. (1980) Milking animals and fermented milks of the Middle East and their contribution to man's welfare, *Journal of Dairy Science* **63**: 1031–8

Gordon, H. M. (1960) Nutrition and helminthosis in sheep, *Proceedings of the Australian Society of Animal Production* **3**: 93–104

Grandin, T. (1980) Livestock behavior as related to handling facilities design, *International Journal for the Study of Animal Problems* **1**: 33–52

Grant, J. L., Naude, J. F. (1977) An investigation of the effect of the implant 'Morebeef' on the performance of lambs in the feed-lot. In *Division of Livestock and Pastures Annual Report for the Year ended 30 September 1976*, pp. 109–10. Department of Research and Specialist Services, Salisbury, Rhodesia

Green, M. L. (1977) Reviews of the progress of dairy science: milk coagulants, *Journal of Dairy Research* **44**: 159–90

Greenhalgh, J. F. D. (1980) Use of straw and cellulosic wastes and methods of improving their value. In Ørskov E. R. (ed.) By-products and wastes in animal feeding, pp. 25–31. *Occasional Publication No. 3.* British Society of Animal Production, Thames Ditton, UK

Greenhalgh, J. F. D., Wainman, F. W. (1972) The nutritive value of processed roughages for feeding cattle and sheep, *Proceedings of the British Society of Animal Production* 61–72

Griffin, L., Allonby, E. W. (1979) Trypanotolerance in breeds of sheep and goats with an experimental infection of *Trypanosoma congolense*, *Veterinary Parasitology* **5**: 97–105

Griffiths, M., Barker, R. (1966) The plants eaten by sheep and by kangaroos grazing together in a paddock in south-western Queensland, *CSIRO Wildlife Research* **11**: 145–67

Guggolz, J., Kohler, G. O., Klopfenstein, T. J. (1971) Composition and improvement of grass straw for ruminant nutrition, *Journal of Animal Science* **33**: 151–6

Guirgis, R. A. (1973) The study of variability in some wool traits in a coarse wool breed of sheep, *Journal of Agricultural Science* (UK) **80**: 233–8

Guirgis, R. A. (1977) Maternal influence on sheep production: a study of some birth traits of lambs, *Journal of Agricultural Science* (UK) **89**: 535–40

Guirgis, R. A. (1980) Response to the use of Merino in improvement of coarse-wool Barki sheep: an analysis of some cross-bred wool traits, *Journal of Agricultural Science* (UK) **95**: 339–47

Guirgis, R. A., Afifi, E. A., Galal, E. S. E. (1982) Estimates of genetic and phenotypic parameters of some weight and fleece traits in a coarse wool breed of sheep, *Journal of Agricultural Science* (UK) **99**: 277–85

Guirgis, R. A., El-Gabbas, H. M. M., Galal, E. S. E., Ghoneim, K. E. (1979) Birthcoat – adult fleece relationship: a study of kemp succession in a coarse wool breed of sheep, *Journal of Agricultural Science* (UK) **92**: 427–36

Guirgis, R. A., Galal, E. S. E. (1972) The association between kemp and some vigour and wool characteristics in Barki and Merino Barki cross, *Journal of Agricultural Science* (UK) **78**: 345–9

Guirgis, R. A., Kazzal, N. T., Haddadine, M. S., Abdallah, R. K. (1978) A study of some wool traits in two coarse wool breeds and their reciprocal crosses, *Journal of Agricultural Science* (UK) **90**: 495–501

Gupta, J. L., Chopra, S. C., Dhamale, S. P. (1974) Effect of mating seasons on growth of Deccani sheep and their crosses with Merino and Rambouillet, *Haryana Agricultural University Journal of Research* **4**: 158–61

Gupta, R. N., Taylor, C. M., Khanna, R. S. (1972) Effect of age and season on staple length, fibre diameter, medullation and fibre density in Bikaneri sheep, *Indian Journal of Animal Production* **3**: 118–22

Hadjipanayiotou, M. (1982) Laboratory evaluation of ensiled poultry litter, *Animal Production* **35**: 157–61

Hafez, E. S. E. (1974) *Reproduction in Farm Animals*, 3rd edn. Lea and Febiger, Philadelphia, USA

Haggar, R. J. (1970) The intake and digestibility of low quality *Andropogon gayanus* hay, supplemented with various nitrogenous feeds, as recorded by sheep, *Nigerian Agricultural Journal* **7**: 70–5

Hall, G. A. B., Savain, J., Figueiro, P. R. P., Lacerda, O., Müller, L. (1977) Zeranol implantation for suckling ram lambs. Weight gain and development of the reproductive tract, *Tropical Animal Production* **2**: 175–9

Hall, H. T. B. (1977) *Diseases and Parasites of Livestock in the Tropics*. Longman, Harlow, Essex, UK

Hansard, S. L. (1983) Microminerals for ruminant animals, *Nutrition Abstracts and Reviews* **B53**: 1–24

Harding, H. P. (1981) Veterinarians willing to get dirty, *Ceres* **14** (1):23–7

Hardjosubroto, W., Astuti, M. (1980) Animal genetic resources in Indonesia. In *Animal Genetic Resources in Asia and Oceania*, pp. 189–204. Tropical Agriculture Research Center, Ministry of Agriculture, Forestry and Fisheries, Tsukuba, Ibaraki, Japan

Hartley, K. T. (1937) An explanation of the effect of farmyard manure in northern Nigeria, *Empire Journal of Experimental Agriculture* **5**: 254–63

Harvey, J. M. (1952a) Chronic endemic fluorosis of Merino sheep in Queensland, *Queensland Journal of Agricultural Science* **9**: 47–141

Harvey, J. M. (1952b) Copper deficiency in ruminants in Queensland, *Australian Veterinary Journal* **28**: 209–15

Haumesser, J. B., Gerbaldi, P. (1980) Observations sur la reproduction et l'élevage du mouton Oudah nigérien [Observations on the reproduction and husbandry of Nigerian Uda sheep], *Revue d'Élevage et de Médecine Vétérinaire des Pays Tropicaux* **33**: 205–13

Hays, V. W., Muir, W. M. (1979) Efficacy and safety of feed additive use of antibacterial drugs in animal production, *Canadian Journal of Animal Science* **59**: 447–56

Healy, W. B. (1972) Ingested soil and animal nutrition, *Proceedings of the New Zealand Grassland Association* **34**: 84–90

Hegarty, M. P. (1982) Deleterious factors in forages affecting animal production. In Hacker, J. B. (ed.) *Nutritional Limits to Animal Production from Pastures*, pp. 133–50. Commonwealth Agricultural Bureaux, Farnham Royal, Slough, UK

Heitzman, R. J. (1980) Growth stimulation in ruminants. In Haresign, W., Lewis, D. (eds) *Recent Advances in Animal Nutrition 1979*, pp. 133–43. Butterworths, London, UK

Helland, J. (1980) An analysis of 'Afar' pastoralism in the northeastern rangelands of Ethiopia. *Working Document 19*. International Livestock Centre for Africa, Addis Ababa, Ethiopia

Hendricksen, R. E., Poppi, D. P., Minson, D. J. (1981) The voluntary intake, digestibility and retention time by cattle and sheep of stem and leaf fractions of a tropical legume (*Lablab purpureus*), *Australian Journal of Agricultural Research* **32**: 389–98

Herlich, H. (1978) The importance of helminth infections in ruminants, *World Animal Review* **26**: 22–6

Hernández, L. J. J. P., Rodríguez, R. O., González Padilla, E. (1976) Evaluación de cuatro métodes para colección de semen en borrego Tabasco o Pelibuey [Evaluation of four methods of collecting semen from Tabasco or Pelibuey rams], *Técnica Pecuaria en México* **30**: 45–51

Hetzel, D. J. S., Yates, N. G., Obst, J. M., Chaniago, T., Bakrie, B. (1982) A preliminary report on the growth of Javanese Thin-tail sheep and their crosses with Suffolk, Polled

Dorset and Wiltshire Horn rams, *Proceedings of the Australian Society of Animal Production* **14**: 629

Heydenrych, H. J. (1977) Study of flock statistics, non-genetic factors, genetic parameters and selection response with regard to the Tygerhoek Merino flock. In *Agricultural Research 1976*, pp. 125–7. Department of Agricultural Technical Services, Pretoria, South Africa

Heydenrych, H. J., Meissenheimer, D. J. B. (1979) Sex differences in the heritabilities of economic traits in South African Merino sheep, *South African Journal of Animal Science* **9**: 69–72

Heydenrych, H. J., Vosloo, L. P., Meissenheimer, D. J. B. (1977) Selection response in South African Merino sheep selected either for high clean fleece mass or for a wider S/P ratio, *Agroanimalia* **9**: 67–73

Hodgson, J. (1982) Influence of sward characteristics on diet selection and herbage intake by the grazing animal. In Hacker, J. B. (ed.) *Nutritional Limits to Animal Production from Pastures*, pp. 153–66. Commonwealth Agricultural Bureaux, Farnham Royal, Slough, UK

Hofmeyer, H. S., Rensburg, W. J. J. van, Kroon, F., Olivier, I. (1976) Verskille in doeltreffendheid van voerverbruik by lammers van drie verskillende skaaprasse: III. Invloed van ras en voedingspeil op doeltreffendheid van energieverbruik [Differences in food conversion efficiency in lambs of three breeds: III. Effect of breed and nutritional level on the efficiency of energy utilisation], *Agroanimalia* **8**: 99–109

Hogarth, P. J. (1978) *Biology of Reproduction*. Blackie, Glasgow, UK

Hohenboken, W. D., Clarke, S. E. (1981) Genetic, environmental and interaction effects on lamb survival, cumulative lamb production and longevity of crossbred ewes, *Journal of Animal Science* **53**: 966–76

Holmes, J. H. G., Leche, T. F. (1977) South East Asian sheep in Papua New Guinea. In *Third International Congress of SABRAO in Association with the Australian Plant Breeding Conference, Australia, February 1977, Animal Breeding Papers*, pp. 1(c)-46 to 1(c)-50. Society for the Advancement of Breeding Researches in Asia and Oceania, Canberra, Australia

Honmode, J., Patil, B. D. (1975) Effect of plane of nutrition on prenatal mortality in Rajasthani ewes, *Indian Journal of Animal Sciences* **45**: 220–3

Honmode, J., Patil, B. D., Pachlag, S. V. (1971) Hormonal induction of superovulatory oestrus in early post-partum indigenous and cross-bred ewes, *Indian Journal of Animal Sciences* **41**: 977–9

Hopkins, P. S., Pratt, M. S. (1976) Some practical considerations for improving the pregnancy rate of tropical Merinos, *Proceedings of the Australian Society of Animal Production* **11**: 153–6

Hughes, G. P., Reid, D. (1952) Studies on the behaviour of cattle and sheep in relation to the utilization of grass, *Journal of Agricultural Science* (UK) **41**: 350–66

Hulet, C. V., Alexander, G., Hafez, E. S. E. (1975) The behaviour of sheep. In Hafez, E. S. E. (ed.) *The Behaviour of Domestic Animals*, pp. 246–94. Ballière Tindall, London, UK

Hulet, C. V., Ercanbrack, S. K., Blackwell, R. L., Price, D. A., Wilson, L. O. (1962) Mating behaviour of the ram in the multi-sire pen, *Journal of Animal Science* **21**: 865–9

Humphreys, L. R. (1978) *Tropical Pastures and Fodder Crops*, Longman, London, UK

Hunt, W. L., Addleman, D., Bogart, R. (1971) Induction of multiovulation in the ewe following synchronisation of estrus, *Journal of Animal Science* **32**: 491–5

Hunter, A. G., Heath, P. J. (1984) Ovine internal parasitism in the Yemen Arab Republic, *Tropical Animal Health and Production* **16**: 95–106

Hunter, R. H. F. (1980) *Physiology and Technology of Reproduction in Female Domestic Animals*. Academic Press, London, UK

Hupp, H., Deller, D. (1983) Virgin Islands white hair sheep. In Fitzhugh, H. A., Bradford, G. E. (eds) *Hair Sheep of Western Africa and the Americas, a Genetic Resource for the Tropics*, pp. 171–5. Westview Press, Boulder, Colorado, USA

Hutagalung, R. I. (1981) The use of tree crops and their by-products for intensive animal production. In Smith, A. J. (ed.) Intensive animal production in developing countries,

pp. 151-84. *Occasional Publication No. 4.* British Society of Animal Production, Thames Ditton, UK
Hyslop, N. St G. (1980) Dermatophilus (streptothricosis) in animals and man, *Comparative Immunology Microbiology and Infectious Diseases* **2**: 389-404
Ibrahim, A. A. E. (1970) Chemical preservation of milk in the Sudan, *Journal of Food Science and Technology in the Sudan* **2**: 31-2
Ibrahim, K. M. (1981) Shrubs for fodder production. In Manassah, J. T., Briskey, E. J. (eds) *Advances in Food-producing Systems for Arid and Semiarid Lands, part A*, pp. 601-42. Academic Press, London, UK
Ibrahim, M. A., Nwude, N., Aliu, Y. O., Ogunsusi, R. A. (1983) Traditional concepts of animal disease and treatment among Fulani herdsmen in Kaduna State of Nigeria. *Pastoral Network Paper 16c.* Overseas Development Institute, London, UK
ICAR (1964) Medical treatment of cattle. In *Agriculture in Ancient India*, pp. 144-51. Indian Council of Agricultural Research, New Delhi, India
ICAR (1981a) *Economics of Mutton Production.* Central Sheep and Wool Research Institute Avikanagar, Rajasthan, India
ICAR (1981b) *Progress Report of Operational Research Project on Sheep and Wool Development 1975-79.* Central Sheep and Wool Research Institute, Avikanagar, Rajasthan, India
ICARDA (1982) *Farming Systems Program Research Report 1982, Project IV: the Role of Livestock in the Farming System.* International Center for Agricultural Research in Dry Areas, Aleppo, Syria
ILCA (1979a) Livestock production in the subhumid zone of West Africa: a regional review. *Systems Study 2.* International Livestock Centre for Africa, Addis Ababa, Ethiopia
ILCA (1979b) Small ruminant production in the humid tropics. *Systems Study 3.* International Livestock Centre for Africa, Addis Ababa, Ethiopia
ILCA (1979c) Towards an economic assessment of veterinary inputs in tropical Africa. *Working Document 1.* International Livestock Centre for Africa, Addis Ababa, Ethiopia
ILCA (1980) Economic trends: small ruminants. *Bulletin 7.* International Livestock Centre for Africa, Addis Ababa, Ethiopia
ILCA (1983) Pastoral systems research. *Bulletin 16.* International Livestock Centre for Africa, Addis Ababa, Ethiopia
ILO (1982) *Yearbook of Labour Statistics.* International Labour Office, Geneva, Switzerland
India, CSWRI (1981) *All-India Coordinated Research Project on Sheep Breeding. Project Coordinator's Report April 1979-March 1981.* Central Sheep and Wool Research Institute, Avikanagar, Rajasthan, India
India, CSWRI (1982) *Annual Report 1981.* Central Sheep and Wool Research Institute, Avikanagar, Rajasthan, India
India, CSWRI (1983) *Annual Report 1982.* Central Sheep and Wool Research Institute, Avikanagar, Rajasthan, India
Iñiguez, L. (1975) Herencia y ambiente en la producción de ovinos Corriedale criados en la altura [Inheritance and environment in the production of Corriedale sheep kept at high altitude]. In *Reunión nacional sobre problemas genéticos en el mejoramiento vegetal y animal, Salta, Cerillos, Mayo 1975*, pp. 33-49. Facultad de Agronomía y Zootecnia, Universidad Nacional de Tucumán, Tucumán, Argentina
Ishaq, S. M., Mumtaz-Ali, S. (1959) A study of the factors affecting performance in Lohi sheep, *Pakistan Journal of Scientific Research* **11**: 113-17
Jackson, M. G. (1978) Treating straw for animal feeding. *Animal Production and Health Paper No. 10.* Food and Agriculture Organization of the United Nations, Rome, Italy
Jahnke, H. E. (1982) *Livestock Production Systems and Livestock Development in Tropical Africa.* Kieler Wissenschaftsverlag Vauk, Kiel, West Germany
Jatsch, O. (1977) Milchfraktionierung beim maschinellen Milchentzug des Schafes [Milk fractionation in machine milking of sheep], *Giessener Schriftenreihe Tierzucht und Haustiergenetik No. 38.* Verlag Paul Parey, Hamburg, West Germany

Jenness, R. (1974) The composition of milk. In Larson, B. L., Smith, V. R. (eds) *Lactation* vol. III, pp. 3–107. Academic Press, New York, USA

Jericho, K. W. F., McCauley, H., Bradley, J., Magwood, S. E., Vance, H. N., Prince, E., Church, T. L. (1974) Economics and disease of farm livestock, *Canadian Veterinary Journal* **15**: 213–18

Jewitt, T. N., Barlow, H. W. B. (1949) Animal excreta in the Sudan Gezira, *Empire Journal of Experimental Agriculture* **17**: 1–17

Johansson, I., Rendel, J. (1968) *Genetics and Animal Breeding*. Oliver and Boyd, Edinburgh, UK

Johari, D. C. (1972) Body weight from birth to one year in Polwarth sheep, *Indian Journal of Heredity* **4**: 38–40

Johari, D. C. (1973) Studies on the semen quality of Polwarth, Rambouillet, Bikaneri and Rampur Bushair rams. (2) Semen quality in summer, rains and autumn seasons, *Indian Journal of Animal Health* **12**: 85–8

Juma, K. H., Karam, H. A., Al-Maali, H. N. A., Al-Barazanji, J. (1973) Effect of docking in Awassi sheep, *Indian Journal of Animal Sciences* **43**: 931–5

Kabbali, A. (1977) Étude de la production laitière et de la croissance des agneaux de brebis Timahdite et Beni-Hsen: influence du niveau énérgetique [Study of milk production and lamb growth in Middle Atlas and Beni Ahsen sheep: effect of energy level], *Hommes, Terre et Eaux* **6**(25): 31–43

Kadry, A. E. H., El-Oksh, H. A., Galal, E. S. E., Ghanem, Y. S. (1980) Heritability and repeatability of some physical wool characteristics of Barki sheep under semi-arid desert conditions, *Beiträge zur Tropischen Landwirtschaft und Veterinärmedizin* **18**: 94–9

Kalla, S. D., Ghosh, P. K. (1974) A note on the relationship between wool quality and blood potassium type in sheep, *Animal Blood Groups and Biochemical Genetics* **5**: 167–9

Kalla, S. D., Ghosh, P. K. (1975) Blood biochemical polymorphic traits in relation to wool production efficiency in Indian sheep, *Journal of Agricultural Science* (UK) **84**: 149–52

Kandasamy, N., Pant, K. P. (1980) Gestational oestrus during pregnancy in sheep, *Indian Veterinary Journal* **57**: 434–5

Kapoor, U. R., Agarwala, O. N., Pachauri, V. C., Nath, K., Narayan, S. (1972) The relationship between diet, the copper and sulphur content of wool, and fibre characteristics, *Journal of Agricultural Science* (UK) **79**: 109–14

Kaschanian, N. (1973) Vergleichende Untersuchungen über die Milchleistung von Awassi-, Schall- und Schall- × Awassi-Schaften im Iran [A comparison of milk production in Awassi, Shal and Shal × Awassi sheep in Iran], *Züchtungskunde* **45**: 270–5

Kategile, J. A., Mgongo, F. O. K., Frederiksen, J. H. (1978) The effect of iodine supplementation on the reproductive rates of goats and sheep, *Nordisk Veterinaermedicin* **30**: 30–6

Kaushish, S. K., Arora, K. L., Dhanda, O. P. (1973) Studies on reproduction in sheep. II. Time of parturition, *Indian Journal of Animal Production* **4**: 56–9

Kaushish, S. K., Sahni, K. L. (1976) Effect of feeding animal protein (egg + milk) and trace elements, and provision of cooler climates on libido, semen quality and certain physiological reactions of Russian Merino rams during summer season, *Indian Journal of Animal Sciences* **46**: 135–9

Kaushish, S. K., Sahni, K. L. (1977) Reproductive performance of Merino sheep during autumn and spring seasons under semi-arid conditions, *Indian Journal of Animal Sciences* **47**: 18–22

Kazmi, M. S., Mathieson, A. R. (1976) Some physical and chemical properties of Nigerian sheep wools, *Tropical Science* **18**: 227–38

Keeler, R. F., Kampen, K. R. van, James, L. F. (eds) (1978) *Effects of Poisonous Plants on Livestock*. Academic Press, London, UK

Keeney, D. R., MacGregor, A. N. (1978) Short-term cycling of 15N-urea in a ryegrass – white clover pasture, *New Zealand Journal of Agricultural Research* **21**: 445–8

Kempster, A. J., Cuthbertson, A., Harrington, G. (1982) *Carcase Evaluation in Livestock Breeding, Production and Marketing*. Granada, London, UK

Kennedy, J. P., Auldist, I. H., Popovic, P. G., Reynolds, J. A. (1976) Reproduction rate of Merino sheep in arid N.S.W., *Proceedings of the Australian Society of Animal Production* **11**: 149–51

Khalil, I. A., Morad, H. M. (1977) Feed intake in Ossimi, Rahmani and Merino rams, *Egyptian Journal of Animal Production* **17**: 87–93

Khattab, A. G. H. (1968) Haemoglobin type and blood potassium and sodium concentrations in Sudan Desert sheep, *Journal of Agricultural Science* (UK) **70**: 95–7

Kim, C. S. (1981) Processing of lignocellulosic materials for animal consumption. In Manassah, J. T., Briskey, E. J. (eds) *Advances in Food-producing Systems for Arid and Semiarid Lands, part A,* pp. 267–302. Academic Press, London, Uk

King, J. M. (1983) Livestock water needs in pastoral Africa in relation to climate and forage. *Research Report No. 7.* International Livestock Centre for Africa, Addis Ababa, Ethiopia

King, J. M., Sayers, A. R., Peacock, C. P., Kontrohr, E. (1984) Maasai herd and flock structures in relation to livestock wealth, climate and development, *Agricultural Systems* **13**: 21–56

Kirton, A. H. (1982) Carcase and meat qualities. In Coop, I. E. (ed.) *Sheep and Goat Production,* pp. 259–74. Elsevier, Amsterdam, Netherlands

Kirton, A. H., Colmer-Rocher, F. (1978) The New Zealand sheep and lamb export carcass classification system, *World Review of Animal Production* **14**(1): 33–41

Kishore, K., Chopra, S. C., Acharya, R. M., Balaine, D. S. (1978) Performance of Rambouillet, Malpura and their crosses for wool traits, *Indian Journal of Animal Sciences* **48**: 737–40

Kishore, K., Gour, D., Rawat, P. S., Arora, C. L. (1980) Note on gestation length in cross-bred sheep, *Indian Journal of Animal Sciences* **50**: 565–7

Kishore, K., Gour, D., Rawat, P. S., Malik, R. C. (1982) A study of some reproductive traits of Avikalin and Avivastra ewes and factors affecting them, *Cheiron* **11**: 20–4

Kishore, K., Sethi, I. C., Basuthakur, A. K. (1983) Study on milk production in Avikalin and Avivastra ewes, *Cheiron* **12**: 257–60

Knipscheer, H. C., De Boer, A. J., Sabrani, M., Soedjana, T. (1983) The economic role of sheep and goats in Indonesia: a case study of West Java, *Bulletin of Indonesian Economic Studies* (Australia) **19**: 74–93

Kolding, K. E., Koford, J. D. (1970) Commercial utilization of milk from nomad sheep flocks in the Near East, *World Review of Animal Production,* special issue **26**: 98–105

Kolff, H. E., Wilson, R. T. (1985) Livestock production in Central Mali. The 'Mouton de Case' system of smallholder sheep fattening. *Agricultural Systems* **16** (4): 217–30

Konczacki, Z. A. (1978) *The Economics of Pastoralism, a Case Study of Sub-Saharan Africa.* Frank Cass, London, UK

Korhonen, H. (1980) A new method for preserving raw milk, the lactoperoxidase antibacterial system, *World Animal Review* **35**: 23–9

Kretchmer, N. (1972) Lactose and lactase, *Scientific American* **227**: 71–8

Krishnamurthy, U. S., Rathnasabapathy, V. (1980) Genetics of haemoglobin in Nilagiri, Merino and their crossbred sheep. Haemoglobin types and their relationship with K types and production and reproduction traits, *Indian Veterinary Journal* **57**: 654–9

Krishnappa, S. B. (1980) Comparative study of Deccani with Corriedale × Deccani crosses of sheep. *Thesis Abstracts, Haryana Agricultural University* **6**: 269

Krishniah, P., Gupta, S.K., Bangaruswamy, S., Rao, J. B. (1980) Studies on the breeding of sheepskins towards improving the quality of shearlings, *Wool and Woollens of India* **17**(2): 35–8

Krueger, W. C., Sharp, L. A. (1978) Management approaches to reduce livestock losses from poisonous plants on rangeland, *Journal of Range Management* **31**: 347–50

Kulshrestha, S. K., Upadahyaya, R. B., Saxena, J. S. (1972) Utility of pangola grass *Digitaria decumbens* for sheep, *Indian Veterinary Journal* **48**: 1152–5

Labban, F. (1973) *A Study on Wool Characteristics of the Barbary Sheep.* Ministry of Agriculture, Tripoli, Libyan Arab Republic

Labban, F. M., Ghali, A. (1969) A study on increasing lambing rate of sheep through management, *Journal of Animal Production of the United Arab Republic* **9**: 285–93
LAC (1978) *Annual Technical Report, Sheep Section July 1977–June 1978.* Lumle Agricultural Centre, Lumle, Pokhara, Nepal
Lahlou-Kassi, A., Marie, M. (1981) A note on ovulation rate and embryonic survival in D'man ewes, *Animal Production* **32**: 227–9
Lall, H. K. (1956) Breeds of sheep in the Indian Union. *Miscellaneous Publication No. 75.* Indian Council of Agricultural Research, New Delhi, India
Lane, I. R. (1981) The use of cultivated pastures for intensive animal production in developing countries. In Smith, A. J., Gunn, R. G. (eds) Intensive animal production in developing countries, pp. 105–43. *Occasional Publication No. 4.* British Society of Animal Production, Thames Ditton, UK
Laredo, M. A., Minson, D. J. (1975) The effect of pelleting on the voluntary intake and digestibility of leaf and stem fractions of three grasses, *British Journal of Nutrition* **33**: 159–70
Lawlor, M. J., Louca, A., Mavrogenis, A. (1974) The effect of three suckling regimes on the lactation performance of Cyprus Fat-tailed, Chios and Awassi sheep and the growth rate of lambs, *Animal Production* **18**: 293–9
Lawrie, R. A. (1979) *Meat Science,* 3rd edn. Pergamon Press, Oxford, UK
Le Gros, G. (1963) £11,700 gain from drought feeding sheep, *Queensland Agricultural Journal* **89**: 647–51
Leibholz, J. (1969) Poultry manure and meat meal as a source of dietary nitrogen for sheep, *Australian Journal of Experimental Agriculture and Animal Husbandry* **9**: 589–93
Lewis, T. R., Suess, G. G., Kauffman, R. G. (1969) Estimation of carcass traits by visual appraisal of market livestock, *Journal of Animal Science* **28**: 601–6
Leyva, V., Henderson, A. E., Sykes, A. R. (1982) Effect of daily infection with *Ostertagia circumcinta* larvae on food intake, milk production and wool growth in sheep, *Journal of Agricultural Science* (UK), **99**: 249–59
Lima, C. R., Araújo, M. R., Souto, S. M. (1972a) Valores nutritivos da silagem de sorgo forrageiro e capins elefante, colonião, pangola e guatemala [Nutritive value of silages of forage sorghum and elephant, colonial, pangola and guatemala grasses], *Pesquisa Agropecuária Brasileira, Zootecnia* **7**: 53–7
Lima, C. R., Souto, S. M., Garcia, J. M. R., Araújo, M. R. (1972b) Valores nutritivos do feno de siratro (*Phaseolus atropurpureus*) em diferentes estádios de crescimento [Nutritive value of siratro (*Phaseolus atropurpureus*) at different stages of growth], *Pesquisa Agropecuária Brasileira, Zootecnia* **7**: 63–6
Lindsay, D. R. (1979) Mating behaviour in sheep, In Tomes, G. L., Robertson, D. E., Lightfoot, R. J. (eds) *Sheep Breeding,* 2nd edn. pp. 473–9. Butterworths, London, UK
Linklater, K. A., Watson, G. A. L. (1983) Sheep housing and health, *Veterinary Record* **113**: 560–4
Littlejohn, A. I. (1964) Foot-rot in feeding sheep. The economic aspect of eradication, *Veterinary Record* **76**: 741–2
Lloyd, D. H., Sellers, K. C. (eds) (1976) *Dermatophilus Infection in Animals and Man.* Academic Press, London, UK
Lonkar, P. S., Uppal, P. K., Belwal, L. M., Mathur, P. B. (1983) Bluetongue in India, *Tropical Animal Health and Production* **15**: 86
Louca, A. (1972) The effect of suckling regime on growth rate and lactation performance of the Cyprus Fat-tailed and Chios sheep, *Animal Production* **15**: 53–9
Lyne-Watt, W. (1942) Stall-feeding of goats and sheep by the Kikuyu tribe, Kenya, *East African Agricultural Journal* **8**: 109–11
Lysandrides, P. (1981) The Chios sheep in Cyprus, *World Animal Review* **39**: 12–16
McArthur, I. D. (1980) Pre-lambing supplementation of Gadic ewes in Western Afghanistan, *Journal of Agricultural Science* (UK) **95**: 39–45
McCorkle, C. M., Jimenez-Zamalloa, L. (1982) Management of animal health and disease in an indigenous Andean community. *Small Ruminant CRSP Publication No. 4.* University of Missouri, USA

References

McCray, C. W. R., Hurwood, I. S. (1963) Selenosis in north-western Queensland associated with a marine cretaceous formation, *Queensland Journal of Agricultural Science* **20**: 475-98

McDonald, I. W. (1981) Detrimental substances in plants consumed by grazing ruminants. In Morley, F. H. W. (ed.) *Grazing Animals,* pp. 349-60. Elsevier, Amsterdam, Netherlands

McDonald, P. Edwards, R. A., Greenhalgh, J. F. D. (1981) *Animal Nutrition,* 3rd edn. Longman, London, UK

MacFarlane, W. V., Howard, B., Haines, H., Kennedy, P. J., Sharpe, C. M. (1971) Hierarchy of water and energy turnover of desert mammals, *Nature* (UK) **234**: 483-4

McGarry, M. G., Stainforth, J. (eds) (1978) *Compost, Fertilizer and Biogas Production from Human and Farm Wastes in the People's Republic of China.* International Development Research Centre, Ottawa, Canada

McIntosh, C. E., Osuji, P. O., Birla, S. C., Hope, C. K. (1976) Economics of supplementary feeding of grazed sheep with ground cottonseed, *Tropical Agriculture* (Trinidad) **53**: 97-103

McIvor, J. G., Williams, W. T., Anning, P., Clem, R. L., Finlay, M. C. (1982) The performance of introduced grasses in seasonally dry tropical environments in northern Australia, *Australian Journal of Experimental Agriculture and Animal Husbandry* **22**: 373-81

Mack, S. (1982) Disease as a constraint to productivity. In Gatenby, R. M., Trail, J. C. M. (eds) *Small Ruminant Breed Productivity in Africa,* pp. 81-3. International Livestock Centre for Africa, Addis Ababa, Ethiopia

McKenzie, J. R. (1981) The effect of growth implants on cattle and sheep production, *Proceedings of the New Zealand Society of Animal Production* **41**: 294-6

McLeod, M. N., Minson, D. J. (1974) Differences in carbohydrate fractions between *Lolium perenne* and two tropical grasses of similar dry-matter digestibility, *Journal of Agricultural Science* (UK) **82**: 449-54

McLeroy, G. B. (1961) The sheep of Sudan (2) Ecotypes and tribal breeds, *Sudan Journal of Veterinary Science and Animal Husbandry* **2**: 101-65

Madani, M. O. K., Williams, H. L. (1983) The effect of an equatorial light environment on reproduction in sheep. 1. The breeding season, *British Veterinary Journal* **139**: 338-48

MAFF (1975) Energy allowances and feeding systems for ruminants. *Technical Bulletin 33.* HMSO, London, UK

Mahajan, J. M., Acharya, R. M., Dhillon, J. S. (1980a) Cross-breeding for the production of fine wool: effect on wool production, *Indian Journal of Animal Sciences* **50**: 553-7

Mahajan, J. M., Acharya, R. M., Dhillon, J. S. (1980b) Cross-breeding for the production of fine wool: effect on wool-quality traits, *Indian Journal of Animal Sciences* **50**: 624-8

Mahajan, J. M., Singh, V. K. (1978) Milk yield of Gaddi ewes and its relationship with lamb growth, *Indian Veterinary Journal* **55**: 550-3

Maheswari, M. L., Talapatra, S. K. (1967) Quantitative nutrition of sheep on pasture, *Indian Journal of Animal Health* **6**: 215-20

Makarechian, M., Farid, A., Sefidbakht, N., Mostafavi, M. S. (1973) The influence of breed and weaning age on feedlot performance of Iranian Fat-tailed sheep, *Iranian Journal of Agricultural Research* **2**: 21-9

Malik, B. S. (1976) Genetic and phenotypic correlations of yearling body weight with yearling wool traits of Chokla sheep, *Indian Veterinary Journal* **53**: 699-701

Malik, B. S., Chaudhary, A. L. (1975) Estimates of genetic and phenotypic parameters of wool traits in Chokla sheep, *Indian Journal of Animal Sciences* **45**: 653-5

Malik, R. C., Acharya, R. M. (1972) Breed differences in pre- and post-weaning body weights of Indian sheep, *Indian Journal of Animal Sciences* **42**: 22-7

Malik, R. C., Acharya, R. M., Mehta, B. S. (1978) Effect of season of birth on growth and subsequent reproduction in crossbred sheep, *Indian Veterinary Journal* **55**: 707-12

Malik, R. C., Singh, R. N., Acharya, R. M., Dutta, O. P. (1980) Factors affecting lamb survival in crossbred sheep, *Tropical Animal Health and Production* **12**: 217-23

Mallikeswaran, K., Mani, V., Raman, K. S. (1980) Milk production in ewes and its relationship with the growth of suckling lambs, *Indian Veterinary Journal* **57**: 235-8

Maloiy, G. M. O. (1973) Water metabolism of East African ruminants in arid and semi-arid regions, *Zeitschrift für Tierzuchtung und Züchtungsbiologie* **90**: 219-28

Mann, I. (1969) *A Handbook on Nigerian Hides and Skins.* Government Printer, Interim Common Services Agency, Kaduna, Nigeria

Mann, I. (1978) Meat and carcase by-products. In Williamson, G., Payne, W. J. A. *Animal Husbandry in the Tropics,* 3rd edn. pp. 679-706. Longman, London, UK

Marai, I. F. M. (1972) A study of twin and triplet lambs in Ossimi fat-tailed sheep, *Rivista di Agricoltura Subtropicale e Tropicale* **66**: 307-16

Marai, I. F. M. (1975) A study of some characteristics of two Egyptian breeds of sheep, *Rivista di Agricoltura Subtropicale e Tropicale* **69**: 157-66

Marais, P. G. (1974) Lammortaliteit by herfsen lentegepaarde ooie [Lamb mortality of autumn- and spring-mated ewes], *Agroanimalia* **6**: 133-7

Marcet, J., Onions, W. J. (1963) Incidence, size and shape of meta-cortex in wool, *Nature (UK)* **199**: 1306-7

Marie, M., Lahlou-Kassi, A. (1977) Étude de quelque paramètres de la reproduction des brebis de race Timahdite [Study of some reproductive parameters in Middle Atlas ewes], *Hommes, Terre et Eaux* **6** (25): 23-9

Maro, M. A. M. (1977) An economic study of goat and sheep production in Tanzania. *RER Paper No. 4.* Department of Rural Economy, University of Dar es Salaam, Morogoro, Tanzania.

Maro, M. A. M. (1978) An economic survey of goat and sheep production in Tanzania, *Beiträge zur Tropischen und Subtropischen Landwirtschaft und Veterinärmedizin* **16**: 371-88

Martin, P. C. (1982) Relationships between nutrient content, digestibility and energy concentration in tropical grasses, *Cuban Journal of Agricultural Science* **16**: 155-60

Martínez, A. (1983) Reproduction and growth of hair sheep in an experimental flock in Venezuela. In Fitzhugh, H. A., Bradford, G. E. (eds) *Hair Sheep of Western Africa and the Americas, a Genetic Resource for the Tropics,* pp. 105-17. Westview Press, Boulder, Colorado, USA

Martínez, A., Herrera, J., Valencia, J., Fernández-Baca, S. (1980) Estudio de la actividad ovárica pos-parto mediante la determinación de progesterona en ovejas Dorset, Suffolk y Tabasco [A study of postpartum ovarian activity by means of determining progesterone levels in Dorset, Suffolk and Tabasco ewes], *Veterinaria* (Mexixo) **11**: 127-31

Mason, I. L. (1969) *A World Dictionary of Livestock Breeds, Types and Varieties,* 2nd edn. Commonwealth Agricultural Bureaux, Farnham Royal, UK

Mason, I. L. (1978) Sheep in Java, *World Animal Review* **27**: 17-22

Mason, I. L. (ed.) (1980a) Prolific tropical sheep. *Animal Production and Health Paper 17.* Food and Agriculture Organization of the United Nations, Rome, Italy

Mason, I. L. (1980b) Sheep and goat production in the drought polygon of Northeast Brazil, *World Animal Review* **34**: 23-8

Mason, I. L., Buvanendran, V. (1982) Breeding plans for ruminant livestock in the tropics. *Animal Production and Health Paper 34.* Food and Agriculture Organization of the United Nations, Rome, Italy

Masud, M., Hussain, S. S. (1961) *Livestock Breeds of West Pakistan.* Pakistan Agricultural Publishers, Lahore, Pakistan

Mathius, W., Pulungan, H., Thomas, N. (1983) A preliminary analysis of results from the nutrition village monitoring programme. *Working Paper 11.* North Carolina State University, Raleigh, USA

Matter, H. E. (1974) Mating and conception during gestation in sheep, *Yearbook, Karakul Breeders Society of South Africa* **16**: 45-55

Matter, H. E. (1975) Zur sexuellen Aktivität innerhalb von Deckzeiten beim Karakul sheep [On sexual activity during the mating season in Karakul sheep], *Deutsche Tierärztliche Wochenschrift* **82**: 368-70

Matthewman, R. W. (1980) Small ruminant production in the humid tropical zone of southern Nigeria, *Tropical Animal Health and Production* **12**: 234-42

Mattner, P. E., Braden, A. W. H., Turnbull, K. E. (1967) Studies in flock mating of sheep. 1. Mating behaviour, *Australian Journal of Experimental Agriculture and Animal Husbandry* **7**: 103–9

Mavrogenis, A. P. (1982) Environmental and genetic factors influencing milk production and lamb output of Chios sheep, *Livestock Production Science* **8**: 519–27

Maxwell, T. J. (1979) Economic techniques for the assessment of sheep systems. In *The Management and Diseases of Sheep,* pp. 400–20. British Council, London, UK

Meinecke-Tillmann, S., Meinecke, B. (1984) Experimental chimaeras – removal of reproductive barrier between sheep and goat, *Nature* (UK) **307**: 637–8

Memon, M. A., Ott, R. S. (1981) Methods of semen preservation and artificial insemination in sheep and goats, *World Review of Animal Production* **17**(1): 19–25

Merrill, L. B., Schuster, J. L. (1978) Grazing management practices affect livestock losses from poisonous plants, *Journal of Range Management* **31**: 351–4

Meyer, W. D. (1951) The Dorper: its characteristics and value. *Farming in South Africa,* **26**: 289–90

Mian, I. A. (1976) Some studies on the clean wool and fat contents of Lohi wool, *Agriculture* (Pakistan) **27**: 173–6

Mignon, J., Diague, G. (1981) Essai du monensin comme additif alimentaire chez le mouton Djallonké [Study of monensin as a feed additive for West African Dwarf sheep], *Annales de Médecine Vétérinaire* **125**: 269–77

Miles, W. H., McDowell, L. R. (1983) Mineral deficiencies in the llanos rangelands of Colombia, *World Animal Review* **46**: 2–10

Milford, R. (1967) Nutritive values and chemical composition of seven tropical legumes and lucerne grown in subtropical south-eastern Queensland, *Australian Journal of Experimental Agriculture and Animal Husbandry* **7**: 540–5

Miller, W. C., Robertson, E. D. S. (1945) *Practical Animal Husbandry,* 4th edn. Oliver and Boyd, Edinburgh, UK

Mills, O. (1982) *Practical Sheep Dairying. The Care and Milking of the Dairy Ewe.* Thorson's Publishers, Wellingborough, UK

Minett, F. C. (1950) Mortality in sheep and goats in India, *Indian Journal of Veterinary Science and Animal Husbandry* **20**: 69–103

Minson, D. J. (1972) The digestibility and voluntary intake by sheep of six tropical grasses, *Australian Journal of Experimental Agriculture and Animal Husbandry* **12**: 21–7

Minson, D. J. (1973) Effect of fertilizer nitrogen on digestibility and voluntary intake of *Chloris gayana, Digitaria decumbens* and *Pennisetum clandestinum, Australian Journal of Experimental Agriculture and Animal Husbandry* **13**: 153–7

Minson, D. J., Stobbs, T. H., Hegarty, M. P., Playne, M. J. (1976) Measuring the nutritive value of pasture plants. In Shaw, N. H., Bryan, W. W. (eds) *Tropical Pasture Research, Principles and Methods,* pp. 308–37. Commonwealth Agricultural Bureaux, Farnham Royal, Slough, UK

Minson, D. J., Ternouth, J. H. (1971) The expected and observed changes in the intake of three hays by sheep after shearing, *British Journal of Nutrition* **26**: 31–9

Mitchell, J. R. (1980) *Guide to Meat Inspection in the Tropics,* 2nd edn. Commonwealth Agricultural Bureaux, Farnham Royal, Slough, UK

Mitidieri, E., Affonso, O. R., Tokarnia, C. H. (1959) Copper and iron in livers of undernourished cattle and sheep from northeastern and northern Brazil, *American Journal of Veterinary Research* **20**: 247–8

Mittal, J. P., Ghosh, P. K. (1980) A note on annual reproductive rhythm in Marwari sheep of the Rajasthan desert in India, *Animal Production* **30**: 153–6

Mittal, J. P., Ghosh, P. K. (1983) Long-term saline drinking and female reproductive performance in Magra and Marwari sheep of the Indian desert, *Journal of Agricultural Science* (UK) **101**: 751–4

Mittal, J. P., Pandey, M. D. (1975) Effect of age and season on wool yield in Bikaneri (Magra) ewes, *Indian Journal of Animal Sciences* **45**: 161–2

Mittendorf, H. J. (1981) Livestock and meat marketing in Asian countries, *World Animal Review* **40**: 34–42

Mlambiti, M. E. (1975) *The Study of Sheep and Goats Marketing in Tanzania.* Department of Rural Economy and Extension, University of Dar es Salaam, Morogoro, Tanzania

MLC (1975) *Planned Crossbreeding and Lamb Carcass Weights.* Meat and Livestock Commission, Bletchley, UK

MLC (1982) Results from recorded commercial flocks selling lambs off grass, summer and autumn 1981. *MLC Sheep Improvement Services Data Sheet 82/1.* Meat and Livestock Commission, Bletchley, UK

Mohan, M., Acharya, R. M. (1980) Heterosis in fleece characteristics, *Indian Journal of Animal Genetics and Breeding* **2**: 35–8

Molokwu, E. C. I., Umunna, N. N. (1980) Reproductive performance of the Y'ankasa sheep of Nigeria, *Theriogenology* **14**: 239–48

Monteith, J. L. (1973) *Principles of Environmental Physics.* Edward Arnold, London, UK

Moore, C. P. (1975) Livestock production specialists – one system for training in the lowland tropics, *World Animal Review* **13**: 38–43.

Moore, R. W. (1967) A comparison of methods of estimating milk intake of lambs and milk yield of ewes at pasture, *Australian Journal of Experimental Agriculture and Animal Husbandry* **7**: 137–40

Moore, R. W., Dolling, C. H. S. (1961) Ability to select diet in Merino sheep fed in pens, *Australian Journal of Experimental Agriculture and Animal Husbandry* **1**: 15–17

Morag, M. (1972) A model for the stratification of dairy and mutton sheep breeds in Middle Eastern deserts, *Journal of Range Management* **25**: 296–300

Morag, M., Degen, A. A., Popliker, F. (1973) The reproductive performance of German Mutton Merino ewes in a hot arid climate, *Zeitschrift für Tierzüchtung und Züchtungsbiologie* **89**: 340–5

Morag, M., Raz, A., Eyal, E. (1970) Mother – offspring relationships in Awassi sheep. IV. The effect of weaning at birth, or after 15 weeks, on lactational performance in the dairy ewe, *Journal of Agricultrual Science* (UK) **75**: 183–7

More, T., Rawat, P. S., Sahni, K. L. (1981) Some observations on high- and low-potassium Chokla sheep given water intermittently under semi-arid conditions, *Journal of Agricultural Science* (UK) **96**: 463–9

More, T., Singh, N. P., Sahni, K. L. (1976) A note on the effect of water deprivation on changes in organ weights, carcass yield and body composition of Chokla sheep under semi-arid conditions, *Indian Veterinary Journal* **53**: 199–201

More, T., Tiwari, S. B., Sahni, K. L. (1980) Some observations on semen quality of rams with genetically determined high and low potassium blood concentrations, *Theriogenology* **13**: 391–5

Moreno Chan, R. (1976) Estado actual y perspectivas de la producción ovina en México [Current position and future prospects of the sheep industry in Mexico], *Veterinaria* (Mexico) **7**: 136–41

Morris, R. S. (1969) Assessing the value of veterinary services to primary industries, *Australian Veterinary Journal* **45**: 295–300

Morris, R. S., Meek, A. H. (1980) Measurement and evaluation of the economic effects of parasitic disease, *Veterinary Parasitology* **6**: 165–84

Moule, G. R. (1951) Drought feeding of sheep, *Australian Veterinary Journal* **27**: 334–8

Moule, G. R., Braden, A. W. H., Lamond, D. R. (1963) The significance of oestrogens in pasture plants in relation to animal production, *Animal Breeding Abstracts* **31**: 139–57

Mount, L. E. (1979) *Adaptation to Thermal Environment. Man and his Productive Animals.* Edward Arnold, London, UK

Mukhamed, I. (1973) Kachestvo shersti sirñskikh ovets porody avassi [Wool quality of Awassi sheep in Syria], *Sbornik Nauchnykh Trudov Moskovskaya Veterinarnaya Akademiya* **71**: 53–6

Mukherjee, T. K. (1980) Animal genetic resources in Malaysia. In *Animal Genetic Resources in Asia and Oceania,* pp. 261–311. Tropical Agriculture Research Center, Ministry of Agriculture, Forestry and Fisheries, Tsukuba, Ibaraki, Japan

Müller, Z. O. (1982) Feed from animal wastes: feeding manual. *Animal Production and Health Paper 28.* Food and Agriculture Organization of the United Nations, Rome, Italy

Murray, R. M. (1978) Annual pattern of ovulation rates in Merinos in semi-arid tropical Queensland, *Proceedings of the Australian Society of Animal Production* **12**: 259

Murugaraj, I., Krishnamurthy, U. S., Rathnasabapathy, V. (1980) Erythrocyte reduced glutathione (GSH) and its relationship with certain biochemical and production traits, *Cheiron* **9**: 102-8

Naga, M. A., El-Shazly, K. (1983) Use of by-products in animal-feeding systems in the delta of Egypt. In Kiflewahid, B., Potts, G. R., Drysdale, R. M. (eds) *By-product Utilization for Animal Production,* pp. 9-15. International Development Research Centre, Ottawa, Canada

Naranjo, Q. A., Sabogal, O. Y. (1978) Comportamiento reproductivo ovino en una región alta de Colombia. 1. Epocas de mayor fertilidad ovina [Reproductive performance of sheep at a high altitude in Colombia. 1. Seasons with the best ewe fertility], *Revista, Instituto Colombiano Agropecuario* **13**: 297-304

Narayan, S. (1960) Skin follicle types, ratios, and population densities in Rajasthan breeds of sheep, *Australian Journal of Agricultural Research* **11**: 408-26

Narayanaswamy, M. (1978) A note on the factors affecting gestation length in Bannur sheep, *Indian Journal of Animal Sciences* **48**: 478-80

Narayanaswamy, M., Balaine, D. S. (1976) A note on the oestrus cycle in Bannur sheep, *Indian Veterinary Journal* **53**: 235-6

Narayanaswamy, M., Balaine, D. S., Chopra, S. C. (1975) A note on gestation length in Mandya sheep, *Indian Journal of Animal Sciences* **45**: 915-17

NAS (1977) *Leucaena, Promising Forage and Tree Crop for the Tropics.* National Academy of Sciences, Washington, DC, USA

NAS (1979) *Tropical Legumes: Resources for the Future.* National Academy of Sciences, Washington, DC, USA

Nash, C. E. (1964) The assessment of N-type fleeces. II. The physical properties and practical testing of the finished carpets, *Journal of the Textile Institute* **55**: T309-23

Naude, J. F., Grant, J. L. (1979) An investigation into an eight-month reproduction cycle for sheep. In *Division of Livestock and Pastures Annual Report for the Year ended 30 September 1978,* pp. 223-4. Department of Research and Specialist Services, Salisbury, Rhodesia

Nel, J. A., Mostert, L., Steyn, M. G. (1960) Karakul breeding and research in South West Africa with special reference to the Neudam Karakul stud, *Animal Breeding Abstracts* **28**: 89-101

Nel, J. E., Allan, J. S., Schalkwyk, D. J. van (1972) Die invloed van ouderdom op prestasie en tempo van genetiese verandering by Merinoskape [The effect of age on performance and rate of genetic change in Merino sheep], *Agroanimalia* **4**: 1-6

Ngere, L. O. (1973) Size and growth rate of the West African Dwarf sheep and a new breed, the Nungua-Black-Head of Ghana, *Ghana Journal of Agricultural Science* **6**: 113-17

Ngere, L. O., Aboagye, G. (1981) Reproductive performance of the West African Dwarf and the Nungua Black Head sheep of Ghana, *Animal Production* **33**: 249-52

Niekerk, C. H. van (1979) Limitations to female reproductive efficiency. In Tomes, G. L., Robertson, D. E., Lightfoot, R. J. (eds) *Sheep Breeding,* 2nd edn. pp. 303-13. Butterworths, London, UK

Nikolaeva, N. (1974) Vliyanie na rodstveniya podbor v"rkhu nyakoi stopanski priznatsi na ovtsete [The effect of inbreeding on some production characters of sheep], *Zhivotnov"dni Nauki* **11**: 51-5

Nivsarkar, A. E., Acharya, R. M. (1980) Selection for maximising productivity (meat and wool) in the Malpura and Sonadi breeds. In *Annual Report 1979,* pp. 63-5. Central Sheep and Wool Research Institute, Avikanagar, Rajasthan, India

Nivsarkar, A. E., Acharya, R. M. (1982) Note on feedlot and carcass characteristics of Malpura sheep, *Indian Journal of Animal Sciences* **52**: 1262-4

Nivsarkar, A. E., Kunzru, O. N., Dwarkanath, P. K. (1971) Studies on the seminal characters of Magra rams, *Annals of Arid Zone* **10**: 58-65

Nogueira Padilha, T. (1980) Observações sobre a classifição das peles caprina e ovina, no Nordeste do Brasil, destinadas à exportação [Observations on the classification of goat and sheep skins produced in north-eastern Brazil and destined for export]. *Pesquisa em Andamento No. 2.* Empresa Brasileira de Pesquisa Agropecuária, Bagé, Brazil

Norman, L. M., Hohenboken, W. (1979) Genetic and environmental effects on internal parasites, foot soundness and attrition in crossbred ewes, *Journal of Animal Sciences* **48**: 1329–37

Nwude, N., Ibrahim, M. A. (1980) Plants used in traditional veterinary medical practice in Nigeria, *Journal of Veterinary Pharmacology and Therapeutics* **3**: 261–73

Obst, J. M., Chaniago, T., Boyes, T. (1980) Survey of sheep and goats slaughtered at Bogor, West Java, Indonesia. *Centre Report No. 10.* Centre for Animal Research and Development, Bogor, Indonesia

Odell, M. J. (1982) Local institutions and management of communal resources lessons from Africa and Asia. *Pastoral Network Paper 14e.* Overseas Development Institute, London, UK

Oji, U. I., Mowat, D. N., Winch, J. E. (1977) Alkali treatments of corn stover to increase nutritive value, *Journal of Animal Science* **44**: 798–802

Okoh, A. E. J., Gadzama, J. N. (1982) Sarcoptic mange of sheep in Plateau State, Nigeria, *Bulletin of Animal Health and Production in Africa* **30**: 61–3

Onions, W. J. (1962) *Wool. An Introduction to its Properties, Varieties, Uses and Production.* Ernest Benn, London, UK

Orji, B. I., Steinbach, J. (1980) Post-partum anoestrous period and lambing interval in the Nigerian Dwarf sheep, *Bulletin of Animal Health and Production in Africa* **28**: 366–71

Ørskov, E. R., Hovell, F. D. de B., Mould, F. (1980) The use of the nylon bag technique for the evaluation of feed stuffs, *Tropical Animal Production* **5**: 195–213

Osman, A. H., Bradford, G. E. (1967) Genotype–environment interaction and compensatory growth in sheep, *Journal of Animal Science* **26**: 1239–43

Osman, A. H., El-Shafie, S. A. (1967) Carcass characteristics of Sudan Desert sheep, *Sudan Journal of Veterinary Science and Animal Husbandry* **8**: 115–19

Osman, A. H, El-Shafie, S. A., Khattab, A. G. H. (1970) Carcass composition of fattened rams and wethers of Sudan Desert sheep, *Journal of Agricultural Science* (UK) **75**: 257–63

Osman, H. E., Fadlalla, B. (1974) The effect of level of water intake on some aspects of digestion and nitrogen metabolism of the 'desert sheep' of the Sudan, *Journal of Agricultural Science* (UK) **82**: 61–9

Osuagwuh, A. I. A., Taiwo, B. B. A., Ngere, L. O. (1980) Crossbreeding in tropical sheep: incidence of dystocia and parturition losses, *Tropical Animal Health and Production* **12**: 85–9

Otchere, E. O., Musah, I. A., Bafi-Yeboa, M. (1983) The digestibility of cocoa husk-based diets fed to sheep, *Tropical Animal Production* **8**: 33–8

Ott, R. S., Memon, M. A. (1980) Breeding soundness examinations of rams and bucks, a review, *Theriogenology* **13**: 155–64

Owen, I. L. (1982) Sheep helminths. In Nunn, M. J. (ed.) *Proceedings of the 1981 Veterinary Officers Conference,* pp. 98–9. Department of Primary Industry, Port Moresby, Papua New Guinea

Owen, J. B. (1976) *Sheep Production.* Baillière Tindall, London, UK

Owen, J. B., Lee, R. F., Lerman, P. M., Miller, E. L. (1980) The effect of reproductive state of ewes on their voluntary intake of diets varying in straw content, *Journal of Agricultural Science* (UK) **94**: 637–44

Owen, J. E., Norman, G. A. (1977) Studies on the meat production characteristics of Botswana goats and sheep – Part II: General body composition, carcase measurements and joint composition, *Meat Science* **1**: 283–306

Oyenuga, V. A., Akinsoyinu, A. O. (1977) Nutrient requirements of sheep and goats of tropical breeds. In Fonnesbeck, P. V., Harris, L. E., Kearl, L. C. (eds) *First International Symposium: Feed Composition, Animal Nutrient Requirements, and Computerization of Diets,* pp. 505–11. Utah Agricultural Experiment Station, Utah State University, Logan, Utah, USA

Pacho, H. N. (1973) Diagnóstico de gestación por laparotomia en ovejas [Pregnancy diagnosis by laparotomy in ewes], *Veterinaria* (Mexico) **4**: 161–5

Pálsson, H. (1955) Conformation and body composition. In Hammond, J. (ed) *Progress in the Physiology of Farm Animals,* vol. 2, pp. 430–542. Butterworths, London, UK

Pandey, J. N., Shukla, D. C., Pal, A. K. (1972a) Studies on resting metabolism of sheep fed akra (*Vicia* species) and concentrate for eight months, *Indian Veterinary Journal* **49**: 48–50

Pandey, M. D., Seth, O. N., Pant, H. C., Roy, A. (1972b) Augmentation of lambing percentage in Bikaneri (Magra) ewes through serum gonadotrophin treatment, *Indian Journal of Animal Sciences* **42**: 28–31

Pant, H. C. (1979) Oestradiol-17β induced LH and FSH surges in the anoestrus and cyclic ewes, *Veterinary Research Bulletin* **2**: 1–6

Pant, K. P., Kulshresta, T. A., Bhatia, R. K. (1980) Studies on the wool quality of Nali and Chokla sheep and their crosses with Rambouillet and Soviet Merino, *Indian Veterinary Journal* **57**: 913–18

Park, R. J., Minson, D. J. (1972) Flavour differences in meat from lambs grazed on tropical legumes, *Journal of Agricultural Science* (UK) **79**: 473–8

Parvaneh, V. (1972) The physical and chemical characteristics of sheep tail fat (Donbeh), *Tropical Science* **14**: 169–71

Pastrana, B. R., Camacho, D. R., Bradford, G. E. (1983) African sheep in Colombia. In Fitzhugh, H. A., Bradford, G. E. (eds) *Hair Sheep of Western Africa and the Americas, a Genetic Resource for the Tropics,* pp. 79–84. Westview Press, Boulder, Colorado, USA

Patterson, H. C. (1976) *The Barbados Black Belly Sheep.* Ministry of Agriculture, Science and Technology, St. Michael, Barbados

Peart, J. N. (1968) Lactation studies with Blackface ewes and their lambs, *Journal of Agricultural Science* (UK) **70**: 87–94

Peart, J. N. (1982) Lactation of suckling ewes and does. In Coop, I. E. (ed.) *Sheep and Goat Production,* pp. 119–34. Elsevier, Amsterdam, Netherlands

Peart, J. N., Doney, J. M., Smith, W. F. (1979) Lactation pattern in Scottish Blackface and East Friesland × Scottish Blackface cross-bred ewes, *Journal of Agricultural Science* (UK) **92**: 133–8

Peart, J. N., Edwards, R. A., Donaldson, E. (1972) The yield and composition of the milk of Finnish Landrace × Blackface ewes. 1. Ewes and lambs maintained indoors, *Journal of Agricultural Science* (UK) **79**: 303–13

Perdomo, J. T., Shirley, R. L., Chicco, C. F. (1977) Availability of nutrient minerals in four tropical forages fed freshly chopped to sheep, *Journal of Animal Science* **45**: 1114–19

Perera, M. E. (1972) Sheep breeding and management under coconut in Ceylon, *Ceylon Coconut Quarterly* **23**: 100–2

Peters, G. H. (1983) The 1982 world development report: a summary, review and conclusions, *World Agricultural Economics and Rural Sociology Abstracts* **25**: xi–xvi

Pfeifer, E. (1953) *Karakul Atlas.* Karakul Breeders Association of South West Africa, Windhoek, South West Africa

Pinstrup-Andersen, P. (1982) *Agricultural Research and Technology in Economic Development,* Longman, London, UK

Piper, L. R., Bindon, B. M., Nethery, R. D. (eds) (1982) *The Booroola Merino.* Commonwealth Scientific and Industrial Research Organization, Division of Animal Production, Melbourne, Australia

Plant, J. W. (1980) Pregnancy diagnosis in the ewe, *World Animal Review* **36**: 44–7

Plucknett, D. L. (1979) *Managing Pastures and Cattle under Coconuts.* Westview Press, Boulder, Colorado, USA

Poggenpoel, D. G., Merwe, C. A. van der (1975) Die gebruik van seleksieindekse by Merinoskape [The use of selection indices for Merino sheep], *South African Journal of Animal Science* **5**: 249–55

Pollott, G. E., Ahmed, F. A. (1978) The effect of weaning age on the growth of Watish (Sudan Desert type) lambs under feedlot conditions. *Research Bulletin No. 2.* Um Banein Livestock Research Station, Singa, Blue Nile Province, Sudan

Pollott, G. E., Ahmed, F. A. (1979a) An investigation into the sex difference in growth rate of Watish lambs (Sudan Desert type) and the related food conversion efficiency. *Research Bulletin No. 10.* Um Banein Livestock Research Station, Singa, Blue Nile Province, Sudan

Pollott, G. E., Ahmed, F. A. (1979b) The preweaning growth of Watish lambs and the influence of sex, time of birth and management on weight-for-age. *Research Bulletin No. 11.* Um Banein Livestock Research Station, Singa, Blue Nile Province, Sudan

Pond, W. G., Ferrell, C. L., Jenkins, T. G., Young, L. D. (1982) Weaning of lambs to a dry diet at ten days of age, *Journal of Animal Science* **55**: 1284-92

Prasad, V. S. S., Nivsarkar, A. E., Singh, R. N., Bohra, S. D. J., Kumar, M. (1983) Body conformation and carcass characteristics of native and crossbred lambs on a high energy ration, *Indian Journal of Animal Sciences* **53**: 156-61

Prasad, V. S. S., Singh, R. N. (1982) Feedlot performance and carcass composition of fat-tailed crossbred lambs, *Cheiron* **11**: 160-4

Prescott, J. H. D. (1979) Growth and development of lambs. In *The Management and Diseases of Sheep*, pp. 358-76. Commonwealth Agricultural Bureaux, Farnham Royal, Slough, UK

Pritchard, C. J. R., Pennell, A. E., Williams, G. L. (1975) A note on the wool characteristics of sheep at the Hofhuf Agricultural Research Centre. *Publication No. 61.* Joint Agricultural Research and Development Project, University College of North Wales, UK, and Ministry of Agriculture and Water, Saudi Arabia

Pritchard, C. J. R., Ruxton, I. B. (1977) A preliminary investigation into the effect of shade on the growth and feed intake of weaned Awassi ewe lambs. *Publication No. 79.* Joint Agricultural Research and Development Project, University College of North Wales, UK, and Ministry of Agriculture and Water, Saudi Arabia

Prucoli, J. O. (1973) Comparação entre épocas de tosquia de ovinos com relação a caracteristicas zootécnicas [Comparing the performance of sheep shorn in different seasons], *Boletim de Indústria Animal* (Brazil) **30**: 323-56

Purohit, G. R., Ghosh, P. K., Taneja, G. C. (1973) Effects of varying degrees of water restriction on the distribution of body water in high- and low-potassium-type Marwari sheep, *Journal of Agricultural Science* (UK) **80**: 177-80

Purushotam, C. H. (1978) Comparative study of the production and reproduction traits of Nellore and Mandya sheep, *Thesis Abstracts, Haryana Agricultural University* **4**: 99

Quartermain, A. R. (1964) Heat tolerance in Southern Rhodesian sheep fed on a maintenance diet, *Journal of Agricultural Science* (UK) **62**: 333-9

Rae, A. L. (1982) Breeding. In Coop, I. E. (ed.) *Sheep and Goat Production*, pp. 15-55. Elsevier, Amsterdam, Netherlands

Ragab, M. T., Sharafeldin, M. A., Imam, S. A., Abdel-Aziz, A. S. (1978) Some non-genetic sources of variation in fleece weights of Fleisch Merino sheep in five commercial flocks in Egypt, *Egyptian Journal of Animal Production* **18**: 13-22

Rahman, M. F., Konuk, T. (1977) A note on transferrin genotypes and their relationship with weight gain in sheep, *Animal Production* **25**: 99-101

Rahman, M. M., Huq, M. A. (1976) A comparative study of gestation period, prolificacy and lambing interval of native and up-graded Lohi sheep, *Bangladesh Veterinary Journal* **10**: 31-35

Rahman, M. M., Rahim, Q. M. F., Hoque, M. (1971) A comparative study of the wool quality of local × Lohi grade I and grade II, *Bangladesh Journal of Animal Science* **4**: 9-12

Ram, S., Singh, B., Balaine, D. S., Chopra, S. C. (1978) A note on fleece characteristics of Nali purebred and its crosses with Corriedale, *Indian Veterinary Journal* **55**: 407-10

Ramadan, M. Y. (1974) A note on the effect of shearing, shade, and water spraying on the heat tolerance of Najdi sheep. *Publication No. 28.* Joint Agricultural Research and Development Project, University College of North wales, UK, and Ministry of Agriculture and Water, Saudi Arabia

Ramadan, M. Y. A., Lucas, I. A. M., Farnworth, J. (1977) Aspects of Najdi ewe and lamb performance in Saudi Arabia: summer and winter lamb production from Najdi ewes.

Publication No. 84. Joint Agricultural Research and Development Project, University College of North Wales, UK, and Ministry of Agriculture and Water, Saudi Arabia
Ramadan M. Y., Robinson, W. I. (1973) The performance of Black Najdi lambs on diets containing date molasses. *Publication No. 23.* Joint Agricultural Research and Development Project, University College of North Wales, UK, and Ministry of Agriculture and Water, Saudi Arabia
Raman, K. S., Sivaraman, T., Mani, V. (1981) Reproductive performance of Nilagiri and Merino sheep, *Indian Veterinary Journal* **58**: 633-5
Rao, L. R., Purushotham, C., Reddy, K. K. (1978) Studies on gestation period in Mandya and Nellore breeds of sheep, *Livestock Adviser* **3** (6): 9-12
Rastogi, R. K., Youssef, F. G., Keens-Dumas, M. J., Davis, D. (1979) Note on early growth rates of lambs of some tropical breeds, *Tropical Agriculture* (Trinidad) **56**: 259-61
Ratan, R., Bhatia, D. R., Singh, M. (1979) A note on chemical composition and nutritive value of wheat bhusa for sheep, *Indian Journal of Animal Sciences* **49**: 676-7
Ravindran, V., Rajamahendran, R., Nadarajah, K., Goonewardene, L. A. (1983) Production characteristics of indigenous sheep under traditional management systems in northern Sri Lanka, *World Review of Animal Production* **19**(3): 47-52
Reddy, K. K., Mehta, B. S., Arora, C. L. (1979) Effect of ewes' weight at service on lamb production in Soviet Merino, *Indian Veterinary Journal* **56**: 487-90
Rees, M. C., Minson, D. J. (1976) Fertilizer calcium as a factor affecting the voluntary intake, digestibility and retention time of pangola grass (*Digitaria decumbens*) by sheep, *British Journal of Nutrition* **36**: 179-87
Rendel, J. M. (1981) Adaptation of livestock to their environment. In Animal genetic resources conservation and management, pp. 190-200. *Animal Production and Health Paper 24.* Food and Agriculture Organization of the United Nations, Rome, Italy
Reyneke, J., Fair, N. J. (1972) Seasonal live body mass changes and wool growth of Döhne Merino sheep in the eastern highveld, *Agroanimalia* **4**: 25-33
Reynolds, S. G. (1980) Grazing cattle under coconuts, *World Animal Review* **35**: 40-5
Richards, J. I. (1981) Methods of alleviating the deleterious effects of tropical climates on animal production. In Smith, A. J., Gunn, R. G. (eds) Intensive animal production in developing countries, pp. 57-87. *Occasional Publication No. 4.* British Society of Animal Production, Thames Ditton, UK
Richards, K. L. (1957) Sheep breeding in Israel, *Agriculture* (UK) **64**: 400-3
Richardson, C. (1972) Pregnancy diagnosis in the ewe: a review, *Veterinary Record* **90**: 264-75
Ricordeau, G. (1982) Major genes in sheep and goats. In *Second World Congress on Genetics applied to Livestock Production, 4-8 October 1982, 6. Round Tables,* pp. 454-61. Editorial Garsi, Madrid, Spain
Robertson, A. (ed) (1976) *Handbook of Animal Diseases in the Tropics,* 3rd edn. British Veterinary Association, London, UK
Robertson, A. M. (1977) *Farm Wastes Handbook.* Scottish Farm Buildings Investigation Unit, Craibstone, Bucksburn, Aberdeen, UK
Robertson, H. A., Sarda, I. R. (1971) A very early pregnancy test for mammals: its application to the cow, ewe and sow, *Journal of Endocrinology* **49**: 407-19
Robinson, J. J. (1978) Response of the lactating ewe to variation in energy and protein intake. In Boyazoglu, J. G., Treacher, T. T. (eds) Milk production in the ewe, pp. 53-65. *Publication No. 23.* European Association for Animal Production, Brussels, Belgium
Rodel, M. G. W. (1972) Effects of different grasses on the incidence of neonatal goitre and skeletal deformities in autumn born lambs, *Rhodesia Agricultural Journal* **69**: 59-60
Rosenweig, N. S. (1969) Adult human milk intolerance and intestinal lactase deficiency. A review, *Journal of Dairy Science* **52**: 585-7
Roux, P. J. le, Barnard, J. P. (1974) The effect of heterosexual contact on libido and mating dexterity in Karakul rams, *South African Journal of Animal Science* **4**: 171-4
Roux, P. J. le, Wyk, L. C. van (1977) Studies on reproduction in Karakul sheep: effect of shortened length of gestation after flumethazone treatment on the quality of Karakul lamb pelts. 1971-1975. In *Agricultural Research 1976,* pp. 86-7. Department of Agricultural Technical Services, Pretoria, South Africa

Roy, A., Datta, I. C., Sahani, K. L., Singh, B., Sen Gupta, B. P. (1962) Studies on certain aspects of sheep and goat husbandry. Artificial breeding, telescoping the breeding season and certain reproductive phenomena in sheep and goats, *Indian Journal of Veterinary Science and Animal Husbandry* **32**: 269-75

Roy, J. H. B. (1980) *The Calf*, 4th edn. Butterworths, London, UK

Rudert, C. P. (1976) Weaning Dorper lambs at 12 or 16 weeks of age, *Rhodesia Agricultural Journal* **73**: 63-4

Russell, E. W. (1973) *Soil Conditions and Plant Growth*, 10th edn. Longman, London, UK

Ruttan, V. W. (1982) *Agricultural Research Policy*. University of Minnesota Press, Minneapolis, USA

Ruxton, I. B. (1975) A preliminary report on the proximate composition of forages grown at the animal production and agricultural research centre, Hofhuf, Saudi Arabia. *Publication No. 59*. Joint Agricultural Research and Development Project, University College of North Wales, UK, and Ministry of Agriculture and Water, Saudi Arabia

Ryder, M. L. (1969) Changes in the fleece of sheep following domestication. In Ucko, P. J., Dimbleby, G. W. (eds) *The Domestication and Exploitation of Plants and Animals*, pp. 495-521. Duckworth, London, UK

Ryder, M. L. (1981) Livestock husbandry and its products with particular reference to skins and fleeces. In Mercer, R. (ed.) *Farming Practice in British Prehistory*, pp. 182-209. Edinburgh University Press, Edinburgh, UK

Ryder, M. L. (1983a) *Sheep and Man*. Duckworth, London, UK

Ryder, M. L. (1983b) Sheep. In Mason, I. L. (ed.) *Evolution of Domesticated Animals*, pp. 63-85. Longman, London, UK

Ryder, M. L., Stephenson, S. K. (1968) *Wool Growth*. Academic Press, London, UK

Saadullah, M., Haque, M., Dolberg, F. (1980) Treating rice straw with urine, *Tropical Animal Production* **5**: 273-7

SABRAO (1980) Animal genetic resources in Asia and Oceania. *Proceedings of a SABRAO Workshop held at University of Tsukuba, 3-7 Sept 1979*. Tropical Agriculture Research Center, Ministry of Agriculture, Forestry and Fisheries, Tsukuba, Ibaraki, Japan

Sacker, G. D., Trail, J. C. M. (1966) The effect of year, suckling, dry season, and type of dam (ewe or gimmer) on milk production in East African Blackheaded sheep as measured by lamb growth, *Journal of Agricultural Science* (UK) **66**: 93-5

Sagi, R. (1978) Udder support as a means for improving milk fractionation in dairy ewes, *Annales de Zootechnie* **27**: 347-53

Sagi, R., Morag, M. (1974) Udder conformation, milk yield and milk fractionation in dairy ewes, *Annales de Zootechnie* **23**: 185-92

Sahani, M. S., Pant, K. P. (1978) Breed differences in the duration of pregnancy in sheep. *Indian Veterinary Journal* **55**: 99-102

Sahani, M. S., Sahni, K. L. (1976) Seasonal variation in oestrus activity and body weight changes in native sheep (Jaiselmeri) under semi-arid conditions, *Indian Veterinary Journal* **53**: 22-7

Sahni, K. L., Roy, A. (1972a) A note on the effect of two storage temperatures on the keeping quality of sheep and goat semen in different diluents, *Indian Journal of Animal Sciences* **42**: 580-3

Sahni, K. L., Roy, A. (1972b) Post-partum conception in Bikaneri sheep, *Indian Journal of Animal Sciences* **42**: 1038-9

Sahni, K. L., Tiwari, S. B. (1973) Artificial insemination of sheep with diluted and stored semen in milk diluents, *Indian Veterinary Journal* **50**: 536-9

Sahni, K. L., Tiwari, S. B. (1974) Effect of early re-breeding on certain aspects of sheep production, *Indian Journal of Animal Sciences* **44**: 767-70

Sahni, K. L., Tiwari, S. B. (1975) Studies on milk production and pre-weaning growth of lambs in different sheep breeds under semi-arid conditions, *Indian Journal of Animal Sciences* **45**: 252-5

Salleh, W. M. B. W., Tan, K. H. (1982) Viability of sheep rearing under rubber. In Jainudeen, M. R., Omar, A. R. (eds) *Animal Production and Health in the Tropics*, pp. 333-5. Penerbit Universiti Pertanian Malaysia, Serdang, Selangor, Malaysia

Sandford, S. (1983) *Management of Pastoral Development in the Third World.* Wiley, Chichester, UK
Santa-María, R., Calderón, R., Campo, C., Rodríguez, M. (1979) Estudio del comportamiento reproductivo de un rebaño de ovejas de la raza Corriedale en la provincia de Pinar del Río. Reporte preliminar [A study on the reproductive performance of a flock of Corriedale ewes in the province of Pinar del Río. Preliminary report], *Ciencia y Técnica en la Agricultura, Ruminantes* **2**: 79–84
Sarson, M. (1972) L'élevage du mouton de la race Barbarine au Centre d'Ousseltia, Tunisie Centrale [Raising Barbary sheep at the Ousseltia Centre in Central Tunisia]. *Documents Techniques No. 55.* Institut National de la Recherche Agronomique de Tunisie, Tunis, Tunisia
Sastry, M. S., Singh, Y. P. (1979) Experimental *Lantana* poisoning and its treatment in livestock, *Indian Veterinary Journal* **56**: 1007–12
Sattar, A., Raizada, B. C., Pandey, M. D. (1978) A note on some wool fibre characteristics of Merino and crossbred lambs, *Veterinary Research Bulletin* **1**: 148–9
Saxena, A. K., Raizada, B. C., Johari, D. K., Pandey, M. D. (1979) Seasonal variation in the semen quality and fertility of Russian Merino sheep raised under semiarid subtropics, *Veterinary Research Bulletin* **2**: 54–9
Saxena, J. S., Kulshrestha, S. K., Johri, C. B. (1971) Studies on exotic legume fodder, siratro (*Phaseolus atropurpureus*) cultivation and nutritive value for sheep, *Indian Veterinary Journal* **48**: 849–53
Schuh, G. E., Tollini, H. (1979) Costs and benefits of agricultural research: the state of the arts. *Staff Working Paper No. 360.* World Bank, Washington, DC, USA
Schumacher, E. F. (1974) *Small is Beautiful.* Sphere Books, London, UK
Scoville, O. J., Sarhan, M. (1978) Objectives and constraints of ruminant livestock production, *World Review of Animal Production* **14**(1): 43–8
Sefidbakht, N., Ghorban, K. (1972) Changes arising from docking of fat-tailed sheep in feedlot performance, *Iranian Journal of Agricultural Research* **1**: 72–7
Sefidbakht, N., Mostafavi, M. S., Farid, A. (1978) Annual reproductive rhythm and ovulation rate in four fat-tailed sheep breeds, *Animal Production* **26**: 177–84
Sehgal, J. P., Singh, M., Rawat, P. S., Singh, R. N. (1983) Feedlot performance in growing lambs for mutton production, *Indian Journal of Animal Sciences* **53**: 715–19
Selaive, A. B., Borba, E. R., Vaz, C. M. S. L., Oliveira, N. R. M. de (1980) Estudo sobre eficiência reprodutiva dos ovinos [Reproductive efficiency in sheep]. In *Pesquisa em ovinos no Brasil 1975–1979,* p. 53. Empresa Brasileira de Pesquisa Agropecuária, Bagé, Brazil
Selim, M. Z., Youssef, A. A. (1971) Body weight as a criterion for selection of rams in a Merino flock, *Agricultural Research Review* **49**: 31–9
Seneviratna, P. (1967) Economic losses due to parasitic diseases in Ceylon, *Ceylon Veterinary Journal* **15**: 5–8
Sengupta, B. P., Mittal, P. C., Mittal, J. P., Pandey, M. D. (1978) Fertility in young and parous Magra ewes superovulated with different combinations of serum and chorionic gonadotropins, *Indian Journal of Animal Sciences* **48**: 889–90
Seth, O. N., Pandey, M. D., Roy, A. (1972) A note on live-weight, growth and mortality in lambs born as singles, twins or triplets, *Indian Journal of Animal Sciences* **42**: 695–8
Seth, O. N., Pandey, M. D., Roy, A (1973) A note on certain economic characters in relation to haemoglobin type in Bikaneri (Magra) sheep, *Indian Journal of Animal Sciences* **43**: 549–52
SFC (1983) *Annual Report 1981 to 1982.* Sugarcane Feeds Centre, Longdenville, Trinidad
Shah, S. I., Müller, Z. O. (1983) Feeding animal wastes to ruminants. In Kiflewahid, B., Potts, G. R., Drysdale, R. M. (eds) *By-product Utilization for Animal Production,* pp. 49–57. International Development Research Centre, Ottawa, Canada
Shah, S. M. A., Fatima, M., Mohsin, A. H. (1971) Incidence of medullation in carpet wools. I. Relationships of medullation characteristics with diameter, *Pakistan Journal of Scientific and Industrial Research* **14**: 550–2

Sharma, K. M., Kidwai, W. A. (1971) Effect of different levels of feeding on the mortality due to parasites, growth and maturity in cross-bred lambs, *Indian Veterinary Journal* **48**: 349–55

Sharma, K. M., Upadhyaya, R. B., Saxena, J. S. (1972) *Setaria sphacelata*– a promising fodder for sheep. 1. Studies on palatability and nutritive value, *Indian Veterinary Journal* **49**: 1137–40

Sharma, V. V., Murdia, P. C. (1974) Utilization of berseem hay by ruminants, *Journal of Agricultural Science* (UK) **83**: 289–93

Sharrow, S. H., Thomas, D. L., Kennick, W. H. (1981) Effect of monensin on the performance of forage-fed lambs, *Journal of Animal Science* **53**: 869–72

Shehata, M. N., El-Shazly, K., Abou-Akkada, A. R. (1978) Studies on the use of diethylstilbestrol (DES) on fattening of lambs. I. Effect of DES implantation at different ages on rate of gain, carcass quality and chemical composition of carcass, *Alexandria Journal of Agricultural Research* **26**: 575–81

Shelton, M., Figueiredo, E. A. P. (1981) Types of sheep and goats in Northeast Brazil, *International Sheep and Goat Research* **1**: 258–68

Shepherd, P. R. (1959) Woollen spinning for the carpet industry, *Wool Technology and Sheep Breeding* **6**: 19–25

Sherrod, L. B., Ishizaki, S. M. (1966) Effects of substituting various levels of feed-grade macadamia nuts for grain upon the nutritive value of concentrate rations for sheep, *Journal of Animal Science* **25**: 439–44

Shevah, Y., Black, W. J. M., Carr, W. R., Land, R. B. (1975) The effects of nutrition on the reproductive performance of Finn × Dorset ewes. I. Plasma progesterone and LH concentrations during late pregnancy, *Journal of Reproduction and Fertility* **45**: 283–8

Shukla, S. P., Misra, B. S., Singh, B. P. (1978) Studies on wool yield and wool quality traits in relation to glutathione (reduced) activity in sheep, *Indian Veterinary Journal* **55**: 291–5

Side, H. A. (1964) Fibre growth rate changes in foetus, lamb and ewe. Ph.D. thesis, University of Leeds, UK

Sidwell, G. M., Ruttle, J. L., Ray, E. E. (1970) Improvement of Navajo sheep. *Agricultural Experiment Station Research Report*. New Mexico State University, Las Cruces, New Mexico, USA

Siebert, B. D., Hunter, R. A. (1982) Supplementary feeding of grazing animals. In Hacker, J. B. (ed.) *Nutritional Limits to Animal Production from Pastures*, pp. 409–26. Commonwealth Agricultural Bureaux, Farnham Royal, Slough, UK

Silva, P. M. da, Quintana, W. N. M., Teixeira, L. H. D., Costanzi, A. R., Rogrigues, C. O. (1980) Efeito da castração em ovinos [Effect of castration in sheep]. In *Pesquisa em ovinos no Brasil 1975–1979*, p. 101. Empresa Brasileira de Pesquisa Agropecuária, Bagé, Brazil

Singh, B., Kalra, S., Ram, S. Chopra, S. C. (1981a) Fleece characteristics of Nali lambs and its crosses with Corriedale and Russian Merino, *Livestock Adviser* **6**(6): 29–32

Singh, G., Singh, V. K., Mishra, R. K. (1980a) Seasonal variation in performance of ewes of Corriedale, Coimbatore and their crosses (1/2 bred, 5/8th and 3/4ths), *Cheiron* **9**: 179–86

Singh, J., Joshi, H. C., Johari, D. K. (1978) Effect of Saxom on libido and semen quality of rams, *Veterinary Research Bulletin* **1**: 82–3

Singh, K. N., Singh, R. C., Jain, P. M., Tomer, V. P. S. (1982) Improving degraded rangelands for sheep rearing, *Indian Farming* **32** (4): 22–3, 28

Singh, L. B., Singh, M., Chaudry, A. L., Dwarkanath, P. K. (1976) Effect of Hb-type and K-type on some production traits in Magra sheep, *Indian Journal of Animal Health* **15**: 139–43

Singh, L. B., Singh, M., Dwarkanath, P. K. (1973a) A note on the significance of haemoglobin and blood potassium types, *Indian Journal of Animal Sciences* **43**: 896–7

Singh, L. B., Singh, M., Dwarkanath, P. K., Lal, A (1979) Effect of some biochemical characters on production traits in sheep. II. Some wool traits, *Indian Journal of Animal Health* **18**: 27–33

Singh, M., More, T., Rai, A. K. (1980b) Heat tolerance of different genetic groups of sheep

exposed to elevated temperature conditions, *Journal of Agricultural Science* (UK) **94**: 63–7
Singh, N., Taneja, G. C. (1979) Effect of prolonged salt intake on the reproduction of ewes and weight of lamb born to Marwari sheep, *Indian Journal of Animal Research* **13**: 71–4
Singh, N. P. (1975) Chemical composition and nutritive value of Anjan *(Cenchrus ciliaris)* grass for sheep, *Indian Veterinary Journal* **52**: 435–8
Singh, N. P. (1980) Sheep nutrition, some suggestions, *Indian Farming* **30**(9): 27–9
Singh, N. P., Pandit, N. N., Sengar, O. P. S. (1968) Influence of feeding various levels of dietary calcium and phosphorus on the metabolism of feed nutrients, *Balwant Vidyapeeth Journal of Agricultural and Scientific Research* **10**: 7–18
Singh, R. N., Arora, C. L., Bohra, S. D. J., Bapna, D. L., Nivsarkar, A. E. (1981b) Note on variation in the pre- and post-weaning body weights in Malpura and Sonadi, and their crosses with Suffolk and Dorset, *Indian Journal of Animal Sciences* **51**: 877–80
Singh, V. K., Mathur, P. B., Bohra, S. D. J. (1973b) A note on carcass traits of Coimbatore male lambs weaned at two ages, *Indian Journal of Animal Sciences* **43**: 783–5
Singh, V. K., Tiwari, S. B., Singh, L. B., Honmode, J. (1973c) Efficiency of milk production and its conversion into lamb weights in Malpura, Chokla and cross-bred ewes, *Indian Veterinary Journal* **50**: 1199–204
Singh, V. P., Bapna, D. L., Mathur, J. P., Arora, R. R., Mahli, R. S. (1980c) Relationship between fibre diameter, fibre length and medullation percentage in Chokla and Nali wools, *Journal of the Textile Association* **41**: 111–14
Sinha, N. K., Wani, G. M., Sahni, K. L. (1980) Note on the incidence of gestational oestrus in Muzaffarnagri ewes, *Indian Journal of Animal Sciences* **50**: 1009
Sirry, I., El-Sokkary, A. M., Hassan, H. A. (1950) Milk production of Egyptian sheep, *Empire Journal of Experimental Agriculture* **18**: 163–8
Sisson, S. (1953) *The Anatomy of Domestic Animals,* revised by J. D. Crossman, 4th edn. Saunders, Philadelphia, USA
Skerman, T. M., Erasmuson, S. K., Morrison, L. M. (1982) Duration of resistance to experimental footrot infection in Romney and Merino sheep vaccinated with *Bacteroides nodosus* oil adjuvant vaccine, *New Zealand Veterinary Journal* **30**: 27–31
Skinner, J. D., Jochle, W., Nel, J. W. (1970) Induction of parturition in Karakul and cross-bred ewes with flumethasone, *Agroanimalia* **2**: 99–100
Slagsvold, P. (1970) Breed difference in water content of faeces from five breeds of sheep, *Acta Veterinaria Scandanavica* **11**: 31–6
Smith, I. D. (1965) The influence of level of nutrition during winter and spring upon oestrus activity in the ewe, *World Review of Animal Production* **4**: 95–102
Smith, I. D., Clarke, W. H. (1972) Observations on the short-tailed sheep of the Malay Peninsula with special reference to their wool follicle characteristics, *Australian Journal of Experimental Agriculture and Animal Husbandry* **12**: 479–84
Smith, O. B., Olubunmi, P. A. (1983) Incidence, prevalence and seasonality of livestock diseases in the hot humid forest zone of Nigeria, *Tropical Animal Production* **8**: 7–14
Soetedjo, R., Beriajaya, D., Henderson, A. W. K., Kelly, J. D. (1980) Use of disophenol for the control of *Haemonchus contortus* in sheep in West Java, Indonesia, *Tropical Animal Health and Production* **12**: 198–202
Sorensen, A. M. (1979) *Animal Reproduction, Principles and Practices* McGraw-Hill, New York, USA
Soulsby, E. J. L. (1982) *Helminths, Arthropods and Protozoa of Domesticated Animals,* 7th edn. Baillière Tindall, London, UK
South Africa, Department of Agricultural Technical Services (1976) Ontwikkeling van 'n dominante wit kleur by die Karakoel [Development of dominant white in the Karakul], *Yearbook, Karakul Breeders Society of South Africa* **18**: 61–7
South Africa, Department of Agricultural Technical Services (1981) Efficiency of growth in Dorper sheep. In *Annual Report of the Secretary for Agricultural Technical Services for the Period 1 July 1979 to 31 March 1980,* pp. 46–7. Pretoria, South Africa
South West Africa, Department of Agricultural Technical Services (1974) Verslag van navorsingsaktiwiteite op die Karakoelnavorsingsstasie te Upington ge durende

1972/1973 [Research report of the Karakul Research Station at Upington for 1972/1973], *Yearbook, Karakul Breeders Society of South Africa* **16**: 13-27

Spencer, G. S. G., Garssen, G. J., Hart, I. C. (1983) A novel approach to growth promotion using auto-immunisation against somatostatin. 1. Effects on growth and hormone levels in lambs, *Livestock Production Science* **10**: 25-37

Spharim, I., Seligman, N. G. (1983) Identification and selection of technology for a specific agricultural region: a case study of sheep husbandry and dryland farming in the northern Negev of Israel, *Agricultural Systems* **10**: 99-125

Spooner, R. L. (1982) Genetics of disease resistance in domestic animals. In *Tropical Animal Production for the Benefit of Man*, pp. 75-81. Prince Leopold Institute of Tropical Medicine, Antwerp, Belgium

Squires, V. (1981) *Livestock Management in the Arid Zone*, Inkata, Melbourne, Australia

Stagnaro, C. G. (1983) Commercial hair sheep production in a semiarid region of Venezuela. In Fitzhugh, H. A., Bradford, G. E. (eds) *Hair Sheep of Western Africa and the Americas, a Genetic Resource for the Tropics*, pp. 85-104. Westview Press, Boulder, Colorado, USA

Stapleton, P. G. (1981) The future for biological improvement of straw digestibility, *Agricultural Progress* **56**: 15-24

Steele, M. (1983a) Khabura Development Project Sultanate of Oman. First results with the pure-bred Chios sheep, November 1981-April 1983. *Preliminary Reports 1B(3)*. Centre for Overseas Research and Development, University of Durham, UK

Steele, M. (1983b) Khabura Development Project Sultanate of Oman. Initial results of sheep cross-breeding programme: Chios × Omani. *Preliminary Reports 1B(4)*. Centre for Overseas Research and Development, University of Durham, UK

Steele, M. (1983c) Khabura Development Project Sultanate of Oman. Synchronised sheep breeding trials at Khabura, November to May 1981/82, and September to February 1982/83. *Preliminary Reports 1B(5)*. Centre for Overseas Research and Development, University of Durham, UK

Stephenson, R. G. A., Hopkins, P. S. (1978) Urea supplementation for ewes in the tropics, *Proceedings of the Australian Society of Animal Production* **12**: 178

Stewart, B. A., Matlock, W. G. (1980) The agricultural marketing system in Zinder, Niger, *Agricultural Systems* **5**: 181-91

Sudiana, D., Mateo, C., Coronel, A. B., Lin, T. C. (1981) Animal quarantine and control of epizootics in three Asian countries. *Extension Bulletin No. 167*. Food and Fertilizer Technology Center, Taiwan

Suliman, A. H., El-Amin, F. M., Osman, A. H. (1978) Reproductive performance of Sudan indigenous sheep under irrigated Gezira conditions, *World Review of Animal Production* **14** (1): 71-9

Sumberg, J. E. (1983) Leuca-fence: living fence for sheep using *Leucaena leucocephala*, *World Animal Review* **47**: 49

Sundstøl, F., Coxworth, E., Mowat, D. N. (1978) Improving the nutritive value of straw and other low-quality roughages by treatment with ammonia, *World Animal Review* **26**: 13-21

Surreaux, P. G., Thompson, D. de M., Corrêa, T. B. (1980) Verficação do desenvolvimento ponderal em cordeiros F1 Ile de France × Corriedale até os 90 dias [Growth of F1 Ile-de-France × Corriedale lambs to 90 days]. In *Pesquisa em ovinos no Brasil 1975-1979*, p. 117. Empresa Brasileira de Pesquisa Agropecuária, Bagé, Brazil

Swain, N. (1982) Note on the grazing behaviour of sheep and goat on natural degraded semi-arid range land, *Indian Journal of Animal Sciences* **52**: 829-30

Symington, R. B. (1959) Light regulation of coat shedding in a tropical breed of hair sheep, *Nature* (UK) **184**: 1076

Taneja, G. C. (1965) Effect of varying the frequency of watering during summer on the temperature, respiration, body weight and packed cell volume of blood of sheep, *Indian Journal of Experimental Biology* **3**: 259-62

Taneja, G. C. (1973) Water economy of sheep. In Mayland, H. F. (ed.) *Water-animal Relations, Symposium held at Twin Falls, Idaho, June 1973*, pp. 95-132. Water-animal Relations Committee, Kimberly, Idaho, USA

Taneja, G. C. (1974) On increasing wool weight through breeding LK-type sheep, *Journal of Agricultural Science* (UK) **83**: 113–16
Taneja, G. C. (1978) *Sheep Husbandry in India.* Orient Longman, New Delhi, India
Taneja, G. C., Narayan, N. L., Ghosh, P. K. (1969) On the relationship between wool quality and the type and concentration of blood potassium in sheep, *Experientia* **25**: 1200–1
Taparia, A. L. (1972) Breeding behaviour in Sonadi sheep, *Indian Journal of Animal Sciences* **42**: 576–9
Teixeira, F. J. L., Bessa, M. N., Souza, A. A. de, Freitas, J. P. de, Mendonça, H. L. (1980) Contribuição ao estudo da eficiência reprodutiva de ovelhas deslanadas Morada Nova – Variedade Branca [Reproductive efficiency of White Morada Nova sheep]. In *Pesquisa em ovinos no Brasil 1975-1979,* p. 55, Empresa Brasileira de Pesquisa Agropecuária, Bagé, Brazil
Tejada, R., Murillo, B., Cabezas, M. T. (1979) Ammonia treated wheat straw as a substitute for maize silage for growing lambs, *Tropical Animal Production* **4**: 172–6
Terrill, C. E. (1974) Sheep. In Hafez, E. S. E. (ed.) *Reproduction in Farm Animals,* 3rd edn, pp. 265–74. Lea and Febiger, Philadelphia, USA
Tešanović, D. (1979) Klimatski faktori i plodnost awassi rase ovaca na Dujaili u Iraku [Climatic factors and reproduction of Awassi sheep at the Dujaila farm in Iraq], *Stočarstvo* **33**: 103–6
Tetteh, A. (1974) Preliminary observations on preference of herbage species by cattle, sheep and goats grazing on range on the Achimota Experimental Farm, *Ghana Journal of Agricultural Science* **7**: 191–4
Thahar, A., Petheram, R. J. (1983) Ruminant feeding systems in West Java, Indonesia, *Agricultural Systems* **10**: 87–97
Thimonier, J. (1981) Practical uses of postaglandins in sheep and goats, *Acta Veterinaria Scandinavica* Suppl. **77**: 193–208
Thomas, D. (1978) Pastures and livestock under tree crops in the humid tropics, *Tropical Agriculture* (Trinidad) **55**: 39–44
Thompson, W. R., Bolsen, K. K., Ilg, H. J. (1982) Mixtures of corn grain and corn silage, nitrogen source and zeranol for feeder lambs, *Journal of Animal Science* **55**: 211–17
Thomson, E. F., Bahhady, F. (1983) Flock composition and fluxes and productivity levels of Awassi flocks in the northwestern Syrian steppe. In *Proceedings of the International Symposium on Production of Sheep and Goat in Mediterranean Area,* pp. 278–89. Ankara University and European Association for Animal Production, Ankara, Turkey
Thornton, R. F., Minson, D. J. (1973) The relationship between apparent retention time in the rumen, voluntary intake and digestibility of legume and grass diets in sheep, *Australian Journal of Agricultural Research* **24**: 889–98
Thwaites, C. J. (1981) Development of ultrasonic techniques for pregnancy diagnosis in the ewe, *Animal Breeding Abstracts* **49**: 427–34
Tiwari, S. B., Sahni, K. L. (1974) A note on the effect of underground housing on semen production of Rambouillet rams under semi-arid conditions, *Indian Journal of Animal Sciences* **44**: 206–8
Tiwari, S. B., Sahni, K. L. (1976a) A note on the effect of cooling rate, temperature of storage and packing material on the livability of ram spermatozoa, *Indian Veterinary Journal* **53**: 232–4
Tiwari, S. B., Sahni, K. L. (1976b) A note on the effect of early weaning on the productivity of Malpura and Chokla lambs under semi-arid conditions, *Indian Veterinary Journal* **53**: 337–40
Tiwari, S. B., Sahni, K. L. (1981) Puberty and reproductive performance in ram lambs of indigenous exotic and cross-bred sheep under semi-arid conditions, *Indian Journal of Animal Sciences* **51**: 634–8
Tomar, S. S. (1978) Relative importance of age and live weight in wool production of Nali sheep, *Indian Veterinary Journal* **55**: 545–9
Tomar, S. S. (1979) Time of parturition in sheep, *Indian Journal of Animal Research* **13**: 68–70

Tomar, S. S., Mahajan, J. M. (1980) Lifetime reproductive performance of Rambouillet ewes under subtemperate conditions, *Indian Journal of Animal Sciences* **50**: 838–43

Topps, J. H. (1963) Some observations on the protein requirements of African 'native' sheep, *Tropical Agriculture* (Trinidad) **40**: 49–54

Topps, J. H. (1967) A note on the selective grazing of sheep in the dry tropics, *Tropical Agriculture* (Trinidad) **44**: 165–7

Torres, R., Hernandez, M., Preston, T. R. (1982) A note on the processing of sugar cane bagasse with alkali, *Tropical Animal Production* **7**: 142–3

Trail, J. C. M., Sacker, G. D. (1966) Lamb mortality in a flock of East African Blackheaded sheep, *Journal of Agricultural Science* (UK) **66**: 97–100

Trail, J. C. M., Sacker, G. D. (1969) Growth of cross-bred Dorset Horn lambs from East African Blackheaded sheep, *Journal of Agricultural Science* (UK) **73**: 239–43

Trapp, M. J., Slyter, A. L. (1983) Pregnancy diagnosis in the ewe, *Journal of Animal Science* **57**: 1–5

Treacher, T. T. (1973) Artificial rearing of lambs: a review, *Veterinary Record* **92**: 311–15

Treacher, T. T. (1978) The effects on milk production of the number of lambs suckled and age, parity and size of ewe. In Boyazoglu, J. G., Treacher, T. T. (eds) *Milk Production of the Ewe*, pp. 31–40, Publication No. 23. European Association for Animal Production, Brussels, Belgium

Tribe, D. E. (1950) Influence of pregnancy and social facilitation on the behaviour of grazing sheep, *Nature* (UK) **166**: 74

Tripathi, S. S. (1977) Synchronization of oestrus in sheep with melengestrol acetate (MGA), *Pantnagar Journal of Research* **2**: 195–6

Trounson, A. O., Rowson, L. E. A. (1976) Research on embryo transfer in domestic animals, *Journal of the Royal Society of England* **137**: 77–85

Tudor, G. D., Morris, J. G. (1971) The effect of the frequency of ingestion of urea on voluntary feed intake, organic matter digestibility and nitrogen balance of sheep, *Australian Journal of Experimental Agriculture and Animal Husbandry* **11**: 483–7

Turner, H. N. (1956) Measurement as an aid to selection in breeding sheep for wool production, *Animal Breeding Abstracts* **24**: 87–118

Turner, H. N. (1969) Genetic improvement of reproduction rate in sheep, *Animal Breeding Abstracts* **37**: 545–63

Turner, H. N. (1972) Genetic interactions between wool, meat and milk production in sheep, *Animal Breeding Abstracts* **40**: 621–34

Turner, H. N. (1977) Australian sheep breeding research, *Animal Breeding Abstracts* **45**: 9–13

Turner, H. N. (1978) Sheep and the smallholder, *World Animal Review* **28**: 4–8

Turner, H. N. (1982) Basic considerations of breeding plans. In Gatenby, R. M., Trail, J. C. M. (eds) *Small Ruminant Breed Productivity in Africa*, pp. 1–6. International Livestock Centre for Africa, Addis Ababa, Ethiopia

Turner, H. N., Young, S. S. Y. (1969) *Quantitative Genetics in Sheep Breeding.* Macmillan Company of Australia, Melbourne, Australia

Tyrrell, R. N., Gleeson, A. R., Peter, D. A., Connell, P. J. (1980) Early identification of non-pregnant and pregnant ewes in the field using circulating progesterone concentration, *Animal Reproduction Science* **3**: 149–53

Ugalde Orta, J. (1978) Análisis de algunos factores genéticos y ambientales que afectan el peso al nacer y crecimiento hasta los tres meses de borregas Romney Marsh [Analysis of some genetic and environmental factors affecting birth weight and growth to 3 months in Romney Marsh lambs], *Veterinaria* (Mexico) **9**: 120

Umunna, N. N., Chineme, C. N., Saror, D. I., Ahmed, A., Abed, S. (1981) Response of Yankasa sheep to various lengths of water deprivation, *Journal of Agricultural Science* (UK) **96**: 619–22

Underwood, E. J. (1977) *Trace Elements in Human and Animal Nutrition,* 4th edn. Academic Press, London, UK

Underwood, E. J. (1981) *The Mineral Nutrition of Livestock,* 2nd edn. Commonwealth Agricultural Bureaux, Farnham Royal, Slough, UK

UNDP/FAO (1976a) Sheep and goat development project, Kenya: Macroeconomic evaluation. *Technical Report 1*. Food and Agriculture Organization of the United Nations, Rome, Italy

UNDP/FAO (1976b) Sheep and goat development project, Kenya: Production economics. *Technical Report 2*. Food and Agriculture Organization of the United Nations, Rome, Italy

Unesco (1979) *Tropical Grazing Land Ecosystems*. United Nations Educational, Scientific and Cultural Organization, Paris, France

Valencia, J., Barrón, C., Fernández-Baca, S. (1977) Pubertad en corderos Tabasco × Dorset [Puberty in Tabasco × Dorset lambs], *Veterinaria* (Mexico) **8:** 127–30

Valencia, J., Mendoza, G., Barrón, C., Fernández-Baca, S. (1978) Manejo y reproducción de ovinos en la región del Ajusco, Mexico, D.F. [Management and reproductive performance of sheep in the district of Ajusco, Mexico], *Veterinaria* (Mexico) **9:** 85–90

Vanselow, B. (1982) Sheep production in a commercial rubber plantation: a Malaysian experience, *Tropical Animal Production* **7:** 50–6

Vassiliades, G. (1981) Parasitisme gastro-intestinal chez le mouton du Sénégal [Gastro-intestinal parasitism in sheep in Senegal], *Revue d'Élevage et de Médecine Vétérinaire des Pays Tropicaux* **34:** 169–77

Veen, T. W. S. van, Buntjer, B. J. (1983) Some marketing aspects of slaughter animals in rural slaughterslabs in Kaduna State of Nigeria, *Revue d'Élevage et de Médecine Vétérinaire des Pays Tropicaux* **36:** 307–12

Veen, T. W. S. van, Folaranmi, D. O. B. (1978) The haemoglobin types of northern Nigerian sheep, *Research in Veterinary Science* **25:** 397–8

Venter, J. J. (1980a) Physical and chemical qualities of wool of good and poor quality. In *Agricultural Research 1979*, p. 89. Division of Agricultural Information, Pretoria, South Africa

Venter, J. J. (1980b) The physical properties of wool of good and poor quality, *Agroanimalia* **12:** 27–38

Venter, J. J., Edwards, W. K. (1977) Damage to wool: influence of time of shearing on weathering. 1970–1974. Final report. In *Agricultural Research 1975*, p. 61. Department of Agricultural Technical Services, Pretoria, South Africa

Venter, J. J., Nel, J. W., Edwards, W. K. (1977) Effect of sheltering on the weathering of wool, *Agroanimalia* **9:** 45–51

Vera, A., Aparicio, F., Echevarria, A. (1978) Estudio de la longevidad y la duración de la vida util en rebaños de ovejas navarras [Longevity and length of useful life in flocks in the Navarre], *Archivos de Zootechnia* **27:** 133–43

Villegas, V., Cruz, E. C. (1960) Ewes for dairy purposes, *Philippine Journal of Animal Industry* **21:** 39–42

Vipond, J. E., Galbraith, H. (1978) Effect of zeranol implantation on the growth performance and some blood characteristics of early-weaned lambs, *Animal Production* **26:** 359

Wallach, E., Eyal, E. (1974) The performance of intensively managed indigenous Iranian sheep and of Awassi sheep imported to Iran from Israel. I. Body and fleece weights of ewes and lambs, *Zeitschrift für Tierzüchtung und Züchtungsbiologie* **91:** 232–9

Wani, G. M. (1981) Ultrasonic pregnancy diagnosis in sheep and goats - a review, *World Review of Animal Production* **17** (4): 43–8

Wani, G. M., Sahni, K. L. (1981) Ultrasonic pregnancy diagnosis in ewes under tropical field conditions, *Indian Journal of Animal Sciences* **51:** 194–7

Ward, H. K. (1959) Some observations on the indigenous ewe, *Rhodesia Agricultural Journal* **56:** 218–23

Wathes, C. M., Jones, C. D. R., Webster, A. J. F. (1983) Ventilation, air hygiene and animal health, *Veterinary Record* **113:** 554–9

Watson, W. A. (1979) Disease and reproduction in sheep, In *The Management and Diseases of Sheep*, pp. 86–99. Commonwealth Agricultural Bureaux, Farnham Royal, Slough, UK

Webster, G. M., Haresign, W. (1981) A note on the use of dexamethasone to induce parturition in the ewe, *Animal Production* **32:** 341–4

Westergaard, J. M. (1972) Trials with a mobile cheesemaking unit, *World Animal Review* **2**: 49–52

Westhuysen, J. M. van der, Merwe, C. A. van der, Bergh, E. C. (1981) The synchronisation of oestrus in sheep: 7. Under dry extensive conditions, *South African Journal of Animal Science* **11**: 53–4

Weston, E. J., Moir, K. W. (1969) Grazing preferences of sheep and nutritive value of plant components in a Mitchell grass association in north-western Queensland, *Queensland Journal of Agriculture and Animal Science* **26**: 639–50

Whiteman, P. C. (1980) *Tropical Pasture Science*. Oxford University Press, Oxford, UK

Whitlock, J. H. (1955) A study of the inheritance of resistance to trichostrongylidosis in sheep, *Cornell Veterinarian* **45**: 422–39

Wiener, G., Purser, A. F. (1957) The influence of four levels of feeding on the position and eruption of incisor teeth in sheep, *Journal of Animal Science* **49**: 51–5

Wiggins, J. P., Wilson, L. L., Ziegler, J. H. (1980) Dosage of zeranol implant: effects on live and carcass traits of lambs, *Veterinary Medicine and Small Animal Clinician* **75**: 121–4

Wilkinson, J. M. (1983) Silages made from tropical and temperate crops. Part 1. The ensiling process and its influence on feed value, *World Animal Review* **45**: 36–42

Willadsen, S. M., Polge, C. (1980) Embryo transplantation in the large domestic species: applications and perspectives in the light of recent experiments with eggs and embryos, *Journal of the Royal Society of England* **141**: 115–26

Williams, H. L. (1975) The reproductive performance of British breeds of sheep in an equatorial environment. III. Further data on imported ewes and their progeny, *British Veterinary Journal* **131**: 23–31

Williams, S. M., Garrigus, U. S., Norton, H. W., Nalbandov, A. V. (1956) The occurrence of estrus in pregnant ewes, *Journal of Animal Science* **15**: 978–83

Williamson, G. (1949) Iraqi livestock, *Empire Journal of Experimental Agriculture* **17**: 48–59

Wilson, A. D. (1969) A review of browse in the nutrition of grazing animals, *Journal of Range Management* **22**: 23–8

Wilson, R. T. (1976) Studies on the livestock of southern Darfur, Sudan. III. Production traits of sheep, *Tropical Animal Health and Production* **8**: 103–14

Wilson, R. T. (1980) Population and production parameters of sheep under traditional management in semi-arid areas of Africa, *Tropical Animal Health and Production* **12**: 243–50

Wilson, R. T. (1981) Livestock production in central Mali: attempts to produce raw materials of animal origin for the French textile industry during the colonial period, *Textile History* **12**: 104–17

Wilson, R. T. (1982) Husbandry, nutrition and productivity of goats and sheep in tropical Africa. In Gatenby, R. M., Trail, J. C. M. (eds) *Small Ruminant Breed Productivity in Africa*, pp. 61–75. International Livestock Centre for Africa, Addis Ababa, Ethiopia

Wilson, R. T. (1983) Livestock production in central Mali: the Macina wool sheep of the Niger inundation zone, *Tropical Animal Health and Production* **15**: 17–31

Wilson, R. T., Clarke, S. E. (1975) Studies on the livestock of southern Darfur, Sudan. I. The ecology and livestock resources of the area, *Tropical Animal Health and Production* **7**: 165–87

Wilson, R. T., De Leeuw, P. N., De Haan, C. (1983) Recherches sur les systèmes des zones arides du Mali. Résultats préliminaires. [Systems research in the arid zones of Mali. Preliminary results]. *Rapport de Recherche No. 5*. International Livestock Centre for Africa, Addis Ababa, Ethiopia

Wilson, R. T., Durkin, J. W. (1983) Livestock production in central Mali: weight at first conception and ages at first and second parturitions in traditionally managed goats and sheep, *Journal of Agricultural Science* (UK) **100**: 625–8

Wilson, R. T., Durkin, J. W. (1984) Age at permanent incisor eruption in indigenous goats and sheep in semi-arid Africa, *Livestock Production Science* **11**: 451–5

Wilson, R. T., Peacock, C. P., Sayers, A. R. (1981) A study of goat and sheep production on the Masai group ranch at Elangata Wuas, Kajiado district, Kenya. *Working Document 24*. International Livestock Centre for Africa, Addis Ababa, Ethiopia

World Bank (1981) Agricultural research. *Sector Policy Paper.* World Bank, Washington, DC, USA

Yenikoye, A., Andre, D., Ravault, J. P., Mariana, J. C. (1981) Étude de quelques caractéristiques de reproduction chez la brebis Peulh, du Niger [Some reproductive characteristics of the Fulani ewe in Niger], *Reproduction, Nutrition, Développement* **21:** 937–51

Yenikoye, A., Pelletier, J., Andre, D., Mariana, J. C. (1982) Anomalies in ovarian function of Peulh ewes, *Theriogenology* **17:** 355–64

Younis, A. A., Afifi, E. A. (1979) On the occurrence of oestrus in pregnant Barki and Merino ewes, *Journal of Agricultural Science* (UK) **92:** 505–6

Younis, A. A., El-Gaboory, I. A. H. (1978) On the diurnal variation in lambing and time for placenta expulsion in Awassi ewes, *Journal of Agricultural Science* (UK) **91:** 757–60

Younis, A. A., El-Gaboory, I. A., El-Tawil, E. A., El-Shobokshy, A. S. (1978) Age at puberty and possibility of early breeding in Awassi ewes, *Journal of Agricultural Science* (UK) **90:** 255–60

Younis, A. A., Kotby, S., Kamar, G. A. R. (1972) Effect of castration on live body weight and certain carcass traits in Ossimi and Rahmani lambs, *Egyptian Journal of Animal Production* **12:** 91–7

Younis, A. A., Seoudy, A. M., Galal, E. S. E., Ghanem, Y. S., Khishin, S. S. (1975) Effect of plane of nutrition on feedlot performance and carcass traits of desert sheep, *Tropical Agriculture* (Trinidad) **52:** 233–42

Zanwar, S. (1981) Embryo transfer in sheep, *Indian Journal of Animal Reproduction* **1:** 102–4

Zeeman, P. J. L., Marais, P. G., Coetsee, M. J. (1983) Nutrient selection by cattle, goats and sheep on natural Karoo pasture. 1. Digestibility of organic matter, *South African Journal of Animal Science* **13:** 236–9

Zemmelink, G. (1980) Effect of selective consumption on voluntary intake and digestibility of tropical forages. *Agricultural Research Report No. 896.* Ahmadu Bello University, Zaria, Nigeria

Zeuner, F. E. (1963) *A History of Domesticated Animals.* Hutchinson, London, UK

Zoby, J. F. F., Holmes, W. (1983) The influence of size of animal and stocking rate on the herbage intake and grazing behaviour of cattle, *Journal of Agricultural Science* (UK) **100:** 139–48

Zyl, G. J. van (1976) Die kleurvariasie in die nageslag van grys Karakoele [Colour variation among the progeny of grey Karakuls], *Yearbook, Karakul Breeders Society of South Africa* **18:** 23–8

Index

abattoir *see* slaughter house
abortion, 48 85, 131, 135
absorption of nutrients, 41, 54
Abyssinian, 5, 133, 230, 269, Fig. 1.2
　see also Menz
adaptation to environment, 1, 4, 13–14, 15, 25, 154, 252–3
Afghanistan, 2, 137, 202, 203
aflatoxin, 85, 113
Africa, 2, 4
　breeds, 4, 5–6
　feeding and nutrition, 70, 72
　growth and meat production, 161, 165
　health, 44, 48
　improvement, 260, 261
　management, 18, 23, 30
　milk production, 202, 203
　reproduction, 127, 129, 136, 141
　skin production, 218
　sociology and economics, 227, 228, 229, 241, 243–7
　wool production, 195
Africander, 5, 269
age
　at culling, 129, 193
　at first mating, 194
　at puberty, 128–9, 138, 142
　at slaughter, 158–9
　at weaning, 29, 31, 149–50, 256
　determination from teeth, 55–6
　effect on diet selection, 83
　effect on growth, 145, 148, 158
　effect on helminthiasis, 41
　effect on milk production, 210–11
　effect on reproduction, 129, 135, 136
　effect on wool production, 184, 190, 193
　structure of flock *see* flock structure
agricultural by-product (as food), 104–15, 158, 249, 251
AI, 34, 138–41, 182, 254

aid programme, 11, 35, 264–5
Algeria, 2, 4, 203, 218, 260
alkali treatment of roughage, 106–8
alkaloids, 85
altitude, 11
America, 202
　see also Central America, North America, South America, West Indies
amino acid, 64, 85
ammonia treatment of roughage, 108–9
anabolic agent, 156
anoestrus, 71, 90, 125–6, 135, 137
ant, 44
anthelmintic, 42, 43, 44
antibiotic (as growth promoter), 157
appetite *see* food intake
apron, 34, 36
Arabi, 180, 191, 196, 270
arid area *see* dry tropics
artificial insemination *see* AI
Asia, 2, 4
　breeds, 4, 6, 8
　growth and meat production, 161
　health, 44
　improvement, 260, 261
　milk production, 202, 203
　pelt production, 219
　reproduction, 141
　sociology and economics, 228, 242–7
Assaf, 8, 142, 206, 207, 208, 270
assessment of adult fleece, 170–4
assessment of lamb's birthcoat, 175–9
assessment of productivity, 10, 228–31, 249
astrakhan *see* Karakul
Australia, 2
　breeds, 10, 11
　feeding and nutrition, 70, 72, 73, 79, 88, 91, 96–7, 114
　growth and meat production, 164, 165

336

health, 44
management, 24
milk production, 210
reproduction, 125, 126, 133, 134, 140
sociology and economics, 15, 226, 239, 245
wool production, 169, 179, 195
Australian Merino, 10, 11, 13, 125, 133, 197, 271
Avikalin, 182, 207, 254, 271
Avivastra, 182, 207, 271
Awassi, 6
growth and meat production, 145, 147, 150, 151, 156, 158, 162, 164
management, 19, 30, 34
milk production, 204–8, 212, 213
reproduction, 121, 128, 133, 140, 142
wool production, 180, 190, 191, 196, 201
see also Israeli Improved Awassi

bagasse, 110, 111
Baggara, 18, 20, 24, 26, 133, 147, 230, 271
Bahrain, 246
banana, 21, 114
Bangladesh (breed), 162, 187, 272
Bangladesh (country), 2, 8, 109, 133, 187
Bapedi, 155, 272
Barbados Blackbelly, 10, 132, 133, 147, 229, 230, 272, Fig. 1.12
Barbary, 6, 272
see also Barki, Libyan Barbary, Tunisian Barbary
barefoot vet, 50
see also stockman
Barki, 6, 272
feeding and nutrition, 66, 73
growth and meat production, 151, 152, 153, 158, 162
milk production, 207
reproduction, 13, 129, 133
wool production, 179, 180, 187–9
barley, 62, 66, 87–8, 104
barrenness, 12, 69, 132, 267
Bedouin, 19, 25, 27, 31
see also Najdi
behaviour, 15, 32, 82–5, 121, 130, 204
Belgium, 247
Beni Ahsen, 207, 212, 272
berseem clover, 102
Bikaneri, 8, 135, 237–9, 273
see also Chokla, Jaisalmeri, Magra, Malpura, Nali, Pugal, Sonadi
biogas, 224, 225
biological control, 44

biological processing of roughage, 106, 109–10, 251
biotin, 73
birth defects, 85
birth weight, 134, 146, 148
birthcoat type, 13, 175–9, 187–8, 189, 192, 199
Blackhead Persian (Africa), 6, 53, 55–6, 138, 167, 254, 273
Blackhead Persian (America), 10, 273
growth and meat production, 147
milk production, 207, 209, 212, 213
reproduction, 133
sociology and economics, 230
bloat, 102
blood, as carcass by-product, 161, 221
blood characteristics, 13, 75, 185–6
bluetongue, 48
body composition, 157–60
body condition, 146, 159–60, 209
body weight *see* liveweight
Bolivia, 3, 4, 133, 218
bone, 70, 74, 157, 158, 161, 221
Bornu, 133, 230, 273
Botswana, 2, 80, 133, 163, 216, 260
bran, 114
brand, 33
Brazil, 3
breeds, 10
feeding and nutrition, 96–7
growth and meat production, 147, 155, 156, 162–3
improvement, 264
management, 20, 23, 25, 32
milk production, 207
reproduction, 134, 140
skin production, 218
sociology and economics, 230, 245
wool production, 194
break crop, 102
breed effect on
compensatory growth, 154
gestation length, 127
grazing behaviour, 84–5
lamb growth, 148, 150, 154–5
milk production, 208
teeth, 55–6
breeding, 12, 34, 130
policy, 34, 181, 253–6
prevention, 34–5, 36
seasonality, 125–6, 131
see also crossbreeding, selection
breeds, 4–11, 17, 269–92
see also individual breed names
bride price, 227

broken mouth, 24, 56, 129, 267
  see also teeth
browse, 32, 82, 83, 85, 267
brucellosis, 49
buffalo, 54, 204, 213
burning, 80–1
bushes, 80, 82
butter, 215
by-product
  agricultural, as food, 104–15, 158, 249, 251
  of carcass see carcass by-products

calcium, 68, 70, 74, 138, 214
calcium hydroxide, 108
calcium oxide, 108
camel, 19, 54, 76, 213
camping site, 84
Canada, 219, 246, 247
canary colouring of wool, 183, 186–7, 191, 201
capital investment, 159, 227, 238–40, 259–60, 261, 266
carbon tetrachloride, 42, 43
carcass
  by-products, 161, 220–2, 232
  composition, 157–8, 159, 161–4
  classification, 161–5
  yield, see dressing percentage
  see also meat
carotene, 73
carpet manufacture, 31, 169, 199–201, 220
carrying capacity, 79–80, 117, 267
caseous lymphadenitis, 49
cash (from sale of sheep), 20, 227, 259
cashew, 20, 116
cassava, 21, 64, 66, 112
castration, 24, 29–30, 155
catch crop, 102
cattle
  feeding and nutrition, 54, 62, 70, 76, 82, 83, 102, 114, 116
  growth and meat production, 158
  health, 42, 48, 49
  improvement, 36
  management, 20, 32, 36
  milk production, 204, 208, 211, 213, 214
  reproduction, 138, 139
  sociology and economics, 228, 231
cellulose, 54
Central America, 3, 4, 10, 20
  production and trade, 203, 219, 244, 245, 246, 247

cereals, 64, 74, 86, 158–9
  see also barley, maize, rice
CGIAR, 264–5
Chad, 2, 18, 23, 24, 133–4, 244
chbâba, 29, 30
cheese, 214, 215
chemical composition of food, 76–7
chemical treatment of roughage, 105–9, 110, 251
Cheviot, 123, 200, 201
children, 26, 27
China, 112, 202, 203, 225
Chios, 274
  feeding and nutrition, 87
  growth and meat production, 145, 147
  improvement, 253
  milk production, 30, 31, 205, 207, 208
chlorine, 68, 70, 214
Chokla, 15, 274
  growth and meat production, 147, 150
  milk production, 207
  wool production, 180, 181, 182–7
choline, 73
chromic oxide, 76
chromosomes, 13
citrus, 110, 113–14, 116
climate, 1, 4–6, 14–16, 17, 79–80, 257–8
  see also genotype-environment interaction
clover, 138
cluster bean, 114
coat, 4, 15, 166
  see also colour, hair coat, pelt, skin
cobalt, 68, 69, 70, 72, 88
coccidiosis, 49
cocoa, 104, 111
coconut, 20, 113, 116, 117
coffee, 111, 116
Coimbatore, 162, 184, 274
cold stress, 15, 63, 135
collar, 20, 21
Colombia, 3, 20, 68, 82, 133, 156
colostrum, 213
colour of coat, 13, 15, 33, 168, 170, 174, 220
communal grazing, 25, 34, 257
communal ownership of land, 227, 260, 261
compensatory growth, 90, 152–4
composition of growth, 157–60
concentrate, 86, 90, 153, 158–9, 257, 267
conception rate, 140, 141, 267

condition scoring, 160
confinement, 27–8
  see also housing
conservation of fodder, 41, 94, 95–101
constraints to sheep production, 250–62
contagious ecthyma, 46
contraception, 34–5, 36
co-operative marketing, 242
copper, 68, 69, 71, 85, 115, 138, 183
corn *see* maize
corpus luteum, 122–3, 127
Corriedale, 11, 274
  growth and meat production, 147, 154, 155, 156
  health, 41
  litter size, 133
  milk production, 207
  sociology and economics, 235–9
  wool production, 184, 194
Costa Rica, 111
cotting of wool, 191
cottonseed, 59, 110, 112, 258
credit, 261, 263
creep feed, 145
Criollo, 10, 133, 162, 194, 274
crop damage, 27, 228
crossbreeding, 11, 12, 38, 253–4
  milk production, 208
  wool production, 181, 183, 186, 188, 196, 197
crude protein *see* protein
Cuba, 3, 96, 133
culling, 24, 129, 193, 234, 235, 236, 267
cultivated pasture *see* pasture
curd, 215
curing (skin), 216–17
cyanogenetic glycosides, 85
Cyprus, 31, 87–8, 145, 147, 207, 208
Cyprus Fat-tailed, 145, 147, 207, 275

dagging, 31, 267
dairy performance *see* milk production
date palm, 110, 113
daylength, 16, 59, 125, 126
death rate *see* mortality
Deccani, 8, 147, 180, 183, 184, 275
degradability of protein, 64, 66
dentition, 55–6
dermatophilus, 48
desert *see* dry tropics
Desert Sudanese, 35, 133, 135, 147, 150, 162, 275
desertification, 252
development of sheep production, 35, 248–66

dexamethasone, 144
diarrhoea, 69
dictyocauliasis, 42–4
diet selection, 82–4, 224
diethylstilboestrol, 156
digestibility, 74, 76–7, 81, 94–7, 107
digestible energy, 60, 76
digestive system, 54–5, 60
dipping, 47, 234, 235, 236
disease, 40–53
  in lactating ewes, 206
  problems, 36, 40, 135, 148, 244, 252
  reproductive, 131
  resistance, 6
  transmission, 15, 16, 257, 258
  see also specific diseases
Djallonké *see* West African Dwarf
D'man, 132, 275
docking, 158, 267
dog, 27, 28, 44
Döhne Merino, 152, 194, 275
domestication, 4
Dongola, 5, 275
donkey, 213
Dorper, 6, 276
  feeding and nutrition, 55, 56, 58, 83
  growth and meat production, 150
  improvement, 254
  reproduction, 133
  sociology and economics, 235, 237–9
Dorset, 11, 85, 123, 148, 151, 182–3, 207, 276, Fig. 1.15
drenching, 43, 69, 234, 235, 236
dressing, 161
dressing percentage, 153, 155, 158–9, 161–3, 267
dried grass, 66, 101
drinking *see* water intake, watering management
drought, 51, 73, 88–91, 244
dry matter intake *see* food intake
dry season, 47, 50, 75, 83, 125
dry tropics, 1, 4
  feeding and nutrition, 75, 79, 82, 84, 85, 86, 88–91, 102
  growth and meat production, 148
  health, 41, 44, 51
  improvement, 251, 252
  management, 17, 18, 22, 25, 38
  pelt production, 220
  reproduction, 125, 136
  sheep adapted to, 4, 5, 13
  sociology and economics, 226, 229, 231
  wool production, 181
Drysdale, 200–1, 276

duck, 44
dystocia, 134, 253, 267

ear, 33
East Africa, 2, 5, 11, 29, 34, 82, 136, 241, 254
East African Blackheaded, 5, 210, 276
East Friesian, 208, 276
East Java Fat-tailed, 8, 133, 276
economics, 231–40
  compensatory growth, 154
  disease control, 40, 42, 50, 52–3, 232
  drought feeding, 90–1
  fencing, 95, 236
  finishing, 119–20, 159, 232, 251, 259
  improvements, 251, 258–62
  manure production, 224, 232
  milk processing, 216
  pasture improvement, 86
  production systems, 36, 159, 231–40
  supplementation of diet, 87–8, 137, 258
  wool production, 232–3, 236
Ecuador, 3, 4
education, 50, 250, 256–7, 262, 263–4
efficiency of conversion
  herbage to meat *see* food conversion efficiency
  milk to meat, 145
Egypt, 2
  feeding and nutrition, 58, 73, 118
  growth and meat production, 150–3, 156, 162–3
  milk production, 207, 209, 212
  reproduction, 12, 127, 129, 133, 138
  sociology and economics, 246, 247
  wool production, 179, 187–90
electric fence, 95, 99
EMBRAPA, 264
embryo, 127, 132, 141
embryo transfer, 141
energy, 60–4
  density of diet (M/D), 56–8
  in food, 62, 63, 107, 110–15
  requirement, 61–3
  system, 60–4
enterotoxaemia, 47, 129, 252, 258
environment, 14–16, 22, 257–8
  see also genotype-environment interaction, housing
enzootic pneumonia, 47–8
  see also pneumonia, respiratory disease
enzymes, 54, 110
epizootic, 51
erosion, 81, 222

erythrocyte GSH, 185–6
estrus *see* oestrus
Ethiopia, 2, 4, Fig. 1.2
  feeding and nutrition, 95, 99
  growth and meat production, 165
  management, 18, 24
  milk production, 203
  reproduction, 133, 136
  skin production, 217, 218
  sociology and economics, 35, 227, 230, 243
  wool production, 31, 32, 200
Europe, 2, 4, 11
  growth and meat production, 154, 161, 164
  management, 17, 24
  milk production, 202, 203, 206
  skin and pelt production, 218, 219
  sociology and economics, 24, 244–7
evaluation
  economic *see* economics
  of breeds, 10
  of foods, 76–7
  of research programmes, 265
evolution, 4
exotic breeds, 12, 48, 90, 127, 131, 182, 196, 253, 254
  see also temperate breeds
export, 15, 160, 218–19, 220, 240, 244–7
extension service, 50, 249–50, 257, 261, 262, 264
extensive systems, 17–20, 25–7, 28, 46, 51, 195
  see also dry tropics
external parasites, 46–7, 195

faeces, 41, 54, 75, 76, 81, 110, 222–5
FAO, 261, 264
farmer
  training for, 250, 256–7, 262
  see also shepherd
farming systems research, 265
fascioliasis, 44
fat
  as sheep food, 55, 60
  consumer preference, 161, 164
  content of milk, 211–14
  content of milk, 211–14
  content of wool, 187
  deposition in carcass, 62, 157–8, 159
  production, 24, 29, 158, 161, 164, 221
fat-rumped sheep, 4, 5, 6, 9
  see also individual breed names
fat-tailed sheep, 4, 5, 6, 8, 9, 19, Fig. 1.3

copulation difficulty, 34, 129, 130
see also individual breed names
see also tail, fatness
fattening *see* finishing
fatty acids, 74
fecundity, 132, 267
feed blocks, 88
feeding, 78–120
feeding trial, 76, 83, 86
feedlot, 52, 150, 261
see also finishing
Fellahi, 6, 277
felt, 168, 169
fence, 27, 95, 236, 260
fertilisation
 of ewe, 25, 127
 of pasture, 59, 69, 70, 71, 92, 94, 251
fertiliser, 70, 221, 222–4
fertility, 69, 132, 267
festivals, 20, 30, 35, 228
fibre *see* hair, wool
fibre-type array, 175–9, 187–8, 192, 199
finishing, 35, 38, 118–20, 153, 158–60, 232, 251, 259, 261
Finnish, 213
fire *see* burning
fish, 44
fish meal, 64, 66, 73
flavour of meat, 165
flaying, 216–17
fleece
 rot, 53
 weight, 31, 172, 180–98
 see also wool
flies, 20, 46
flock size, 18, 23, 25, 32, 33
flock structure, 17, 24–5, 129, 136
flood, 20
fluke, 43, 44
fluorine, 70, 72, 85
flushing, 137, 267
fodder conservation *see* conservation of fodder
fodder crop, 102–3, 104, 251
fodder tree *see* trees
foetus, 59, 127–8, 132
folic acid, 73
food
 conversion efficiency, 41, 54, 71, 75, 118–20, 150, 154, 155, 159
 evaluation, 76–7
 intake, 14, 41, 56–9, 67, 69, 71, 74, 81, 87–8, 94–5, 137, 145, 152, 209
 preference, 82–4, 224
 supplements *see* supplementary feeding

footrot, 45–6
forage conservation *see* conservation of fodder
fostering of lambs, 29, 203, 220
fractionation, 205–6
France, 140, 203, 215, 247
freezing semen, 139–40
frequent lambing, 141–2
frozen meat, 165, 240
FSH, 122–3, 130
fuel, 225
Fulani, 5, 277
 management, 18, 20, 24
 milk production, 216
 reproduction, 121, 123, 127, 133
 see also Bornu, Uda, Yankasa
fungus, 85, 109–10, 113
fur *see* pelt

gastrointestinal roundworms, 40–2, 43, 82, 104, 206
gelatine, 221
generation interval, 12, 129
genes, 11–14
genetic improvement, 12, 34, 253–6
 of milk production, 208
 of pelt production, 220
 of wool production, 196–9
 see also crossbreeding, selection
genetic resistance to disease, 13, 53
genetics, 11–14, 252–6
genotype–environment interaction, 12, 13–14, 255
geophagia, 68
German Mutton Merino, 133, 150, 162, 188–90, 195, 277
gestation length, 127
Ghana, 2, 82, 133–4, 137, 142, 167, 231, 233, 237–9, Fig. 1.11
ghee, 215
gid, 44
gliricidia, 20, 95, 99, 102
glucose, 55
goat
 crossing with sheep, 14
 feeding and nutrition, 54, 59, 75–6, 82, 83, 85, 116, 118
 health, 42, 46, 47, 49
 management, 19, 32–3
 milk production, 202, 204, 206, 213, 214
 sociology and economics, 228, 244–7, 257
goitre, 71, 85
government
 intervention, 242, 248, 251, 254, 262
 objectives, 35, 36, 260

gram, 104
grass, 59, 62, 64, 66, 78–101
grass staggers, 206
grassland, 79
grazing
  behaviour, 32, 82–5
  duration, 25–9, 84
  management, 25–9, 41, 84, 92, 251, 257
  preference, 82–4, 224
Greece, 203, 247
gross energy, 60, 62, 76
groundnut, 85, 113
growth, 145–60
  composition, 157–60
  curve, 145–6, 152
  nutrient requirement, 61–2
  post-weaning, 150–7
  potential, 145, 150, 154
  pre-weaning, 145–9
  rate, 14, 22, 64, 69, 75, 145–57, 158–60, 253
growth hormone, 155
growth promoter, 156–7
guar, 114
guts, 220–1

haematuria, 85
haemoglobin type, 13–14, 185
Haemonchus, 41
haemorrhagic enterotoxaemia, 47
hair breeds, 4–10, 216
  see also individual breed names
hair coat, 69, 166–7
hair production, 24, 136, 167, 220
Hamdani, 180, 191, 201, 277
Hampshire, 194, 278
hay, 62, 66, 90, 95, 98–100, 110
haylage, 101
health, 40–53, 206, 252
  see also disease
heartwater, 49
heat *see* oestrus
heat production, 60, 186
heat stress, 14–16, 257
  feeding and nutrition, 59, 63, 75
  health, 40
  reproduction, 125, 135
  wool production, 186, 195
hedge, 20, 21, 95
helminthiasis, 13, 40–2, 43, 69, 82, 117, 206
hema, 260
herbal medicines, 51
herding, 23, 25–7, 84, 85
heritability, 12, 13, 132, 183, 188, 190, 193, 194, 199

growth rate, 12
milk yield, 12, 208
reproductive traits, 12, 132, 255
udder conformation, 205
wool production, 183, 188, 190, 193, 194, 199
heterosis, 12, 190, 254, 267
Hissardale, 182, 278
hoof, 45, 222, 232
hormone levels, 123–5, 126, 127, 143–4
hormones, 71, 122–5, 126, 127–8, 155
horns, 13, 222
horse, 213
household scraps, 21
housing, 27–9, 106, 131, 132, 223–4, 252, 257–8
  see also shade, shelter
humid topics, 1, 13
  feeding and nutrition, 79, 85, 91
  health, 42, 44, 252
  management, 20, 21, 22
  reproduction, 125, 132
  sociology and economics, 226, 231, 232
  wool production, 195, 201
humidity, 13, 15, 47, 50, 186, 257
Hungarian Combing Wool Merino, 188–9, 278
hybrid vigour *see* heterosis
hydrocyanic acid, 112
hypocalcaemia, 206
hypomagnesaemia, 206

ICAR, 264
ICARDA, 264–5
identification, 33
ILCA, 229, 264
Ille-de-France, 147, 278
ILRAD, 264
improved pasture *see* pasture
improvement, 22, 34, 136–7, 138–44, 196, 248–66
  see also genetic improvement
inbreeding, 34, 254
incisor teeth, 55–6
income, 226, 227, 232–9
India, 2, 4
  breeds, 8, 10, 11, 15, Fig. 1.6, Fig. 1.7
  feeding and nutrition, 63, 66, 74, 75, 81, 96–7, 102, 117
  growth and meat production, 145,
  hair production, 220
  health, 40
  improvement, 249, 250, 261, 262–3, 264

management, 17, 19, 24, 26, 27, 31, 34
manure production, 223, 224, 225
milk production, 30, 207, 208, 209
reproduction, 12, 125, 127–9, 131, 133, 135, 138, 139, 141
skin and pelt production, 218, 219
sociology and economics, 226–7, 231–3, 237–9, 240, 242, 245
wool production, 31, 169, 179, 181–7, 200, 201
indigenous breeds, 11, 12, 131, 196, 252–3, 254–5
Indonesia, 2, 8
feeding and nutrition, 22
growth and meat production, 147, 151, 162
health, 41
management, 21, 22, 28
milk production, 207
reproduction, 125, 129, 133
induction
of lambing, 144
of oestrus, 126, 142
infertility, 125, 130, 131, 138
see also anoestrus
insulin, 155
intake see food intake
integration
sheep with field crops, 17, 19, 20, 33, 98, 102, 103–15, 226, 259
sheep with tree crops see tree crops
intensive system, 19, 35–9, 69, 85, 88, 266
diseases of, 48, 52
milk production, 203, 206
intercrop, 102
international aid, 12, 34, 264–5
international trade, 15, 218–19, 219, 243, 244–7
intestines, 54, 157, 221
investment, 159, 227, 238–40, 259–60, 261
iodine, 69, 71, 138
Iran, 2
growth and meat production, 145, 147, 150
milk production, 202, 203, 207
reproduction, 125
sociology and economics, 246, 247
Iraq, 2
growth and meat production, 147, 151, 158, 162
milk production, 202, 203, 212
reproduction, 133
skin production, 218

sociology and economics, 246, 247
wool production, 190–1
Ireland, 140
iron, 68, 69, 76, 138
irrigation, 22, 95
isoflavins, 85
Israel, 2, 6
feeding and nutrition, 72, 209
reproduction, 133, 140, 142
wool production, 191
Israeli Improved Awassi, 6, 180, 208, 278
Italy, 203, 207, 208, 245, 246, 247
Ivory Coast, 2, 20, 34, 134, 157, 245, 246

Jaffna, 8, 147, 151, 279
Jaisalmeri, 8, 279
growth and meat production, 147
milk production, 207, 208, 209
reproduction, 125
wool production, 180, 181, 183–5
Japan, 246, 247
Javanese Thin-tailed, 8, 125, 133, 147, 151, 162, 279
Jordan, 244, 246, 247
Junin, 194, 279

kangaroo, 76, 82
kapok, 116
Karadi, 180, 190, 191, 201, 279
Karakul, 11, 279
feeding and nutrition, 75
growth and meat production, 145, 147, 151
pelt production, 36, 219–20
reproduction, 126, 144
Kashmir Valley, 182, 279
kata, 49
Kelantan, 8, 162, 279
kemp, 166–7, 172–3, 178–9, 187–92, 199, 200
Kenya, 2, 4
feeding and nutrition, 85
growth and meat production, 147, 160, 164
management, 18, 21, 23, 24, 36
milk production, 207
reproduction, 133, 136
skin production, 218
sociology and economics, 227, 230, 231, 234–9, 260
wool production, 36
ketosis, 48
kinship, 23, 227
kjeldahl, 77
knife, 31

knitting, 168
kola, 85, 116
kunan, 34, 35
Kuwait, 218, 245, 246, 247

laban, 215
labour, 34, 90, 226, 232, 236–7
   see also children, herding, women
lactation
   curve, 31, 208, 209, 211
   effect on wool production, 191
   length, 31
   nutrient requirements, 59, 62, 66, 70, 73, 75
   see also milk
lactose, 55, 213, 216
lamb
   digestive system, 54–5
   dysentery, 47
   see also growth
lambing
   dystocia, 134, 253
   food intake at, 137
   hormonal control, 127–8
   induction, 144
   interval, 24, 129, 131–2, 135–6, 137, 141–2, 150, 255
   rate, 132, 220, 232
   season, 125–6
   sickness, 206
   synchronisation, 144
lameness, 45, 48, 129
Latin America *see* Central America, South America
lead, 85
leather, 216–19
Lebanon, 19, 31, 207, 212, 246
leeches, 19
legumes, 59, 64, 80, 82, 92, 93–4, 102, 251
lending, 23, 227
leucaena, 20, 95, 102, 103
LH, 122–4, 126, 130
libido, 71, 130, 131, 132
Libya, 2, 191, 218, 245, 246, 257
Libyan Barbary, 180, 191
lice, 46
light
   response of pasture, 78–9, 92, 117
   response of sheep, 15, 59, 74, 125, 189, 192–3
   see also shade
limarin, 112
Lincoln Longwool, 194, 280
linoleic acid, 74
litter size, 11, 12, 131–4
   effect on growth rate, 146
   effect on lamb mortality, 134, 135, 136, 255
   effect on milk production, 210
   effect on pelt quality, 220
   effect on wool production, 190, 193
   nutrient requirement, 62
liver fluke, 43, 44
liveweight, 91, 125, 145–60
liveweight gain *see* growth rate
llanos, 82
loans (to farmers), 261, 263
Lohi, 133, 151, 187, 280
longevity, 128–30
lotaustralin, 112
lucerne, 66, 102
lumpy wool disease, 48
lung worms, 42–4

macadamia nut, 114
Macina, 11, 20, 30, 31, 36, 37, 281
macro-minerals, 70–1, 86, 214, 251
Madagascar, 2, 5, 281
Madras Red, 147, 151, 281, Fig. 1.7
magnesium, 68, 69, 70, 71, 138
Magra, 76, 180, 181–5, 281
Maintenance requirement, 57, 58, 61, 63, 91
maize, 21, 59, 66, 90
Malawi, 2, 5
Malaysia, 2, 8, 20, 63, 117
Mali, 2, 11, 55
   management, 18, 22, 24, 31, 32, 33
   reproduction, 36, 133, 136
   skin production, 218
   sociology and economics, 230, 244
   wool production, 31, 32
Malpura, 15, 281
   feeding and nutrition, 81
   growth and meat production, 147, 150, 151, 162
   milk production, 207
   reproduction, 131
   wool production, 180, 181, 182–7
management, 17–39, 256–8
   cooling, 131, 257
   grazing, 27, 41, 84, 251, 257
   milking, 27, 31, 202–6
   of newborn lambs, 16, 134
   of sick animals, 27
   presence of ram, 126
   reproductive, 29–30, 126, 135
   shade, 187, 258
Mandya, 8, 135, 147, 151, 163, 281, Fig. 1.6
manganese, 68, 69
mange, 47

mango, 116, 117
manure
  as sheep food, 114–15
  from sheep, 20, 82, 104, 222–5, 226, 232
mark *see* identification
market information, 243
marketing, 240–4, 256, 266
Marwari, 75, 76, 125, 182–5, 282
Masai, 5, 282
  growth and meat production, 145, 146, 147
  health, 53
  improvement, 254, 260
  management, 18, 24, 36
  reproduction, 36, 133, 136
  sociology and economics, 230, 234–9
mastitis, 129, 206
mating, 25, 34, 130
Maure, 5, 133, 282
Mauritania, 2, 18, 24, 29, 30, 203, 244, 245
Mayo-Kebbi, 133, 282
ME *see* metabolisable energy
meat
  preservation, 165
  production, 22, 38, 145–65, 226, 256
  quality, 161, 164–5
meat and bone meal, 66
Menz, 31, 282
Merino, 11, 282
  feeding and nutrition, 58, 73, 75, 83, 85, 126
  growth and meat production, 147, 151, 163, 210
  health, 53
  milk production, 207, 210
  reproduction, 12, 125, 134
  wool production, 36, 169, 172, 179, 182–3, 186–95, 197–9, 201
  see also Australian Merino, German Mutton Merino, Hungarian Combing Wool Merino, South African Merino, South African Mutton Merino, Soviet Merino
metabolic disorder, 206
metabolic rate, 14, 63, 152, 155, 186
metabolisable energy, 57, 58, 60–7, 76, 86–8, 90
Mexico, 3, Fig. 1.15
  feeding and nutrition, 63, 115
  growth and meat production, 147
  management, 20
  milk production, 207, 212, 215
  reproduction, 123, 131, 133, 134, 138, 139

skin production, 218
sociology and economics, 245, 246
wool production, 194
micro-minerals *see* trace minerals
micro-organisms
  antibiotic resistance, 157
  in cheese, 214–15
  in meat, 165
  in poultry manure, 115
  in rumen, 54–5, 59, 60, 71, 73
  in silage, 100
Middle Atlas, 207, 212, 282
Middle East *see* Near East
midges, 48
migration, 5, 19, 23, 51
  see also nomadism, transhumance
milk, 202–16
  composition, 211–14
  FCE of production, 118
  genetic improvement of yield, 256
  intolerance, 216
  let-down, 208–9
  nutrient requirement for, 62, 66, 70, 73, 75
  production system, 25, 26, 30–1, 33, 38, 202–6, 226
  products, 214–16
  worldwide distribution of production, 202–3
  yield, 12, 14, 145, 206–11
  see also lactation
milking, 27, 31, 204–6
  machine, 205
millet, 66, 104
mimosine, 102
mineral licks, 69, 88
minerals, 68–72, 86, 214, 251
mites, 46, 47
molasses, 86, 110, 115
molybdenum, 69, 71, 85
monensin, 157
Mongolia, 202, 203
Morada Nova, 10, 135, 147, 154, 207, 283
morbidity, 50, 268
Morocco, 2, 4, 125, 132, 207, 212, 218
mortality, 11, 12, 50, 51, 232, 253
  adult, 16, 50, 129
  lamb, 16, 38, 50, 131–2, 134–5, 149
  prenatal, 131, 132
mothering ability, 148
moulting, 166
mud, 20, 46
Muzaffarnagari, 147, 162, 283

Najdi, 283
  growth and meat production, 151, 158, 163
  milk production, 207, 212
  reproduction, 133
  wool production, 180, 191
Nali, 147, 149, 151, 180, 181–5, 283
Namibia, 11, 36, 125, 219–20, 244
Near East, 2, 4, 6, 8
  feeding and nutrition, 19
  health, 15, 44, 48
  management, 19, 23, 25, 28
  milk production, 30, 31, 202, 204, 208, 215
  sociology and economics, 15, 245, 246, 261
  wool production, 187–91, 201
Nellore, 6, 117, 284
nematodes, 40–2
Nepal, 45, 100, 161, 164, 171, 241
net energy, 60–3
New Zealand, 72, 164, 201, 210, 239, 245
Nguni, 5, 284
  see also Bapedi
nickel, 69
nicotinic acid, 73
Niger, 2
  management, 18, 20, 24
  reproduction, 121, 123, 125, 126, 127, 133, 136
  sociology and economics, 244
Nigeria, 2, 4, 13
  feeding and nutrition, 58, 75, 85, 97
  growth and meat production, 147–8, 151, 159
  health, 51, 252
  manure production, 223
  milk production, 207, 212, 213, 216
  reproduction, 134
  skin production, 218
  sociology and economics, 35, 228, 241, 245, 246, 261
  wool production, 191–2
night grazing, 26, 27, 187
night management, 18, 21, 27–8, 261
Nilgiri, 11, 133, 181, 182–5, 201, 207, 284
nitrate, 85
nitrogen see protein
nitrogen fertiliser, 59, 92
nomadism, 17–22, 26, 41, 51, 68
  see also migration, transhumance
North America, 24, 203, 218, 244, 245, 246, 247
  see also Canada, USA

Northern Sudanese, 5, 284
  see also Desert Sudanese
Nungua Blackhead, 133, 142, 284
nutrition, 25, 40, 54–77
  as a constraint to sheep production, 251–2
  effect on carcass composition, 157–8
  effect on milk production, 209–10
  effect on reproduction, 126, 129, 137–8
nutritive requirements, 56–76
nylon bag, 76

oats, 102
Oceania, 10, 203, 218, 244, 245, 246, 247
  see also Australia, New Zealand, Pacific islands
oesophagus, 54
oestrogen, 122–4, 126, 127–8, 137, 138
oestrous cycle, 121–5
oestrus, 121–2, 125–6, 128
  detection, 121, 140
  synchronisation, 140, 141
offal, 161, 220–1, 232
offtake, 24, 129, 231, 268
oil-palm, 20, 85, 113, 116
oil-seed meal, 66
Oman, 247, 253
oral dosing, 69
  see also drenching
orange, 113
orf, 46
oryx, 76
Ossimi, 6, 285
  feeding and nutrition, 58
  milk production, 207, 212
  reproduction, 12, 121, 129, 133
  wool production, 181, 187–9
ovary, 122–3, 125
overgrazing, 51, 75, 82, 86, 251, 252, 260
overstocking, 51, 89, 252, 260
ovulation, 123
ovulation rate, 132, 134, 137, 141
ovum transfer, 141
ownership of sheep, 23, 227–8
oxalates, 85
oxytocin, 208

Pacific islands, 114
Pakistan, 2, 8, 133, 187, 218, 220
pantothenic acid, 73
Papua New Guinea, 2, 41, 133
parasitism, 40–5, 46–7, 82, 104, 195, 206

parity of ewe
  effect on milk production, 210–11
  effect on reproduction, 135, 136
partition of nutrients, 78, 146, 154, 196
parturition *see* lambing
pastoral systems research, 265
pasture
  improved, 91–5, 251
  in worm cycle, 41, 44
  natural, 78–82
Patanwadi, 181, 285
Pelibuey, 10, 285
  feeding and nutrition, 63, 115
  milk production, 207, 212, 213
  reproduction, 123, 130, 133, 139
  sociology and economics, 230
pelt, 36, 219–20
penis, 130
performance testing, 255
Permer, 253
Peru, 3, 4, 194, 218
perverse selling, 24, 227–8
phalaris staggers, 72
phenotypic variation, 11, 12, 253
Philippines, 206, 207, 212
phosphorus, 68, 79, 70, 74, 88, 107, 138, 214
photoperiod *see* daylength, light
photosensitisation, 16, 85
physical treatment of straw, 105, 106, 110
pig, 114, 221, 265
pig manure, 115
pineal gland, 125
pining, 72
pituitary, 122–3, 155
placenta, 127–8
plantain, 114
pneumonia, 47–8, 129, 252
  see also respiratory disease
poisonous plants, 16, 19, 85–6
Polwarth, 131, 147, 155, 162, 285
population
  human, 86, 248
  sheep, 1–4, 86
  see also overstocking
potassium, 68, 70, 214
potassium hydroxide, 108
potassium type, 13, 14, 75, 185
poultry, 221, 265
poultry litter, 113, 114–15
pox, 49, 51
PPR, 49, 252, 258
predation, 26, 261
pregnancy, 127–8
  diagnosis, 125, 143–4
  duration, 127
  grazing behaviour, 84
  hormones, 127–8
  nutrient requirements during, 59, 62, 73, 75, 137–8
  oestrus during, 127
  toxaemia, 48
preservation of meat, 165
Priangan, 8, 41, 132, 133, 207, 285
productivity index, 10, 228–31
progeny testing, 255
progesterone, 122–4, 126, 127–8, 143
prolactin, 122–3, 126
prolific breeds, 12, 132, 134
prolificacy, 132, 268
proportion of reproductive females, 24–5, 129, 132, 136
prostaglandins, 122–3, 127–8
protein
  in food for sheep, 60, 77, 81, 82, 86, 94, 96–8, 107, 110–15
  in milk, 55, 211–14
  requirements, 64–7, 210
puberty, 128–30, 132, 138, 142
Pugal, 185, 285
pulpy kidney, 47

Q fever, 131
Qatar, 246
quality count (of wool), 169, 170
quarantine, 53

Rabo Largo, 10, 286
raddle, 143
Rahmani, 6, 133, 163, 187–9, 207, 212, 286
rain, 15, 19, 98, 258
rainfall, 79–80, 135, 148
rainy season
  feeding and nutrition, 80, 98
  growth and meat production, 148
  health, 45, 47, 49, 50, 135
  management, 15
  wool production, 186
Ralgro *see* growth promoter
Rambouillet, 11, 15, 286
  growth and meat production, 147
  milk production, 207
  reproduction, 128, 129, 131, 133
  wool production, 182–7, 195
Rampur Bushair, 185
ranching, 18, 28, 41, 52, 84, 90, 251, 260
rangeland, 20
  see also pasture
ratio of ewes to rams, 24–5, 121, 130

religion, 29, 30, 160
repeatability, 12, 134, 188, 190, 193, 199
reproduction, 121–44
  ewe physiology, 121–30
  ram physiology, 130–1
reproductive diseases, 131
reproductive rate, 14, 131–8
research, 249–51, 263–5
  wool production, 179–95
respiratory disease, 47–8, 257
  see also pneumonia
restricted suckling, 29–30, 203, 208
retinol, 73
return on capital investment, 238–40
rhizobia, 82
riboflavin, 73
rice, 66
rice bran, 114
rice straw *see* straw
rickets, 74
rinderpest, 49
risk, 17, 38, 227, 239, 259, 260
roadsides, 21, 103
Romanov, 182
Romney Marsh, 11, 75
  growth and meat production, 147, 156
  reproduction, 133
  sociology and economics, 235–9
  wool production, 176, 194, 200
roughage, 54, 77
roundworms, 40–2, 43
rubber, 20, 116, 117
rumen, 54–5, 59
  see also micro-organisms
rumen degradable protein (RDP), 64–7
rural development projects, 264

SABRAO, 10
Sahara, 5, 25, 79
Sahel (area), 33
Sahel (sheep) *see* West African Long-legged
salmonellosis, 15, 131, 257
salt, 69, 70–1, 76, 138
salted meat, 165
Santa Inês, 10, 147, 163, 207, 287
Sardinian, 207, 208, 287
Saudi Arabia, 2
  feeding and nutrition, 96
  growth and meat production, 151, 163, 165
  milk production, 202, 203, 207, 212
  reproduction, 133
  skin production, 218

  sociology and economics, 165, 245, 246, 247, 259, 260
  wool production, 191
savannah, 79, 82
sawdust, 115
scavenging, 20, 226
scissors, 31
Scottish Blackface, 123, 211, 213, 287
seasonal effect on
  birth, 126
  disease, 15, 47, 50
  lamb growth, 148, 152
  lamb mortality, 135
  reproduction, 121, 125–6, 127, 131, 136
  vegetation, 20, 22, 79–80, 81, 125
  wool production, 184–5, 189, 190, 194
sedentary system, 17–22, 23, 26, 51–2
selection
  breeding, 13, 129, 132, 181, 193, 196, 197, 208, 220, 253, 254–6
  of diet, 82–4, 224
selenium, 68, 69, 72, 74, 85, 88, 138
semen, 130–1, 138–40
semi-arid area *see* dry tropics
Senegal, 2, 4, 134, 136, 148, 229, 230, 245, 246
sewage, 115
sex effect on
  lamb growth, 62, 148, 150, 155
  wool production, 190
sex ratio, 34
shade, effect on
  pasture, 92
  sheep, 10, 15, 117, 187, 258
  wool, 189, 193
sharing, 23
shearing, 27, 31–2, 59, 187, 196
shearing interval, 32
shears, 31–2
sheep scab, 47
shelter, 48, 117
  see also housing, shade
shepherd, 23, 84, 256
  assistance in mating, 130
  training for, 50, 250, 256–7, 262
shepherding, 23, 85, 226
Shropshire, 206, 207, 212, 287
shrubs, 80, 82
silage, 59, 66, 100–1
sisal, 112
skin, 46, 49, 161, 216–19, 232, 240
  see also pelt
slats, 224, 258
slaughter, 158–61

slaughter house, 41, 125, 160–1, 240
smallholder, 51, 130, 256, 261
smoked meat, 165
snail, 44
sociological factors, 50, 226–8, 258–62
sodium, 68, 70, 88, 214
sodium hydroxide, 108, 115
soil, 68, 76, 222–3
soil erosion, 81, 222
solar radiation, 16, 74, 79, 92, 117, 189, 192–3, 258
Somali, 6, 10, 75, 163, 287, Fig. 1.4
Somalia, 2, 4, 203, 244, Fig. 1.4
somatotrophic hormone, 155
Sonadi, 151, 163, 181–4, 288
sorghum, 66, 104
South Africa, 2, 5–6, 11
  feeding, 72
  growth and meat production, 150, 152, 155, 163
  pelt production, 36, 219
  reproduction, 126, 127, 133, 134, 144
  skin production, 218
  sociology and economics, 245, 246
  wool production, 167, 192–4
South African Merino, 5, 11, 133, 155, 163, 288
South African Mutton Merino, 155, 163
South America, 3, 4, 8, 11
  health, 44
  management, 20, 28
  milk production, 203
  reproduction, 141
  skin production, 218
  sociology and economics, 226, 244, 245, 246, 247
  wool production, 194–5
Southdown, 85
Southeast Asia, 2, 4, 23, 130, 241, 245, 247
Southern Sudanese, 5, 288
Soviet Merino, 11, 133, 185, 195, 288
soya bean, 73, 86, 87–8, 112
sperm, 121, 130–1, 138–40
spinning, 168–70
sponge, 141
Sri Lanka, 8
  feeding, 117
  growth and meat production, 147, 151
  management, 20, 23, 25, 28, 136
stall feeding, 21, 22, 41
  see also feedlot, housing
starch, 60
state marketing enterprise, 242

steppe, 79
stiff lamb disease, 74
stocking rate, 41
  see also carrying capacity, overgrazing
stockman, 250
stomach, 54
stratification of the sheep industry, 35–9, 254, 261
straw, 64, 66, 70, 73, 74, 87–8, 104–10, 251
streptothricosis, 46, 48–9
strippings, 205
struck, 47
stubble, 19, 20, 28, 42, 104
sturdy, 44
stylo, 82
subsidy, 259
suckling, 29, 203–4
suckling prevention, 29–30, 203, 208
Sudan, 2, 4, 13
  growth and meat production, 147, 162
  management, 18, 20, 24, 26
  manure production, 222
  milk production, 203, 215
  reproduction, 133, 136
  skin production, 218
  sociology and economics, 230, 241, 244, 245
Suffolk, 11, 288
  growth and meat production, 147, 151, 154, 163
  reproduction, 123, 133
  wool production, 182–3, 194
sugar-cane, 21, 104, 110–11
sulphur, 68, 71, 88, 183
sunlight, 16, 74, 79, 92, 117, 189, 192–3, 258
superovulation, 141
superstition, 51, 262
supplementary feeding, 21, 69, 86–8, 256, 258
  effect on growth and meat production, 150–1, 158–9
  effect on reproduction, 131, 137–8
swayback, 71
sweat, 15, 186
synchronisation of oestrus, 140, 141
Syria, 2
  feeding and nutrition, 104
  management, 19, 31, 136, 261
  milk production, 202, 203
  skin production, 218
  sociology and economics, 228, 246
  wool production, 191

systems of sheep production, 17–22, 51–2
  milk, 30–1, 202–6
  see also nomadism, ranching, scavenging, sedentary system, smallholder, transhumance

tail
  fatness, 19, 34, 129, 130, 158, 164
  docking, 158
tanning, 216–18
Tanzania, 2, 20, 138, 145, 146, 227
Tanzania Long-tailed, 5, 289
tapeworm, 43, 44–5
Targhee, 133, 154, 289
tassels, 13
teaser rams, 126, 143
Teeswater, 196
teeth, 55–6, 129
  see also broken mouth
temperate breeds, 11, 14, 125, 158
  see also exotic breeds
temperature
  air, 14, 15, 19, 75, 186
  body, 48
  effect on sperm, 139–40
  see also cold stress, heat stress
tenderness of meat, 164
terminal sire, 154, 254
testes, 130
testosterone, 130
tethering, 27, 41, 84
textile production, 167–70
Thailand, 2, 8, 27, 162
theft, 27, 28, 228, 261
thermal stress *see* cold stress, heat stress
thiamin, 73
thin-tailed sheep, 4, 9
  see also individual breed names
thyroid, 71, 155
thyoprotein, 257
ticks, 27, 46, 131
Togo, 2, 134
total solids (in milk), 211–14
toxin, 85, 113
  see also poisonous plants
trace minerals, 68, 69, 71–2, 88, 214
trade, 240–7
trader, 215, 240–3
traditional medicines, 51
training, 50, 250, 257, 262, 263–4
transferrin, 13
transhumance, 17–22, 26, 41, 68
  see also migration, nomadism

transport
  of food, 90
  of sheep and products, 15, 160, 165, 244
tree crops, integration with sheep, 20, 92, 116–17
trees, 80, 82, 256, 261
trekking *see* walking
Trinidad and Tobago, 63, 99, 103, 116, 147, 207, 212
trypanosomiasis, 1, 53
Tswana, 5, 133, 163, 289
Tuareg 5, 18, 24, 25, 290
Tunisia, 2, 4, 207, 212, 218
Tunisian Barbary, 207, 212, 290
Turkey, 202, 203, 205, 261
twin lamb disease, 48
twinning, 134, 137, 262

Uda, 7, 125, 126, 133, 147, 151, 167, 191–2, 253, 290
udder, 29, 204, 205–6
Uganda, 2, 4, 135
UK, 55–6, 60, 63, 138, 140, 154, 159, 247, 260
Ultrasonic scanning, 143, 160
undergradable protein (UDP), 64–7, 87
underground housing, 28
undernutrition, 45, 50, 51, 69, 135, 137, 152, 158, 196, 251
United Arab Emirates, 246, 247
university, 50, 250, 256, 264
Upper Volta, 2, 4, 24, 28, 244, 250
urea, 59, 87–8, 109, 110, 210
urine, 82, 109, 222–4
USA, 11, 55, 72, 112, 114, 129, 138, 140, 154, 157, 164, 219
USSR, 140, 203, 218, 219, 244, 245, 246, 247
uterus, 122–3

vaccination, 45, 46, 47, 48, 53, 252, 258
Venezuela, 3, 58, 63, 82, 91, 96, 133–4, 147–8, 151, 163, 207
veterinary assistant, 50
veterinary service, 50–2, 252
vibriosis, 131
Virgin Island, 230, 290
vitamins, 34, 70, 72–4, 138, 214
voluntary food intake *see* food intake

walking, 19–20, 25, 52, 61, 160, 244
water intake, 15, 75–6
watering interval, 19, 25, 74–5
watering management, 26, 28, 39, 74–6, 89, 187, 251

watering place, 25, 51, 75, 84, 89
weaning, 29, 30, 31, 42, 149–50, 203, 206
weaning rate, 12, 132, 234, 237
weathering of fleece, 192–3
weaving, 168–70
weight gain *see* growth rate
Wensleydale, 189, 192, 196, 291
West Africa, 2, 5, 10
　management, 18, 20, 23
　reproduction, 134
　sociology and economics, 226, 230, 233–4, 244, 245, 247
West African, 10, 291
　growth and meat production, 147–8, 151, 163
　milk production, 207
　reproduction, 133–4
　sociology and economics, 230
West African Dwarf, 6, 10, 13, 167, 291, Fig. 1.11
　feeding and nutrition, 112
　growth and meat production, 146, 148, 149, 159
　improvement, 253, 254
　milk production, 207, 212, 213
　reproduction, 132, 134, 136, 137, 142
　sociology and economics, 229, 230, 237–9
West African Long-legged, 55–6, 167, 291
　see also Fulani, Maure, Tuareg
West Germany, 247
West Indies, 10, 28, 230, Fig. 1.12
wet area *see* humid tropics
wet season *see* rainy season
Wiltshire Horn, 166
women, 27, 31
wool, 166–201
　assessment of adult fleece, 170–4
　assessment of lamb's birthcoat, 175–9
　breeds, 4, 8, 9, 11, 166
　canary colouring, 183, 186–7, 191, 201
　carpet, 169, 170, 172, 181, 187, 194, 195, 199–201
　chemical composition, 71, 167
　coloured, 168, 170, 171, 174, 181
　cotting, 191
　crimp frequency, 183, 185, 193, 194
　dyeing, 170
　fibre density, 182, 184, 198
　fibre diameter, 13, 168–200
　fibre length, 171, 173–4, 180–1, 188, 190–4, 198, 201
　fibre-type array, 175–9, 187–8, 189, 192, 199
　fleece weight, 31, 172, 180–98
　follicle density, 190
　halo hair grade, 175, 187–8, 192
　kemp percentage, 187, 189, 190, 191, 199, 200
　marketing, 240–2
　medullation percentage, 168, 171, 180–90
　milling, 169
　pigmentation, 170, 174
　processing, 167–70, 199–201, 240
　production, 11, 12, 24, 31–2, 36, 118, 256
　quality, 13, 168, 194, 197, 249
　quality count, 169, 170
　soundness, 174
　S/P ratio, 166, 176, 183, 190, 192, 193, 194, 201
　spinning, 168–70
　staple length, 32, 168, 173, 180–95, 199
　textiles, 167–70
　woollen process, 169
　worsted process, 169
　yield, 182, 188
World Bank, 228, 264
worms *see* helminthiasis

Yankasa, 7, 167, 191–2, 292
　feeding and nutrition, 75
　growth and meat production, 148, 151
　reproduction, 134, 253
Yemen, 246, Fig. 1.3
yoghurt, 214, 215

Zaghawa, 5, 18, 292
Zaïre, 5
Zambia, 58, 59, 75, 96
zeranol, 156
Zimbabwe (breed), 5, 125, 292
Zimbabwe (country), 2, 55, 66, 82, 156
zinc, 68, 69, 138